AF601570

Handbook of Expert Systems Applications in Manufacturing

Intelligent Manufacturing Series

Series Editor: Andrew Kusiak
Department of Industrial Engineering
The University of Iowa, USA

Manufacturing has been issued a great challenge – the challenge of Artificial Intelligence (AI). We are witnessing the proliferation of applications of AI in industry, ranging from finance and marketing to design and manufacturing processes. AI tools have been incorporated into computer-aided design and shop-floor operations software, as well as entering use in logistics systems.

The success of AI in manufacturing can be measured by its growing number of applications, releases of new software products and in the many conferences and new publications. This series on Intelligent Manufacturing has been established in response to these developments, and will include books on topics such as:

- design for manufacturing
- concurrent engineering
- process planning
- production planning and scheduling
- programming languages and environments
- design, operations and management of intelligent systems

Some of the titles are more theoretical in nature, while others emphasize an industrial perspective. Books dealing with the most recent developments will be edited by leaders in the particular fields. In areas that are more established, books written by recognized authors are planned.

We are confident that the titles in the series will be appreciated by students entering the field of intelligent manufacturing, academics, design and manufacturing managers, system engineers, analysts and programmers.

Titles available

Object-oriented Software for Manufacturing Systems
Edited by S. Adiga

Integrated Distributed Intelligence Systems in Manufacturing
M. Rao, Q. Wang and J. Cha

Artificial Neural Networks for Intelligent Manufacturing
Edited by C.H. Dagli

Handbook of Expert Systems Applications in Manufacturing
Structures and rules
Edited by A. Mital and S. Anand

Handbook of Expert Systems Applications in Manufacturing Structures and rules

Edited by

A. Mital

Industrial Engineering, University of Cincinnati, Cincinnati, USA

and

S. Anand

Industrial Engineering, University of Cincinnati, Cincinnati, USA

SPRINGER-SCIENCE+BUSINESS MEDIA, B.V.

First edition 1994

Originally published by Chapman & Hall in 1994

Typeset in Times 10/12 by Interprint Limited, Malta

ISBN 978-0-412-46670-0 ISBN 978-94-011-0703-7 (eBook)
DOI 10.1007/978-94-011-0703-7

A catalogue record for this book is available from the British Library

Library of Congress Cataloging-in-Publication data available

Printed on permanent acid-free text paper, manufactured in accordance with the proposed ANSI/NISO Z 39.48-199X and ANSI Z 39.48-1984

To
Little Aashi and Growing Anubhav for their
Undemanding Love
And
Our Wives, Chetna and Meera, for their
Inspiration and Encouragement

Contents

Contributors

Sam Anand, Industrial Engineering, University of Cincinnati, Cincinnati, OH 45221-0116, USA.

P. Banerjee, Department of Mechanical Engineering, The University of Illinois at Chicago, Chicago, IL 60680, USA.

Nina M. Berry, Intelligent Design and Diagnostic Research Laboratory, Department of Industrial and Management Systems Engineering, The Pennsylvania State University, University Park, PA 16802, USA.

B. Bibanda, Department of Industrial Engineering, University of Pittsburgh, Pittsburgh, PA 15261, USA.

James C. Chen, Department of Industrial Engineering, University of Wisconsin-Madison, Madison, WI 53706, USA.

C.N. Chu, School of Industrial Engineering, Purdue University, West Lafayette, IN 47907, USA.

P.H. Cohen, Department of I & MSE, The Pennsylvania State University, University Park, PA 16802, USA.

Cihan H. Dagli, Engineering Management Department, University of Missouri-Rolla, Rolla, MI 65401, USA.

Suranjan De, Department of Decision and Information Sciences, Santa Clara University, Santa Clara, CA 95053, USA.

C.R. Emerson, Department of Mechanical and Industrial Engineering, T.J. Watson School of Engineering and Applied Science, State University of New York, NY 13902-6000, USA.

Jiebo Guan, Sun Japan Corporation, MMBS Otowa Building 9F, 1-20-14 Otowa Bunkyu-ku, Tokyo 112, Japan.

Reuven Karni, Faculty of Industrial Engineering and Management, Technion – Israel Institute of Technology, Haifa, Israel 32000.

R.L. Kashyap, School of Electrical Engineering, Purdue University, West Lafayette, IN 47907, USA.

Soundar R.T. Kumara, Intelligent Design and Diagnostic Research Laboratory, Department of Industrial and Management Systems Engineering, The Pennsylvania State University, University Park, PA 16802, USA.

Tsuang Kuo, Industrial Engineering, University of Cincinnati, Cincinnati, OH 45221-0116, USA.

Anita Lee, Department of DSIS, University of Kentucky, Lexington, KY 40506, USA.

Chung-Yu Liu, Department of Mechanical and Industrial Engineering, T.J. Watson School of Engineering and Applied Science, State University of New York, NY 13902-6000, USA.

M. Marefat, AI-Simulation Group, Department of Electrical & Computer Engineering, The University of Arizona, Tucson, AZ 85721, USA.

Anil Mital, Industrial Engineering, University of Cincinnati, Cincinnati, OH 45221-0116, USA.

Gary P. Moynihan, The University of Alabama, Department of Industrial Engineering, Box 870288, Tuscaloosa, AL 35487-0288, USA.

Setsuo Ohsuga, Research Center for Advanced Science and Technology, The University of Tokyo, 4-6-1 Komaba Meguro-ku, Tokyo 153, Japan.

Ming Rao, Intelligence Engineering Laboratory, Department of Chemical Engineering, University of Alberta, Edmonton, Canada T6G 2G6.

Jacob Rubinovitz, Faculty of Industrial Engineering and Management, Technion – Israel Institute of Technology, Haifa, Israel 32000.

Chana S. Syan, Department of Mechanical and Process Engineering, University of Sheffield, P.O. Box 600, Mappin Street, Sheffield SI 4DU.

Alice E. Smith, Department of Industrial Engineering, University of Pittsburgh, Pittsburgh, PA 15261, USA.

K. Srihari, Department of Mechanical and Industrial Engineering, T.J. Watson School of Engineering and Applied Science, State University of New York, NY 13902-6000, USA.

Gürsel A. Süer, Industrial Engineering Department, University of Puerto Rico-Mayaguez, Mayaguez, Puerto Rico 00681.

Arne Thesen, Department of Industrial Engineering, University of Wisconsin-Madison, Madison, WI 53706, USA.

C. Tunasr, Department of Industrial Engineering, University of Pittsburgh, Pittsburgh, PA 15261, USA.

Mario van Vliet, Econometric Institute, Erasmus University, Rotterdam, The Netherlands.

Qun Wang, Intelligence Engineering Laboratory, Department of Chemical Engineering, University of Alberta, Edmonton, Canada T6G 2G6.

I.C. You, School of Electrical Engineering, Purdue University, West Lafayette, IN 47907, USA.

Ji Zhou, Department of Mechanical Engineering, Huazhong University of Science and Technology, Wuhan, People's Republic of China 430074.

Preface

Artificial intelligence (AI) is playing an increasingly larger role in production and manufacturing engineering. Much of this growth is the result of special-purpose computer controlled machines that are dominating modern manufacturing operations, such as computer numerically controlled machines and robots, and production activities, such as materials handling and process planning. Since a great deal of production and manufacturing engineering knowledge can be put in the form of rules, expert systems have emerged as a promising practical tool of AI in solving manufacturing and production engineering problems. The expert systems allow knowledge to be used for constructing human–machine systems that have specialized methods and techniques for solving problems in a particular application area.

Over the years, many expert systems have been developed for applications in manufacturing and production engineering. Most of these expert systems, however, have been of little use to practitioners at large. The primary reason for this limited utility is that in most cases the developers do not divulge the knowledge base and inference mechanism that form the backbone of an expert system. Without the knowledge base, users can only derive a very limited benefit from an expert system and, for all practical purposes, a technical publication describing the expert system for the reader merely becomes a publicity brochure. The reader must either develop his own knowledge base or purchase the system from the developer, often at a substantial cost.

Our and our colleagues' frustration with such publications provided us the impetus for undertaking the development of this book, which is aimed at both researchers and practitioners, and provides a collection of expert systems in manufacturing and production engineering along with their knowledge base and rules. We believe that inclusion of the knowledge base and associated rules is essential if practitioners are to derive full benefit from these expert systems. This unique book is the result of our belief and the efforts of our distinguished colleagues who subscribe to this philosophy. We are confident that the expert systems included in this volume will be well received and readers will appreciate the efforts of all those who so generously agreed to contribute to this effort.

A total of 15 different expert systems are included in this book. These expert systems are preceded by an introductory chapter written by Kuo,

Mital and Anand. The expert system rules are included on a floppy disk in ASCII and can be easily accessed. These rules and the description of the expert system's structure should assist the users in customizing these systems.

In the introductory chapter, Kuo *et al.* provide a brief overview of expert systems basics, outline the advantages and limitations of expert systems, describe the basic expert systems development process, and review the various expert systems in production and manufacturing engineering.

The first expert system included in the book, developed by van Vliet, presents optimization rules for two flexible manufacturing system design problems: the machine allocation problem and the traffic allocation problem. This system should be particularly useful to automobile and electronics manufacturers in designing logistically sound manufacturing systems. In the following expert system, Marefat and Banerjee provide a reusable framework to generate solutions to manufacturing problems such as those encountered in laying out a facility and integrating process planning in designing manufacturing systems. The framework consists of hierarchical knowledge representation, preliminary design, iterative modification, four distinct information flow and reasoning paths, and solution validation.

Ohsuga and Guan provide a general purpose knowledge-based system that can be applied to a variety of design problems including mechanical design, chemical compound design and feedback control system design. The authors first describe the outline of their knowledge-based system and then demonstrate its application by designing a feedback control system. Solution to multi-model line balancing is provided by Suer and Dagli through their knowledge-based expert system. This expert system assigns different models to a varying number of lines with a varying number of workstations over a period of time such that the total number of resources, manpower and robots are not exceeded.

The knowledge regarding the production supervisor's informational needs and decision-making processes is modeled in the expert system developed by Moynihan. The system utilizes data commonly found in external manufacturing systems and files. Wang, Rao and Zhou provide an expert system that assists in the development of conceptual product designs. It's focus is on minimizing energy consumption, increasing raw material utilization and profits, and reducing environmental effects of effluents. The expert system also ensures flexibility, operability, controllability and safety of the manufacturing process. The expert system developed by Syan, on the other hand, assists in the selection of surface treatments and coatings at the product design stage.

The knowledge-based system developed by You, Chu and Kashyap assists in evaluating three-dimensional casting designs for manufacturability by incorporating rules on tolerancing, rounding, shrinkage, junction rules, wall thickness, parting line and solidification. The next system, developed by Rubinovitz and Karni, provides a qualitative approach to selecting materials handling and transfer equipment. It also serves as a design tool.

The next two expert systems primarily deal with scheduling. The knowledge-based system developed by De and Lee generates detailed production schedules for a flexible manufacturing system, and considers both assembly and non-assembly jobs. Chen and Thesen's expert system, on the other hand, assists in determining when a specific scheduling rule should be used. The goal of this system is to increase the production capacity.

The expert system developed by Liu, Emerson and Srihari is a multi-attribute expert system which utilizes fuzzy reasoning in selecting a pick-and-place machine for a surface mount technology printed circuit board assembly line. The goal is to enhance productivity and profitability of the manufacturing facility. Bidanda, Cohen and Tunasar's expert system is also an equipment selection system. The application, however, is limited to workholding devices for rotational parts.

The expert system developed by Berry and Kumara is intended to assist users in developing improved techniques for robot task planning. It combines machine learning with robotic planning systems to improve decision and reactive processes. Finally, Smith and Dagli provide an expert system for selecting appropriate quality control charts for a variety of manufacturing processes. The system can also assist users in interpreting results (trends) and selecting acceptance sampling plans.

Overall, the expert systems included in this volume cover a fairly wide variety of manufacturing and production engineering topics. We realize that the coverage leaves out many important areas but we hope our readers will appreciate the difficulty in compiling such a volume. Perhaps this book will act as a catalyst, inspiring further volumes at some point in the future.

Any book of this magnitude requires input from a number of individuals. In this case, these are individuals who contributed to this volume and those who assisted us in the review process. To all these people, our sincere thanks. We would also like to thank Mark Hammond, our editor at Chapman & Hall, for his constant encouragement and assistance.

Anil Mital
Sundararaman Anand

CHAPTER 1

An introduction to expert systems in production and manufacturing engineering: the structure, development process and applications

Tsuang Kuo, Anil Mital and Sam Anand

1.1 INTRODUCTION

Due to the global competition, manufacturing is facing several challenges: short product life cycle, frequent design changes and small in-process inventory. The performance of a manufacturing system is affected by numerous needs, such as material requirement planning (MRP), capacity planning, facility and material handling device planning, inventory control, tool management, scheduling, quality control, and manufacturing information system management (Chang *et al.*, 1991). These necessities have led to the implementation of computer technologies, such as automated test facilities; MRP; computer-aided design (CAD); computer-aided manufacturing (CAM); computer-aided process planning (CAPP); computer-aided quality control (CAQ); digital computers simulation; and data collection, storage and analysis. While these operations and primarily digital manipulations that take advantage of speedy data storage, retrieval and computational capabilities of computers, other kinds of expertise, such as knowledge that is based on past experience and cannot easily be cast into mathematical formulae for conventional algorithmic programs, have yet to be fully exploited (Grimson and Patil, 1987). Expert systems, a technology falling into this last category, improve the quality of information provided to the responsible experts and users for operating modern automated and integrated systems, and assist them in reliable operation of their systems (Christie, 1990; Braun, 1990).

Expert systems, also known as knowledge-based systems, are the most widely used of all artificial intelligence (AI) techniques (Michie, 1982). Expert systems have been developed for a variety of applications. Among the many successful expert systems reported are: DENDRAL, developed for interpreting mass spectrograms to identify chemical constituents (Buchanan and Feigenbaum, 1978); PROSPECTOR, developed for interpreting geological structure (Duda *et al.*, 1979); MYCIN, developed for diagnosing infectious disease (Davis *et al.*, 1977); and XCON/R1, developed for designing computer system configuration (McDermott and Bachant, 1984). These expert systems not only preserve valuable knowledge for future use, they have significant economic impacts. For example, PROSPECTOR has discovered a mineral deposit worth $100 million and XCON/R1 is saving Digital Equipment Corporation (DEC) more than a million dollars a year (Giarratano and Riley, 1989).

Expert systems range from analysis systems to synthesis systems. Analysis systems perform interpretation, prediction and diagnostic tasks. Synthesis systems are developed for design, planning, monitoring, debugging, repair, instruction and control, and involve more complex reasoning tasks (Interrante and Biegel, 1990).

An expert system is a computer program which uses explicitly represented knowledge and computational inference techniques to mimic a human expert in a specific domain (Grimson and Patil, 1987; Winston, 1984). Therefore, expert systems usually have two major components: a knowledge base and an inference strategy. The knowledge base is a collection of facts and rules (Grimson and Patil, 1987; Rowe, 1988). Facts are used to describe the current task domain and the goal. Rules are used for solving problems in that domain and sometimes generate new facts. The inference strategy of expert systems navigates through the knowledge base in order to satisfy a goal, i.e. what rule to apply and where to apply it (Miller *et al.*, 1988). Two command inference strategies are forward chaining and backward chaining. The application of rules to facts in the knowledge base to produce a modified situation is often called forward chaining. Backward chaining does not apply rules to current task-domain situations, but to the goal.

The development of expert system shells leads to easier modeling of complex problems for expert systems (Wolfgram *et al.*, 1987). An expert system shell is a software package that provides a high level, user friendly programming language, text editing, file manipulation and an interpreter, or compiler, to run the expert system.

Expert systems differ from conventional programs in several ways. While most conventional programs deal with numerical data with known algorithm, expert systems are concentrated in symbolic data with no known algorithm to solve the problem. Expert systems usually handle problems which have multiple acceptable answers while conventional programs are best suited for problems with one precise answer. Conventional programs are computation intensive. Expert systems are search intensive programs (Wolfgram *et al.*, 1987). The major difference between the two is the degree of separation between the knowledge and knowledge application mechanism. In other words, the difference is between data and control (Linn and Wysk, 1990).

The objectives of this introductory chapter are to provide an understanding of expert systems' concepts, terminologies, tools, and their capabilities and limitations. Following that, the expert system building process and latest applications are presented and key factors for successful implementation of expert systems in real world are discussed.

1.2 EXPERT SYSTEM BASICS

Expert systems are typically domain-specific and tend to operate by utilizing heuristic problem solving approaches (Badiru, 1988). Much of the knowledge in expert systems is heuristic in nature – it consists of rules of thumb that are often, but not always, true (Grimson and Patil, 1987). Several representation schemes and data structures have been proposed for heuristic knowledge. As mentioned earlier, the reasoning process is controlled by the inference engine.

1.2.1 Knowledge representation

Knowledge representation schemes describe the various architectures used to represent the experts' knowledge. Major approaches to knowledge representation include logic representations, semantic networks, procedural representations, logic programming formalisms, frames, production rules and predicate calculus (Barr and Feigenbaum, 1981; Cercone, 1987). However, production rules and frames are most frequently used in commercial expert system shells.

(*a*) *Production rules*

Production rules were first developed by Newell and Simon (1972). Rules capture informal, often judgmental or heuristic, information and are typically chained together to form a line of reasoning leading to the answer. The term 'production rule' is used to describe one very general, underlying idea – the notion of 'IF–THEN' (condition–action) pairs, i.e. 'If this condition holds, then this action is appropriate'. The IF part of the production rule states the condition(s) that must be present for the production to be applicable. The THEN part is the appropriate action to take. During the execution, a production rule whose condition part is satisfied can be fired or executed by the interpreter. The resulting action will usually generate new fact(s) to be added to the knowledge base.

(b) Frames

A frame is a partition of knowledge which consists of a collection of slots to describe an object (Minsky, 1975). Frames can be treated as a prototypical instance, i.e. explicit common sense knowledge, assumptions, what things to look for and in what way to look for them. Each slot has one or more declarative and/or procedural information in the form of predefined internal relations. A slot describes its individual properties. For instance:

AUTOMOBILE Frame

Parent:	Vehicle
Type:	sedan, wagon, jeep, ...
Number_of_Wheels:	an integer (default 4)
Type_of_engine:	4_cylinders, 6_cylinders, ...
Color:	yellow, red, ...
Trim:	economy, deluxe, ...

the automobile frame has six slots (i.e. Parent, Type, Number_of_Wheels, Type_of_engine, Color, and Trim). Reasoning with frames is a process of matching prototypes with specific individual slot. The basic task is to match the set of frames with the collection of facts about a specific instance, to determine which frame(s) best matches.

Frames are often organized in a hierarchical manner, providing an economy of knowledge presentation and reasoning mechanism. The concept of inheritance is based on the principle that attributes are true of a general category are true of the more specific subcategories as well. For example, all the properties of Vehicle frame will pass to Automobile frame, such as all vehicles have wheels, then all automobiles will have wheels too. The economy of reasoning mechanism is accomplished by successive refinement. The system begins by using general category frames near the top of the hierarchy, then refines this by using subcategory frames at the next level. This focusing effort is very important in a large knowledge base (Rowe, 1988; Wolfgram *et al.*, 1987).

1.2.2 Inference techniques

The inference engine is the backbone of an expert system. An inference engine interprets knowledge in the knowledge base and makes reasoning among the knowledge to accomplish conclusions for the domain (Barr and Feigenbaum, 1981; Rowe, 1988; Winston, 1984).

(a) Backward chaining

Backward chaining reasons from goals back towards more primitive data. It will retrieve all the rules that are relevant to the goals, i.e. one can draw conclusion(s) in a rule if each of the clauses in its 'IF' part are true. How do we determine the truth of each of these clauses? We start all over again doing exactly the same thing: retrieve all the rules that are relevant to the first clause and try each of those rules in turn. Eventually, the system encounters topics for which there are no rules. At that point it requests specific facts from the user in order to determine the truth or falsity of the premise clause under consideration.

(*b*) *Forward chaining*

Rules can also be run forward, deducting all logical conclusions from a basic set of input data. A new fact asserted to the system can cause any rule whose premises match that fact to fire. Then, if all the other premises are also matched among facts that have been asserted, the rule's conclusion is in turn asserted as a new fact.

1.2.3 Development tools (shells)

Expert systems may be developed in any computer language. While use of conventional languages (e.g. C, Fortran or Pascal) to develop expert systems will eliminate integration problems, these languages are poorly adapted to symbolic expressions. AI languages (e.g. Lisp or Prolog) are very capable in symbolic reasoning. However, they suffer from the lack of programming tools and established standards to communicate with conventional computer programs. For example, the delivery of LISP-based expert system applications by NASA has proven difficult as a result of: (1) low availability of LISP on conventional computers; (2) high cost of state-of-the-art LISP tools and hardware; and (3) poor integration of LISP with other languages, making embedded application difficult (Culbert *et al.*, 1988). Recently, numerous commercial expert system development tools (shells) have become available. These shells provide natural language knowledge representation and a built-in inference engine to aid the development of an expert system [see Badiru (1988) for a collection of commercial expert system shells].

An expert system shell is a software package that provides a high level, user friendly programming language, text editing, file manipulation and interpreter, or compiler, to run the expert system. Thus, an expert system shell is a development environment containing all the functions needed to write an expert system without having to use other software (Huber and Huber, 1988). Shells can be classified into several types: inductive shells, in which rules are constructed by the shell from case examples; rule-based shells, in which a knowledge base of rules is constructed by the developer from analysis of the problem domain; and high-end 'hybrid' shells, which contain multiple knowledge representation and inference methods (Miller *et al.*, 1988). The following checklist provides very useful hints for selecting the right shell for expert system development (Dubas, 1990).

1. **Ease of use.** The expression of knowledge should be as close as possible to a natural language. The function of the inference engine should be easy to understand.
2. **Documentation and support.** Because shells are not standardized software (Mackerle, 1989), a clear and complete manual, and experienced individuals in the vicinity become very helpful in solving programming problems.
3. **Maturity and reliability of the software product.** Expert systems require a great deal of work before they are complete and should remain in use over many years. Therefore, the basic software tool (shell) should remain available for a long time and new releases should replace old ones without compatibility problems.
4. **A good user interface.** In order to produce a program users can cope with easily, many features, like menus or forms for the input, help texts on request, reports in any prescribed form and so on, must be provided.
5. **Well developed and working interfaces to other programs or languages.** Expert systems often need to access a data base, a spreadsheet program or a conventional language, so communication with them must be effective.
6. **Several methods of representing knowledge and inferencing.** In this way, knowledge can be expressed and structured in a form which is near to that used by human experts.

Although expert system tools greatly decrease the effort involved in developing an expert system, tool's users must still undergo some training and must have programming skills (Culbert, 1986). Transfer of human knowledge into an expert system and knowledge acquisition are very time consuming and labor intensive tasks. The use of shells might allow domain experts to develop expert systems based on their own expertise and eventually eliminate the knowledge acquisition process which has been the big bottleneck in the construction of expert systems.

1.2.4 Advantages and limitations

Expert system applications have been quite diverse in most respects because of many good features, such as modularity, uniformity, naturalness (Barr and Feigenbaum, 1981), availability, reliability and cost reduction (Giarratano and Riley, 1989).

1. **Modularity.** Individual knowledge in the expert systems' knowledge base can be added, deleted or changed independently. Although changing rules may change the behavior of the system, it has no direct effect on other rules.
2. **Uniformity.** Since all information must be encoded within the rigid structure of knowledge presentation scheme, the information can be easily understood by another person.

3. **Naturalness.** Production rules are most frequently used by human experts to explain how they do their jobs. Therefore, production rules encode statements about what to do in predetermined situations, naturally. In addition, the inheritance property of frames, which applies attributes of general objects to more specific objects, is a natural reasoning mechanism of humans.
4. **Availability.** Expert systems are available on any suitable computer hardware. The expertise becomes permanent. The system is steady, unemotional and provides complete response at all times.
5. **Reliability.** Due to the explicit organization of knowledge, expert systems provide the capability to obtain consistent and reliable results over time. DEC's XCON/R1 currently configures 98% of company hardware orders correctly. Prior to implementation of XCON/R1, human experts were configuring orders correctly only 65% of the time (Byrd and Hauser, 1991).
6. **Cost reduction.** The cost of providing expertise per user is greatly lowered.

Although expert systems have been shown to be a very useful tool for complex problems, several problems exist and need to be attended to.

1. **Expectations.** Expert systems conjure up a wide variety of expectations about their performance and role. An expert system can sometimes function as an assistant to a human who is still responsible for most of the problem solving. An expert system can be a colleague to humans and may be programmed as a real expert. Therefore, the expectation to replace all human experts is unrealistic. For example, expert systems do not have the same sense humans have about knowing when they might be wrong (Biegel and Gupta, 1990; Kaewert and Frost, 1990). Typical expert systems cannot generalize their knowledge by using analogy to reason about new situations the way people do. It is easier to program the empirical/heuristic knowledge than to have the system understand the underlying relationship of causes and effects. For example, it is easier to have a rule:

IF temperature is below 0°C
THEN water freezes

than to make the system understand why water will freeze (Boley, 1990).

2. **Limited scope of problem.** The scope of problem has to be defined narrowly enough to make the knowledge base construction task feasible, yet ensure that we do not define away the difficult part of problem.
3. **Inexact reasoning.** Substantial difficulties surround the problem of inexact reasoning. One fundamental problem is to specify what is meant when a rule says 'probably', 'maybe' or 'very likely'. A second difficulty lies in defining the 'arithmetic' to use in combining them. For example, what does 'probably + maybe' add up to (Grimson and Patil, 1987)?

4. **Inefficiency.** The strong modularity and uniformity of the knowledge representation result in high overhead in their use in problem solving. For example, since production systems perform every action by means of match-action cycle and convey all information by means of the context data structure, it is difficult to make them efficiently responsive to predetermined sequences of situations or to take larger steps in their reasoning when the situation demands it (Barr and Feigenbaum, 1981).
5. **Opacity.** It is hard to follow the flow of control in problem solving algorithms which are less apparent than they would be if they were expressed in a programming language. The flow of execution is not sequential in an expert system, so that it is impossible to read the code line-by-line and understand how the system operates. In other words, although situation-action knowledge can be expressed naturally in production systems, algorithmic knowledge is not expressed naturally (Barr and Feigenbaum, 1981). It also makes the development of adequate test procedure for expert systems very difficult. The lack of knowledge and techniques on how to test an expert system cause the failure of many expert systems to be integrated into operational environment (Biegel and Gupta, 1990).

1.3 DEVELOPMENT PROCESS OF EXPERT SYSTEMS

Some of the practical concerns of developing knowledge-based expert systems are: (1) when should the knowledge based expert systems be used, and (2) how to identify potential application areas and plan for the development of knowledge-based expert systems considering staffing, funding, time and hardware requirements (Wentworth, 1990). Expert systems development can be divided into four stages: problem selection, prototype development, migration from prototype to expert system and deployment of the expert system (Biegel and Gupta, 1990; Kaewert and Frost, 1990).

1.3.1 Problem selection

Selected problems for potential expert system development should be carefully evaluated from business, and users' and technical points of view. Unfortunately, the technical characteristics normally receive most of the attention; the business and end users' perspectives are often overlooked. Lippolt (1990) reported that most expert systems in The Netherlands were rated positive on technical grounds, but the organizational and user evaluations were considerably less positive. In order to gain resources and support for the development and implementation of expert systems, the potential system must produce measurable, recognizable benefits to the business organization. The problem should have direct impact on the success of the company and also meet the management's long-range plans for the organization. Important issues that need to be addressed are such as does the task have direct payback, is scarce expertise captured, is expertise preserved, is expertise decentralized and will the need continue for years.

Because the implementation of an expert system will change the work environment of the end users, the characteristics of the user organization should be evaluated (e.g. end users' commitment to deploying the solution, users' understanding of the need for expert system to solve the problem, and users' understanding of the limitations and capabilities of the expert systems). Changes to the work environment should be carefully planned and implemented as part of the overall project.

From the technical perspective, the following issues should be addressed: can a subset of the task be prototyped; how user requirements can be clearly defined; has the problem been taught to novice; how the problem should be solved with the current state of the art technology; is 100% correctness required; does the problem require primarily symbolic, not numeric, reasoning; is the problem well structured; and have the traditional techniques failed in the past?

1.3.2 Prototype development

A prototype is a model of the planned expert system, built to address a subset of the problem. Divide and conquer is a very useful strategy for rapid prototyping. Prototyping is used to demonstrate a subset of the system's capability to end users and management, as well as to detect flaws in the strategy planned for the expert system. Test results from the initial implementation will reveal more about the problem area (Badiru, 1988; Kaewert and Frost, 1990).

1.3.3 Migration from prototype to expert system

The project team should plan ahead to target the expert system for the appropriate delivery environment, such as hardware requirements and user supports. Both users and developers should participate in determining the standard for correct performance and success. The system should be refined by moving from artificial to real data, enhancing the knowledge base, testing the application's results for correctness and documentation. The project team should also observe any user resistance and management changes.

1.3.4 Deployment of the expert system

An implementation plan should be created to achieve a smooth and orderly transition of the expert system from development to daily operation within the user organization. During this stage, the development team needs to follow subjects closely: (1) reconfirm business commitment to use the system, (2) track the performance of the system, (3) address jobs affected by the system, (4) change reward systems necessary, (5) provide user training, (6) test within the user environment, and (7) refine and complete user documentation.

1.4 EXPERT SYSTEMS IN PRODUCTION AND MANUFACTURING

Expert systems have potential use in a wide range of applications (Badiru, 1988) and are being pursued by nearly every major industry to increase automation and productivity (Byrd and Hauser, 1991; Culbert, 1986).

1.4.1 Scheduling

Scheduling is a decision process to allocate resources (generally equipment and personnel) with limited availability to all the jobs to be accomplished in the factory. The need for a job to be rescheduled emerges when the availability of workers, tools and machines changes unpredictably over time. Suer and Dagli (1988) demonstrated a knowledge-based system which generates relevant scheduling papers based on the information pertaining to the user's input. Sarin and Salgame (1990) developed an interactive, real-time, knowledge-based approach for dynamic scheduling. The blackboard concept has been utilized to organize and maintain the dynamic data base.

Arinze and Partovi (1990) proposed to use three knowledge-based systems to assist line balancing tasks for groups of similar products. These systems include one for generating precedence networks, one for selecting a procedural job allocation and one for implementing solutions generated by the models. Among all three systems, the development of a precedence network is of primary importance to increase the overall effectiveness of line balancing, because it requires a great deal of time and professional expertise. These systems will reduce the time and skill requirements substantially and increase the efficiency of the line balancing process.

In a flow shop, the task of sequencing is to control *N* jobs on *M* machines. Each job can have a maximum of *M* operations and every job has the same order of machining process. Chengalvarayan and Parker (1990) integrated linear programming, heuristics and expert system approaches for job sequencing in a flow shop. Results show that a flexible and fast solution emerges.

The production activity control (PAC) for assembly systems is considered as short-term planning and control which should be operated as close to real time as possible. Therefore, the capability of analyzing the manufacturing system and generating a schedule promptly is very important. Copas and Browne (1990) combine heuristic rules with the expert system technique to develop an expert system for manufacturing at PAC level. Knowledge is gained from the application of just-in-time, optimized production technology and manufacturing resource planning (MRP II) approaches to production management.

Intelligent scheduler (IS) is a multiple criteria knowledge-based scheduler for batch manufacturing systems (job shops). A heuristic algorithm is embedded within two knowledge bases, one for job scheduling and the other for selecting a suitable schedule based on the user provided criteria. Many important factors have been taken into account, such as multiple machines, multiple fixtures, multiple tools, alternate processing routes, machine set-up time, machine process time, due date job arrival time, initial shop loading and hot jobs (Jiang, 1991).

1.4.2 Engineering design

Conceptual design is part of the engineering design process which converts design requirements into an acceptable design solution. Nowadays, engineering design problems have become more complex than ever before. The increasing complexity has made synthesis a very difficult process. It is during this stage of the design process that the creativity and experience of an engineer are most needed (Hartley *et al.*, 1986).

Maher and Longinos (1987) describe the development of an expert system shell for the preliminary phase (synthesis) of engineering design in a structured and organized fashion. The proposed shell adopts principles of the morphological approach to design, incorporating heuristics in the form of design constraints.

Inappropriate component fixturing requirements will significantly reduce the capability of a flexible manufacturing system to meet the demands for small to medium size batch production. Darvishi and Gill (1990) use a knowledge-based system to demonstrate an optimum solution for fixture design problem in a real, although simple, prismatic engineering component manufacturing environment.

Many factors need to be considered for a good design of a precision forging die, such as allowance, draught angle, radius of fillet, parting surface from the forging drawing and selection of forging machines. Unfortunately, no analytical methods are available for this process. The engineer's experience has been the only resource for many years. Wang and Chen (1990) proposed a structured method of knowledge extraction in manufacturing (SMKEM) for building an expert system for precision forging dies. Pre-characters are extracted from the engineering drawing of the forging component and further divided into two sets: primary characters and secondary characters. The development of the object is mainly determined by primary characters. Knowledge of three forging shapes resides in the system: rotation shape, plate shape and vane shape.

Traditional geometric-information-only CAD systems simply cannot meet the future demands of CAD systems for two reasons: (1) geometric form is not the only type of information which has been considered during engineering design, and (2) logic and hierarchical structures of the concept representing the design object are the prevailing features at the upper level of design (especially at the conceptual level). A production rules systems has been proposed to assist engineers in the design process of component synthesis. The intelligent CAD system is developed using an object oriented programming paradigm because it simulates the designer's way of thinking during conceptual design (Kusiak *et al.*, 1991).

Jacobs *et al.* (1991) developed an expert system to generate numerically controlled machine programs from engineering drawings for two-dimensional punched objects. This prototype rule-based system applies manufacturing rules to accomplish two objectives: (1) object recognition and (2) alternative processes selection. The approach used here is different from most of numerically controlled (NC) code generating programs in the following ways: (1) it requires no operator intervention, (2) it uses a standard CAD exchange file as input (not the detail part code or a special description file) and (3) it uses a rule-based language.

1.4.3 Plant layout

Facilities layout planning is a decision process which requires human experience and subjective judgments to choose a suitable facility layout from a mix of specialized models. Basically, the layout problems involve the arrangement of machines on a factory floor so that the total time required to transfer material between each pair of machines is minimized. Traditionally, facility layout is treated as a single criterion problem, and is solved via a linear and quadratic assignment model to minimize the total material handling cost. However, the facility layout should be considered as a multi-objective problem: (1) minimizing material handling cost, (2) maximizing safety level, (3) maximizing the level of satisfaction of special requirements subject to certain hard and soft constraints, and (4) minimizing noise level hazard (Kumara *et al.*, 1987).

Malakoot and Tsurushima (1989) developed a multiple criteria knowledge-based facility layout whose knowledge base consists of two parts: (1) construction of a layout based on a set of rules and restrictions, and (2) improvement of the layout based on interaction with the decision maker. The expert system functions in the following way: for construction puposes in (1), rule priorities and adjacency of departments are used; for improvement purposes in (2), material handling cost, flexibility and materials handling time for paired comparison of generated layout are used.

Arinze *et al.* (1989) proposed a knowledge-based decision support system for the selection of an appropriate model for facility layout problems. Heuristics knowledge is elicited from experienced layout planners. Heragu and Kusiak (1990) developed an expert system for machine layout problems in an automated manufacturing system. The system combines the optimization and expert system approaches, and takes both quantitative as well as qualitative factors into account (e.g. width of material handling carrier path, clearance between machines, etc.) Under specific combinations of manufacturing and materials handling systems, Abdou and Dutta (1990) incorporated six factors into a knowledge-based system. These factors are product variety and quantity, degree of flexibility, level of automation, material handling system, work in process, and environmental considerations. The program interfaces with algorithms to optimize the selection of materials handling equipment and generates appropriate layouts.

1.4.4 Flexible manufacturing systems (FMSs)

FMS is a manufacturing system for processing a wide variety of different parts with low to medium demand volume. It consists of a group of NC machines and workstations connected by an automated materials handling system under the control of one or more computers (Byrd and Hauser, 1991). The major advantage of FMSs is to increase flexibility and productivity of discrete part manufacturing (Sabuncuoglu and Hommertzheim, 1989).

FMSs are usually classified into several classes based on the number of NC machines and their arrangement in association with a materials handling system. Two significant costs of FMSs are capital investment and design. Mellichamp *et al.* (1990) developed an expert system to evaluate the capital investment and design of FMSs. By analyzing the output from an FMS simulation model, the expert system determines whether operational and financial objectives are met. It then identifies design deficiencies or opportunities for improvement. Finally, the system processes designs to overcome deficiencies or exploit improvement opportunities.

Sabuncuoglu and Hommertzheim (1989) have reviewed recent developments in simulation and discussed of the role of expert systems simulation in FMS.

1.4.5 Quality control

KIMS (knowledge-based image management system) is a generic prototype expert system for image analysis and interpretation tasks. It oversees the entire process of image processing, segmentation, feature extraction and knowledge representation along with an expandable capability for image understanding tasks. It shows the usefulness of photo-to-image inspection in a manufacturing environment (Ntuen *et al.*, 1989).

NEC Corporation has an expert system for diagnosing the malfunctions of a chip mounting machine used to mount chips on integrated circuit boards. Because of the operator's inability to efficiently diagnose many malfunctions of the chip mounting machine, experts were frequently forced to leave their tasks to help operators troubleshoot malfunctions. The associated cost of production delays and the disruption of the human expert's regular tasks triggered this development. Knowledge was elicited from the machine designer. Using a hierarchical classification of malfunction symptoms and causes, 15 flow diagrams and a matrix were obtained. These explained the designer's troubleshooting procedures for particular malfunction symptoms. In 1988–1989, the expert system successfully diagnosed 92% of the chip mounting machine's malfunctions (Naruo *et al.*, 1990).

To achieve computer-aided inspection, Pham *et al.* (1991) developed a knowledge-based system which provides a general link for CAD and a coordinate measuring machine (CMM). This system is a customized generic preprocessor for transferring CAD files to neutral data file (NDF) format. Function definition and program synthesis are two major modules of the system. While the expert system is implemented in the function definition module, a uniform storage format is adopted for the program synthesis module.

Kuo and Mital (1992) give a more complete discussion of quality control expert systems.

1.4.6 Process planning

Process planning is a systematic procedure which translates design information from an engineering drawing into a feasible technological sequence of manufacturing instructions. In other words, it is the process of converting a piece part from its initial form to a final form. Therefore, knowledge-based process planning systems usually consist of three modules: a part design input scheme, a knowledge base and a control scheme (Gupta, 1990).

Control of an automated manufacturing facility is usually divided into two hierarchical levels: the cell level and the workstation level. Assigning jobs to machines is the primary goal of process planning at the cell level. The process plans are affected by global factors, such as production goals and resource availability. It is critical that speeds of various processes are coordinated so as to avoid collisions or deadlocks or to maximize some criterion at the workstation level. Two knowledge-based controls were developed by Ben-Arieh *et al.* (1989), one for the cell level and one for the workstation level. The variant approach for the workstation level uses a coding scheme and a look-up table to create the process plan, while the generative approach provides the solution using global knowledge of the domain with no specific tables. The evaluation was done by simulation which changes variables at different levels and behavior of the system was monitored.

The electronic manufacturing has become a very dynamic industry due to frequent and rapid technology advances. The effectiveness of execution and generation of process plans totally depends on the consistency of the process planner. Irizarry-Gaskins and Chang (1990) present a knowledge-based process planning system for electronic assembly environment. The system uses network representation of plans that allows for flexibility and modularity in the planning system. A hybrid architecture consisting of production rules, procedural algorithms and a hierarchical data base has been implemented.

To produce a machining plan, a subarea of process planning, a large amount of rather subjective and specialized knowledge is required. As technology continuously changes, the regeneration of machining plans becomes increasely complex and time consuming. Yeo *et al.* (1991) developed a frame-based expert system for the generation of machining tasks for turned parts in a machining-based manufacturing environment. Knowledge is extracted from an experienced process planner and NC programmer.

Gupta (1990) further discusses knowledge-based process planning systems.

1.4.7 Production management

An expert system for manufacturing system simulation (XMAS) allows the user to model a proposed or existing flexible manufacturing system in a few hours, without having a working knowledge of a simulation language. XMAS is an integrated system which includes the simulation software MAST and the knowledge base of FMS modeling experts (Wichmann, 1986).

Parlar (1989) presents a knowledge-based expert system (EXPIM) which can identify and recommend up to 30 production inventory models ranging from the simple extensions of the classical economic ordering quantity (EOQ) model to complex multi-level control systems.

The use of computational (OR) models to support decision making in production systems is quite involved. For example, the selection of appropriate models, the formulation of inputs, the skill for solving the model and the interpretation of model results using OR models is quite complex. Knowledge required for a successful application includes familiarity with the available models for solving the problem, their use and limitations, and the data input requirements. For example, if a regression model is needed, it is important to know what data is required and in what format. Knowledge for efficient model execution includes problem solving skills, such as manual computational procedures and ability to use statistical software. The interpretation of model results requires both experience and knowledge of derived statistics. Burd and Kassicieh (1990) demonstrate the use of expert systems in selecting, formulating and solving mathematical models for production systems.

Linn and Wysk (1990) developed an expert system for an automated storage and retrieval system (AS/RS) control decision process using a hierarchical control structure which partitions control procedures into strategic, tactical and process control levels. In addition, a multipass simulation technique is employed to adopt control policies to system changes. The system demonstrated that the control flexibility is capable of including fluctuations in demand and maintaining quality performance.

For the rectangular trim loss or cutting-stock problem, the number of possible configurations increase exponentially with the number of different pieces used in the problem; it is very time consuming to find the cutting pattern with the least amount of stock sheet waste. Dietrich and Yakowitz (1991) present a rule-based system to find the most effective heuristic rules for rectangular trim loss problem in terms of stock sheet utilization.

1.5 CONCLUDING REMARKS

Technical features of information system are not the only factors which are relevant to the adoption or rejection of an expert system. The successful implementation of information systems is largely affected by behavioral factors such as individuals' attitude and the degree of management support as well as organizational features such as the size and complexity of the department (Robey, *et al.*, 1978). Therefore, the success of an expert system should be evaluated in terms of these three aspects: technical, business and end users. An expert system must be targeted at a real business need. Business people will keep the project focused on the constantly changing needs of the business organization to stay ahead in a fast-paced, competitive environment. End users will remind both the management and technical experts of the realities of the job, and what is both necessary and possible in everyday practice. In evaluating the suitability of expert systems technology for a particular problem, it is important to consider the expectations of the group in which the system will reside. Very often the expectations of managers and users will make or break the system before it is begun.

An expert system shell is an ideal development tool for prompt prototypes of expert systems (Rajaram, 1987). The shell should have a knowledge representation as close to natural language as possible. Therefore, the knowledge base can be easily examined by experts, end users, managers and knowledge engineers for correctness. This might eliminate the painstaking knowledge acquisition process and gain confidence among all parties. As long as the efficiency of the shell remains tolerable, traditional languages should be excluded from final expert system implementation tools. All expert system tools have strengths and weaknesses, and no single tool is dominant for a wide spectrum of applications or over a wide range of functionality (Mettrey, 1991). The expert system itself should be carefully evaluated in terms of: (1) speed performance, (2) memory occupation, (3) reliability of the results, (4) consistency of the knowledge base, (5) user friendliness and (6) updating capabilities (Pau, 1987).

An expert system is usually a small component of a large computer application, and is integrated with conventional programs so that it can use existing applications and data. Most expert system solutions consist of both procedural and non-procedural tasks, programs or procedures written in conventional programming languages should be available to be called by the expert system. A study has shown that the lack of integration of expert systems in the traditional data processing environment is a major criterion for project failure (Lippolt, 1990).

Finally, the system should be transferred to an information system organization as a daily operational program. Continuing support from knowledge engineers and experts should be focused on ongoing enchancement to make the expert system smarter.

REFERENCES

Abdou, G. and Dutta, S.P. (1990) An integrated approach to facilities layout using expert systems. *International Journal of Production Research*, **28**(4), 685–708.

Arinze, B. and Partivo, F. (1990) A knowledge based method for designing precedence networks and performing job allocation in line balancing. *Computers and Industrial Engineering*, **18**(3), 351–64.

Arinze, B., Banerjee, S. and Sylla, C. (1989) A methodology for knowledge-based decision support for facilities layout planning. *Computers and Industrial Engineering*, **17**(1–4), 31–6.

Badiru, A.B. (1988) Expert systems and industrial engineers: a practical guide to a successful partnership. *Computers and Industrial Engineering*, **14**(1), 1–13.

Barr, A. and Feigenbaum, E.A. (1981) *The Handbook of Artificial Intelligence – 1*, William Kaufmann, San Mateo, CA.

Ben-Arieh, D., Moodie, C.L. and Chu, C.C. (1989). Knowledge-based scheduling under unpredicted conditions: two approaches. *International Journal of Production Research*, **27**(5), 869–82.

Biegel, J. and Gupta, U.G. (1990) Expert systems in manufacturing: promises and perils. *Computers and Industrial Engineering*, **19**(1–4), 127–30.

Boley, H. (1990) Expert system shells: very-high-level languages for artificial Intelligence. *Expert Systems*, **7**(1), 2–8.

Braun, R.J. (1990) Turning computer into experts. *Quality Progress*, **23**(2), 71–5.

Buchanan, B.G. and Feigenbaum, E.A. (1978) DENDRAL and Meta-DENDRAL: their applications dimension. *Artificial Intelligence*, **11**, 5–24.

Burd, S.D. and Kassicieh, S.K. (1990) The use of AI methodologies in production system modeling. *Computers and Industrial Engineering*, **18**(4), 559–70.

Byrd, T.A. and Hauser, R.D. (1991) Expert systems in production and operations management: research directions in assessing overall impact. *International Journal of Production Research*, **29**(12), 2471–82.

Cercone, N. (1987) Knowledge representation: an overview. *Indian Journal of Technology*, **25**(12), 521–43.

Chang, T.C., Wysk, R.A. and Wang, H.P. (1991) *Computer-Aided Manufacturing*, Prentice-Hall, New York.

Chengalvarayan, G. and Parker, S.C. (1990) A knowledge based system for flow shop scheduling. *Computer and Industrial Engineering*, **19**(1–4), 17–21.

Copas, C. and Browne, J. (1990) A rules-based scheduling system for flow type assembly. *International Journal of Production Research*, **28**(5), 981–1005.

Cristie, R.D. (1990) *Impact of Artificial Intelligence on Plant and System Operations.* Proceedings of the 33rd Power Instrumentation Symposium. Proc. Venue. Publisher, pp. 193–7.

Culbert, C. (1986) *State of The Art: Expert System Tools*, Southcon Conference Record.

Culbert, C., Giarratano, J., Riley, G. and Savely, R.T. (1988) *Solution to the Expert System Delivery Problem.* Proceedings of the ISA/88 International Conference and Exhibit. Proc. venue. Publisher,

Darvishi, A.R. and Gill, K.F. (1990) Expert system rules for fixture design. *International Journal of Production Research*, **28**(10), 1901–20.

Davis, R., Buchanan, B.G. and Shortliffe, E.H. (1977) Production systems as a representation for knowledge-based consultation program. *Artificial Intelligence*, **8**, 15–45.

Dietrich, R.D. and Yakowitz, S.J. (1991) A rule-based approach to the trim loss problem. *International Journal of Production Research*, **29**(2), 401–15.

Dubas, M. (1990) Expert systems in industrial practice: advantages and drawbacks. *Expert Systems*, **7**(3), 150–6.

Duda, R., Gaschnig, J. and Hart, P. (1979) Model design in the PROSPECTOR consultant system for mineral exploration, in *Expert Systems in the Micro-Electronic Age*, (ed. D. Michie) Edinburgh University Press, Edinburgh, pp. 153–167.

Giarratano, J.C. and Riley, G. (1989) *Expert Systems: Principles and Programming*, PWS, Kent.

Grimson, E.L. and Patil, R.S. (1987) *AI in the 1980s and Beyond*, MIT Press, Cambridge, MA.

Gupta, T. (1990) An expert system approach in process planning: current development and its future. *Computers and Industrial Engineering*, **18**(1), 69–80.

Hartley, P., Sturgess, C.E.N. and Rowe, (1986) *Emerging Trends in Manufacturing*. Proceeding of the 12th All India Machine Tool Design and Research Conference. Proc. venue. Publisher.

Heragu, S.S. and Kusiak, A. (1990) Machine layout: an optimization and knowledge-based approach. *International Journal of Production Research*, **28**(4), 615–35.

Huber, L.E. and Huber, R.C. (1988) Artificial intelligence series, part 9: cracking researchers' monopoly on expert systems expertise. *Industrial Engineering*, **20**(1), 58–65.

Interrante, L.D. and Biegel, J.E. (1990) Design knowledge-based systems: matching representations with application requirements. *Computers and Industrial Engineering*, **19**(1–4), 92–6.

Irizarry-Gaskins, V.M. and Chang, T.C. (1990) A knowledge based approach for automatic process plan generation in an electronic assembly environment. *International Journal of Production Research*, **28**(9), 1673–93.

Jacobs, F.R., Mathieson, K., Muth, J.F. and Hancock, T.M. (1991) A rule-based system to generate NC programs from CAD exchange files. *Computers and Industrial Engineering*, **20**(2), 167–76.

Jiang, J.C. (1991) IS: an intelligent scheduler for batch manufacturing systems. *Computers and Industrial Engineering*, **21**(1–4), 319–23.

Kaewert, J.W. and Frost, J.M. (1990) *Developing Expert Systems for Manufacturing: A Case Study Approach*, McGraw-Hill, New York.

Kumara, S.R.T., Kashyap, R.L. and Moodie, C.L. (1987) Expert system for industrial facilities layout planning and analysis. *Computers and Industrial Engineering*, **12**(2), 143–52.

Kuo, T. and Mital, A. (1992) *The Development of Quality Control Expert Systems: Past, Present, and Future Trends*. Proceeding of the Second International FAIM Conference, 179–88. CRC Press, Boca Raton, FL.

Kusiak, A., Szczerbicki, E. and Vujosevic, R. (1991) Intelligent design synthesis: an object-orient approach. *International Journal of Production Research*, **29**(7), 1291–308.

Linn, R.J. and Wysk, R.A. (1990) An expert system framework for automated storage and retrieval system control. *Computers and Industrial Engineering*, **18**(1), 37–48.

Lippolt, B.J. (1990) Failures and successes of expert systems. *VTT Symposium*, **1**(116), 115–26.

Mackerle, J. (1989) Review of expert systems development tools. *Engineering Computations*, **6**(1), 2–17.

Maher, M.L. and Longinos, P. (1987) Development of an expert system shell for engineering design. *International Journal of Applied Engineering Education*, **3**(3), 279–86.

Malakoot, B. and Tsurushima, A. (1989) An expert system using priorities for solving multiple-criteria facility layout problems. *International Journal of Production Research* **27**(5), 793–808.

McDermott, J. and Bachant. J. (1984) R1 revisited: four years in the trenches. *AI Magazine*, **3**, 21–32.

Mellichamp, J.M., Kwon, O.J. and Wahab, A.F.A. (1990) FMS designer: an expert system for flexible manufacturing system design. *International Journal of Production Research*, **28**(11), 2013–24.

Mettrey, W. (1991) A comparative evaluation of expert system tools. *Computer*, **24**(2), 19–31.

Michie, D. (1982) *Introductory Readings in Expert Systems*, Gordon and Breach, London.

Miller, E.J. Wilson, K.D., and Lewis, C.R. (1988) Expert system shells: do they deliver what they promise? *Chemical Engineering Progress*, **84**(10), 37–44.

Minsky, M. (1975) A framework for representing knowledge, in *The Psychology of Computer Vision* (ed. P. Winston), McGraw-Hill, New York, pp. 211–77.

Naruo, N., Lehto, M., and Salvendy, G. (1990) Development of a knowledge-based decision support system for diagnosing malfunctions of advanced production equipment. *International Journal of Production Research*, **28**(12), 2259–76.

Newell, A. and Simon, H.A. (1972) *Human Problem Solving*, Prentice-Hall, New York.

Ntuen, C.A., Park, E.H. and Kim, J.H. (1989) KIMS – a knowledge-based computer vision system for production line inspection. *Computers and Industrial Engineering*, **16**(4), 491–508.

Parlar, M. (1989) EXPIM: a knowledge-based expert system for production/inventory modelling. *International Journal of Production Research*, **27**(1), 101–18.

Pau, L.F. (1987) *Prototyping, Validation and Maintenance of Knowledge-based Systems Software*. Proceedings of The Third Annual Expert Systems in Government Conference. Proc. venue. Publisher, pp. 248–53.

Pham, D.T., Martin, K.F. and Khoo, L.P. (1991) A knowledge-based preprocessor generator for coordinate measuring machines. *International Journal of Production Research*, **29**(4), 677–94.

Rajaram, N.S. (1987) Tools and Technologies for Expert Systems. A Human Factors Perspective. *NASA Contractor Rep. 172009*, 26.1–26.18.

Rowe, N.C. (1988) *Artificial Intelligence through Prolog*, Prentice-Hall, New York.

Sabuncuoglu, I. and Hommertzheim, D.L. (1989) Expert simulation systems – recent developments and applications in flexible manufacturing systems. *Computers and Industrial Engineering*, **16**(4), 575–85.

Sarin, S.C. and Salgame, R.R. (1990) Development of a knowledge-based system for dynamic scheduling. *International Journal of Production Research*, **28**(8), 1499–512.

Sür, G. and Dagli, C. (1988) Knowledge-based system for the selection of production scheduling papers. *Proceedings of Computers in Engineering*, 231–6.

Wang, S.C. and Chen, J. (1990) An expert system for designing precision forging dies. *International Journal of Production Research*, **28**(5), 943–51.

Wentworth, J.A. (1990) Developing knowledge-based expert systems. *VTT Symposium*, **1**(116), 36–77.

Wichmann, K.E. (1986) *Intelligent Simulation Environment for the Design and Operation of FMS*. Proceedings of the 2nd International Conference on Simulation in Manufacturing. Proc. venue. Publisher. pp. 1–11.

Winston, P.H. (1984) *Artificial Intelligence*, Addison-Wesley, Reading, MA.

Wolfgram, D.D., Dear, T.J., and Galbraith, C.S. (1987) *Expert Systems for Technical Professional*, Wiley, New York.

Yeo, S.H., Wong, Y.S. and Rahman, M. (1991) Integrated knowledge-based machining system for rotational parts. *International Journal of Production Research*, **29**(7), 1325–37.

CHAPTER 2

Operations research/artificial intelligence rules for the optimal design of manufacturing systems: machine and traffic allocation

Mario Van Vliet

2.1 INTRODUCTION

To gain a competitive edge in world markets, many automobile and electronic manufacturers have diversified their products in order to gain/maintain their market shares. In order to produce several types of products simultaneously in an effective way, many manufacturers have installed flexible manufacturing systems that consist of numerically controlled machines. These machines are flexible in the sense that negligible set-up time is incurred when a machine switches from processing one product to another. One of the most important reasons for a manufacturer to introduce a flexible manufacturing system is the possibility to combine optimal logistic performance of the manufacturing system [e.g. work-in-process (WIP) levels or product leadtimes] with a high flexibility of production, which is today's prerequisite to survive in a competitive consumer market. To ensure this it becomes more and more important to properly design a flexible manufacturing system such that the required logistic performance is achieved within the given investment constraints. Logistic performance measures that have received much attention lately are, for example, the average leadtimes of the products, the average WIP levels and the average throughputs. What becomes important then is to investigate the quantitative aspects of these logistic performance measures related with a given design. Expert systems using operations research/artificial intelligence (OR/AI) rules can play a very important role in the optimal tuning of these quantitative aspects.

In this paper we discuss OR/AI rules for two flexible manufacturing system design problems: *machine allocation* and *traffic allocation*. In machine allocation one tries to find an optimal allocation of machines over workstations such that a certain logistic performance criterion is optimal. In traffic allocation the optimal allocation of incoming traffic of products over workstations is subject of study. For the above problems we develop OR/AI rules that can be used within expert systems which help the practitioner to design a flexible manufacturing system in a sound way.

The organization of this paper is as follows. In Section 2.2 we discuss the general class of models in which the OR/AI rules fall that we use to solve the design problems. Section 2.3 deals with the two design problems: Section 2.3.2 with machine allocation and Section 2.3.3 with traffic allocation. For both problems, we discuss the problems addressed, the rules used and illustrate the quality of the rules by means of numerical results for several real life manufacturing systems. Section 2.4 gives some concluding remarks.

2.2 OR/AI MODELS FOR MANUFACTURING SYSTEM DESIGN

2.2.1 Modeling framework

The models we propose for optimizing the design of a manufacturing system consist of two phases. The first phase is a *performance analysis phase*. In this phase steady-state queueing analysis is used to analyze the performance of a given manufacturing system design in terms of, for example, product leadtimes, throughput and WIP. This first phase supplies us with analytical expressions that relate a certain system design to performance indicators like average queueing time and queue lengths. For example, by using steady-state queueing analysis, we can evaluate the relation between the number of machines at a given workstation and the average queueing time at that workstation. The expressions from the performance analysis phase are, together with their mathematical characteristics, e.g. convexity, used in the second phase, the so-called *optimization phase*. In this phase we use the expressions from the performance analysis phase in OR/AI optimization rules to arrive at a (near-) optimal design of the manufacturing system. This paper will focus upon the optimization phase. The techniques used in the performance analysis phase are briefly discussed; however, in describing the OR/AI models, the performance analysis phase is referred to as a 'black box' which is used in the OR/AI rules of the optimization phase.

2.2.2 Use of queueing models: the performance analysis phase

Queueing is a phenomenon that frequently occurs in manufacturing. Products form queues when they have to wait in order to be processed by a workstation. The resulting queueing time that a product encounters during production can form an extensive part of the total time it takes to convert the raw materials into a finished product (the so-called product leadtime). Recent research has shown that for some manufacturing environments queueing time can consume as much as 80% of the product leadtime! Hence, it is safe to say that reducing the queueing time of a product will significantly improve the product leadtime and hence the logistic performance of a manufacturing system. Therefore, queueing analysis of manufacturing systems is an important tool for manufacturing system design.

In general, queueing occurs due to the fact that both the inter-arrival time of products into the system and the service times at workstations are not fixed but fluctuate. A common way to model these fluctuating inter-arrival and service times is by means of a stochastic variable. This also implies that performance measures, like average queueing times or queue lengths, are stochastic variables. Therefore, if we analyze a manufacturing system by means of a queueing model, we assume the system to behave as a stochastic process. Products arrive into the manufacturing system according to an inter-arrival time distribution and are processed by machines (servers) according to a service time distribution. When several identical machines operate in parallel with one common buffer we shall refer to the whole processing unit as a workstation. Since products normally undergo several processing stages during manufacturing, a product may encounter several queues during the whole manufacturing process. To analyze this, we model the manufacturing system as a system of queues interconnected by a network through which the products follow their own specific routes. The methodology to analyze this kind of network is called queueing network theory.

In the sequel the focus shall be on steady-state queueing analysis. Although queueing analysis can also be used to analyze the transient behavior of a manufacturing system, the long-term character of design issues makes steady state analysis a much more appropriate tool. The steady state performance indicators that shall be used in the sequel are logistic performance measures like average time in queue, WIP, throughput and utilization. The average time in queue, or queueing time, is the time a product spends on average in queues during production. This measure is, for example, used to calculate the average product leadtime. The WIP is the total value of all the products that are on average in the system. Note that this includes products that are either in queue or in process. The throughput of a manufacturing system is the average number of products that are produced per unit of time. Utilization of a workstation (or traffic intensity) is the average proportion of time a workstation is busy processing.

2.2.3 Use of OR/AI rules: the optimization phase

The problems we consider are optimization problems in the sense that we try to allocate either machines or incoming traffic over workstations such that the logistic performance of the system is optimal. In the optimization phase, the expressions from the performance analysis phase are used as relations between variables (like number of machines) and performance indicators (like average queue lengths). In the optimization phase first an optimization problem is formulated and then rules are developed to solve this optimization problem. In this paper we focus upon the following two aspects of these rules.

1. The quality of the rules in solving the optimization problems.
2. The applicability of the rules for use in practice and numerical results of the rules for some practical environments.

2.3 ALLOCATION PROBLEMS

2.3.1 General class of allocation problems

As in any optimization problem, we have to decide on what our decision variables shall be. In our context this means that we have to choose the 'knobs' to turn in order to improve the performance of a manufacturing system. In terms of manufacturing system design one often refers to these 'knobs' as system design parameters. In the problems we address we assume the product routings, throughput, location of workstations and technology (e.g. processing times) to be specified. The system design parameters we consider are related to the utilization of the workstations. The two design parameters that we treat in the sequel are the number of machines at workstations and the incoming traffic of products at workstations. The central performance criterion is the average WIP (or inventory) in the system. The number of machines and the traffic offered at a workstation affect the workstation's utilization and therefore influence the average queue length at the workstation. The optimization problems we consider concern the allocation of machines over the workstations (machine allocation) and the allocation of total incoming traffic over the workstations (traffic allocation). The objective is to allocate these design parameters in such a way that a WIP-related objective function is optimal.

2.3.2 Machine allocation

(a) *Introduction*

The problems we consider represent two important manufacturing system design issues. The first issue concerns problems where a fixed amount of machinery has to be distributed across the system. Here the aim is to make optimal (in terms of system performance) use of machinery. Problems like this often emerge in designing flexible manufacturing systems (FMSs). Imagine, for example, the situation in which a fixed number of homogeneous machines is available. Each machine can be made suitable for any kind of operation by assigning different tools to the machine. The problem now is to distribute the available machines among the system so as to optimize the performance of the system.

The second issue concerns the design of a manufacturing system such that the system satisfies a WIP performance level. The costs associated with the design must be as small as possible. One can think, for example, of a plant manager who wants the average WIP not to exceed a given level. The problem is then to design the system such that this target WIP is met, while keeping costs as low as possible.

We consider a production process that can be modeled as an open queueing network with different product types. In our context this means that the production process consists of workstations through which each product follows its own individual deterministic route. An example of such a system is a production process where printed circuit boards are made. A certain type of circuit board will only visit a workstation if that particular workstation prints a component that belongs to that type of circuit board. Furthermore, if different types of circuit boards use the same type of component, they will all visit the workstation that prints that component.

We aim to optimally allocate servers such that either

1. The WIP level of the system satisfies a certain target WIP level and the costs of the configuration are minimal [this problem shall be called the 'server allocation problem' (SA)].
2. The WIP level is minimized while keeping the total number of servers fixed [the 'server reallocation problem' (SR)].

Since WIP and leadtimes are linearly related through Little's law (Little, 1961), the problems we discuss and the set of rules to solve them also relate to the latter performance measure.

The manufacturing system we consider consists of J workstations. Each workstation j has m_j identical parallel servers with independent distributed service times with mean $1/\mu_j$. N product types are produced by the network. Products of type i arrive at the first workstation they visit according to a stochastic process with parameter λ^i and then follow a deterministic route through a subset of the set of workstations. Products waiting to be processed by a workstation are served according to a first-come-first-served (FCFS) discipline. Furthermore, it is assumed that the arrival processes and service processes are independent.

For further analysis we can treat the different product types as one aggregate product with an aggregate arrival rate λ_j at each workstation j. In the sequel we assume that the arrival and service rates are given, while the numbers of machines (servers) at workstations are decision variables. We can model such a manufacturing system as a queueing network and analyze its steady-state behavior. This steady state analysis is the performance analysis phase of the machine allocation models One of the functional relations that come out of this steady state analysis is the relation between the average queue length at workstation j (L_j) and the number of machines at the workstation (m_j). We shall denote this functional relation by $L_j(m_j)$. Important for our analysis are the mathematical characteristics of this relation. Obviously, $L_j(m_j)$ is a decreasing function in m_j. Furthermore, Boxma, Rinnooy Kan and Van Vliet (1990) have shown that for the above manufacturing system $L_j(m_j)$ is a convex function in m_j if the arrival processes are Poisson and the service times are exponentially distributed. Van Vliet and Rinnooy Kan (1991) show that one can also assume such a property for more general manufacturing systems. Therefore, in the sequel we assume $L_j(m_j)$ to be a convex, decreasing, function in m_j.

We shall measure the steady state performance of the network by the WIP. The WIP is the total value of all the products that are in the network. Without loss of generality, we make the assumption that the value of a product at workstation j, either in queue or in process, is independent of the type of product and equal to v_j. The formulation for WIP then becomes

$$\text{WIP}(m_1, \ldots, m_J) = \sum_{j=1}^{J} v_j L_j(m_j).$$

Furthermore, we assume that the allocation of m_j servers at workstation j generates investment costs of $F_j(m_j)$ with $F_j(m_j)$ a convex and increasing function in m_j. Such functions are of interest since they can model situations where capacity increments are achieved by using cheaper options initially. Bitran and Tirupati (1989) also mention that with regard to over-time wage structure, convex investment functions are a proper representation of these decision problems.

In order to prevent the system from becoming instable, we have to require that the traffic intensity at a workstation j $(=\lambda_j/m_j\mu_j)$ is less than one. This results in requiring that $m_j \geq m_j^L = \lfloor \lambda_j/\mu_j \rfloor + 1$, where $\lfloor \cdot \rfloor$ represents the integer rounddown operation.

For convenience we use the following notation.

$$m = (m_1, \ldots, m_J)$$

$$S = \{m \mid m_j \geq m_j^L\}$$

$$F(m) = \sum_{j=1}^{J} F_j(m_j)$$

$$L(m) = \sum_{j=1}^{J} v_j L_j(m_j)$$

$$\Delta F_j(m_j) = F_j(m_j) - F_j(m_j - 1) \quad (m_j > m_j^L)$$

$$\Delta L_j(m_j) = v_j(L_j(m_j - 1) - L_j(m_j)) \quad (m_j > m_j^L).$$

(b) The server reallocation problem

The server reallocation problem is the problem of allocating servers to workstations such that the WIP is minimized. The total number of servers that can be allocated is fixed and equal to M. The allocation of servers can be regarded as the distribution of these M servers over the queueing network. The mathematical formulation is as follows.

Problem (SR)

$$\min_{m} L(m)$$

$$\text{s.t.} \sum_{j=1}^{J} m_j = M$$

$$m_j \geqslant m_j^L, \quad m_j \text{ integer} \quad (j \in \{1, \ldots, J\})$$

The OR/AI rules to solve (SR) are very natural rules and can very easily be applied in practice. For each workstation we first calculate the minimal number of servers necessary ($= m_j^L$ for workstation j). Then for each workstation we calculate the decrease in average queue length at that workstation if a new server is added to that workstation. The workstation for which this decrease is maximal gets an additional server. This procedure is repeated until all the servers from the pool of M servers have been allocated. In mathematical terms these allocation rules can be stated by the following set of rules.

Set of rules SR

Step 1 Start with c^0 where $c_j^0 = m_j^L$.
Step 2 $k := 1$.
Step 3 Set $c^k := c^{k-1} + e_i$, where e_i is the ith unit vector and

$$i = \text{Arg} \max_{j \in \{1, \ldots, J\}} \Delta L_j(c_j^{k-1} + 1)$$

with Argmax denoting the index at which the maximum is achieved.
Step 4 If $k = M - \Sigma_{j=1}^{J} m_j^L$, stop; else $k := k+1$, go to step 3.

The rules as stated by the set of rules SR belong to the class of greedy rules in the sense that at each step we allocate servers to the workstation where locally the greatest decrease in queue length is achieved. Despite the fact that the rules as given by set of rules SR are simple and straightforward, one can prove that the rules provide us with an optimal solution of problem (SR). For this, we refer to Boxma, Rinnooy Kan and Van Vliet (1990).

(c) The server allocation problem

In this problem we want to allocate servers in such a way that the WIP is below a target WIP level W_T. The configuration we are looking for is a minimal cost configuration: the investment costs associated with allocating the servers have to be minimal. The mathematical formulation is as follows.

Problem (SA)

$$\min_m F(m)$$

$$\text{s.t. } L(m) \leqslant W_T$$

$$m_j \geqslant m_j^L, \quad m_j \text{ integer} \quad (j \in \{1, \dots, J\})$$

Problem (SA) belongs to the class of NP-hard problems. Therefore, our aim is to come up with rules that provide us with near-optimal solutions. For this problem we present two set of rules. The first set of rules is similar to the rules as presented for problem (SR). Again we start with the minimal number of servers required at each workstation ($=m_j^L$). We then look at whether this allocation is feasible in terms of the required WIP target. If not, we have to add servers to workstation in order to decrease the WIP. The condition under which to add servers to the workstations is different from the condition we used for problem (SR). In problem (SR) we only had to consider the decrease in queue length, since WIP was the only performance measure we looked at. Here, however, we have two performance measures to consider: WIP and investment cost. Looking at the problem formulation, adding a server to a workstation has to give a large decrease in queue length, in order to satisfy the WIP target as fast as possible. However, the investment costs of adding a server should also not be too large. Considering these observations, intuitively we would add a server to that workstation where the ratio of the increase in investment costs and the decrease in WIP is minimal. This is exactly what the first set of rules does. It adds a server to the workstation where the above mentioned ratio is minimal, and it keeps on repeating this procedure until the WIP target is achieved. Again, this set f rules belongs to the class of greedy rules in the sense that a server is added to the workstation where the best local improvement in terms of the ratio is achieved.

In mathematical terms this set of rules is stated as follows.

Set of rules SA1

Step 1 Start with c^0 where $c_j^0 = m_j^L$.

Step 2 $k := 1$.

Step 3 Set $c^k := c^{k-1} + e_i$, where e_i is the ith unit vector and

$$i = \operatorname*{Arg\,min}_{j \in \{1,\dots,J\}} \frac{\Delta F_j(c_j^{k-1}+1)}{\Delta L_j(c_j^{k-1}+1)}$$

Step 4 If $L(c^k) \leqslant W_T$, $c^{SA1} := c^k$, stop; else $k := k+1$, go to step 3.

The above set of rules does not provide us with an optimal solution to problem (SA). However, one can prove the following theorem.

Theorem 2.1 If $c^0, \ldots, c^p$ are the allocations generated by set of rules SA1 and c^* is an optimal allocation for (SA) then it holds that

$$F(c^{p-1}) < F(c^*) \leqslant F(c^p)$$

Proof: See Boxma, Rinnooy Kan and Van Vliet (1990).

Theorem 2.1 shows that the solution generated by the set of rules SA1 provides us with bounds to check whether the allocation found by set of rules SA1 is sufficiently close to the optimal allocation. The difference (if any) between the heuristic and the optimal solution is due to the greedy character of the set of rules. One can prove (see Boxma, Rinnooy Kan and Van Vliet, 1990) that the difference between the optimal solution and the heuristic solution is created by the last server added. The adding of the last server could result in a WIP which is substantially smaller than the target WIP. In the greedy manner we only look at the workstation where adding a server causes the minimal ratio. There could be another workstation, however, where adding a server causes a higher ratio, but this addition also satisfies the WIP target but with lower investment costs. This observation is used in the next set of rules. It starts with the allocation generated by the set of rules SA1 and then tries to improve upon this allocation. The set of rules SA2 generates J allocations. The first allocation is the allocation as given by SA1. To get the second allocation, the set of rules SA2 takes the number of servers at each workstation equal to those as given in the first allocation except for the workstation where the last server is added in the previous allocation. At this workstation (say j'), the number of servers is decreased by one and kept fixed at this level in all of the following steps. Note that this allocation is not feasible. Given this allocation, servers are added in the same greedy manner as in SA1. The procedure stops as soon as a feasible allocation is found. The resulting allocation is the second of the J allocations. The third allocation is found by the same procedure, with now j' equal to the workstation at which the last server is added to get the previous allocation. This procedure is repeated until J allocations have been generated. The allocation with the lowest objective function value is the heuristic allocation as given by set of rules SA2.

In mathematical terms the set of rules are stated as follows.

Set of rules SA2

Initialization Set $c^{H_1} := c^{SA1}$; $C := \{c^{H_1}\}$; $A := \{1, \ldots, J\}$; $k := 1$.

Step 1 Let j_k be the index of the workstation at which a server is added in the final iteration to obtain heuristic H_k. $A := A \backslash \{j_k\}$; $c^{H_{k+1}} = (c_1^{H_k}, \ldots, c_{j_k}^{H_k} - 1, \ldots, c_J^{H_k})$; $k := k+1$.

Step 2 Set $c^{H_k} := c^{H_k} + e_i$, where e_i is the ith unit vector and

$$i = \underset{j \in A}{\mathrm{Argmin}} \frac{\Delta F_j(c_j^{H_k}+1)}{\Delta L_j(c_j^{H_k}+1)}$$

Step 3 If $L(c^{H_k}) \leqslant W_T$
then $C := C \cup \{c^{H_k}\}$. If $k=J$ go to step 4, else go to step 1.
else go to step 2.

Step 4 Choose $c^{SA2} := \underset{c \in C}{\mathrm{Argmin}}\, F(c)$.

One can prove the following theorem.

Theorem 2.2 Let c^* be the optimal allocation for (SA), then it holds that

$$F(c^{SA2}) \leqslant 2F(c^*)$$

and this bound is tight.

Proof: See Frenk *et al.* (1991).

From Theorem 2.2 we see that we can give a quality performance guarantee on the allocation as generated by the set of rules SA2. The objective function value of the allocation generated by SA2 will in the worst case be twice as large as the optimal objective function value. In Frenk *et al.* (1991) a set of rules is presented that improves this bound to 3/2. However, in all the practical examples where we tested the set of rules on (Section 2.3.2.e) the results generated by the set of rules SA2 where identical to the allocations generated by this improved set of rules. This is due to the fact that the worst-case bounds as presented here are theoretical worst-case bounds, where in practice, the set of rules might show a far better performance. Furthermore, Frenk *et al.* (1991) show that for a large class of problems the set of rules SA2 will produce the same solution as the improved set of rules.

(*d*) *Comments on the allocation rules*

The class of rules as presented for the server reallocation and server allocation problems can very easily be applied in practice. In principle one only needs a mechanism to estimate the decrease in queue length if a server is added to a workstation. In this paper we use state-of-the-art queueing network analysis techniques to provide for this 'black box'. In practice however, even 'on the floor' estimates or experience from other manufacturing systems can be used to come up with an idea about the relation between number of servers and queue lengths. Note, however, that then the results on optimality or worst-case performance might not hold. The experience we have obtained in applying these rules tells us that even in such situations, these kind of optimization rules give the designer a very good rough-cut indication on the effects that a design (in terms of machine allocation) has on the logistic performance of the manufacturing system.

(*e*) *Numerical results*

The three sets of rules as described in Sections 2.3.2.b and 2.3.2.c have been implemented to obtain results for two manufacturing systems. The computer code is written in Pascal and it runs on an IBM-ps/2 model 30 with a 8087 math coprocessor.

The first system example comes from a paper by Bitran and Tirupati (1989). The manufacturing system produces custom specific semiconductor devices and consists of 13 workstations. The system can produce up to 10 different semiconductor devices (product types). The total number of operations which is performed by the system equals 93.

The second manufacturing system is a fictitious system. The system consists of 11 workstations and produces two product types. A total of 64 operations is performed by the system. Although this system also performs a large number of operations, the number of products is small. Therefore, it serves as a good complement to the first system to study the performance of the set of rules for two complex, but in essence different, manufacturing systems. For the relevant data on the two manufacturing systems, we refer to Van Vliet and Rinnooy Kan (1991).

We applied the set of rules SR for a wide variety of M values. Figures 2.1 and 2.2 show the trade-off curves we obtained for the two systems. For each

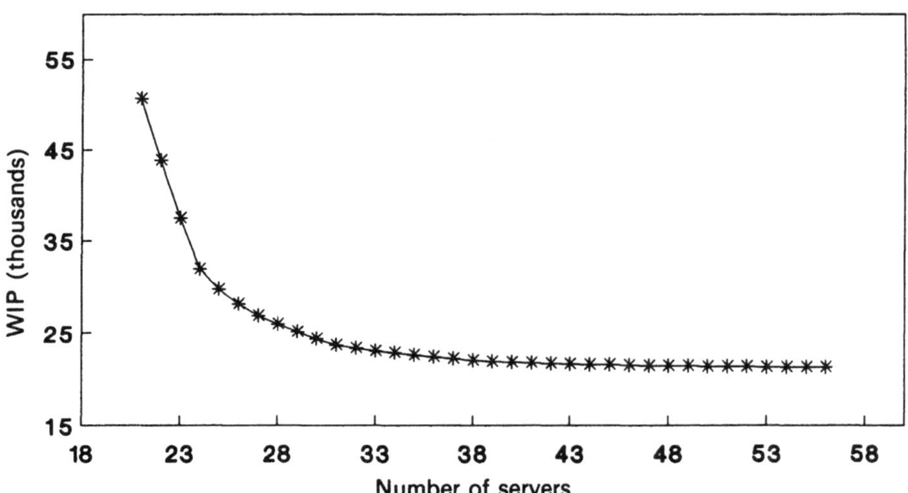

Fig. 2.1. System MA1 – trade-off curve server reallocation. *, Set of rules SR.

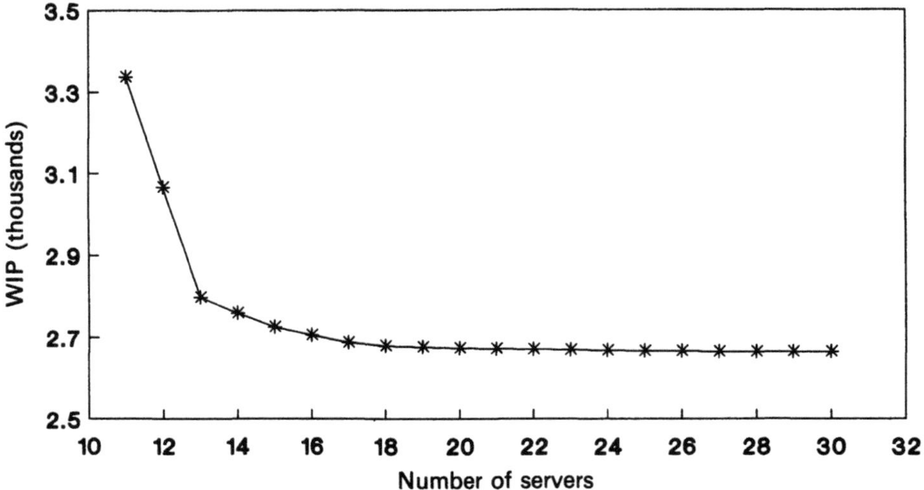

Fig. 2.2. System MA2 – trade-off curve server reallocation. ∗, Set of rules SR.

value of M (number of servers) we show the minimal WIP obtained. The left-most point on the curve illustrates the minimal number of servers one can use before getting instabilities (traffic intensities greater than or equal to one) at one or more workstations. The right-most M value is chosen such that beyond that point the WIP does not substantially decrease anymore and the costs keep increasing in a similar manner as before. The marks illustrate all the possible machine allocations that lie on the trade-off curve.

For the (SA) problem we present results obtained by the sets of rules SA1 and SA2. We first compare the performance of the two sets of rules in terms of deviations from the optimal allocation. The best set of rules is then used to obtain trade-off curves in the same nature as Figs 2.1 and 2.2.

To compare the performance of heuristics SA1 and SA2 we use a relative error indicator. For each heuristic SAi we use an upperbound (UBi) and the lowerbound (LBi) on the optimal solution value. The lowerbound is given by Theorem 1.1, while the objective function value of the heuristic solution as given by each set of rules is an obvious upperbound on the optimal solution value. The relative error is then calculated by

$$\text{Relative error of SAi} = \frac{\text{UBi} - \text{LBi}}{\frac{\text{UBi} + \text{LBi}}{2}}. \tag{2.1}$$

We applied the heuristics for a wide range of WIP values. Figures 2.3 and 2.4 show the results of heuristics SA1 and SA2 for system 1. From Figs 2.3 and 2.4 we see that the relative errors decrease substantially when the set of rules SA2 is used. Especially when the relative error for SA1 is large (up to

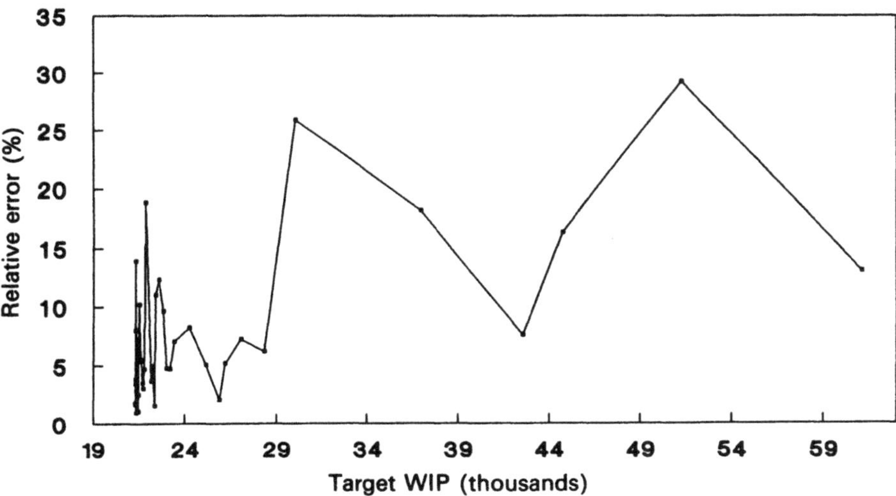

Fig. 2.3. System MA1 – relative error set of rules SA1. ·, Set of rules SA1.

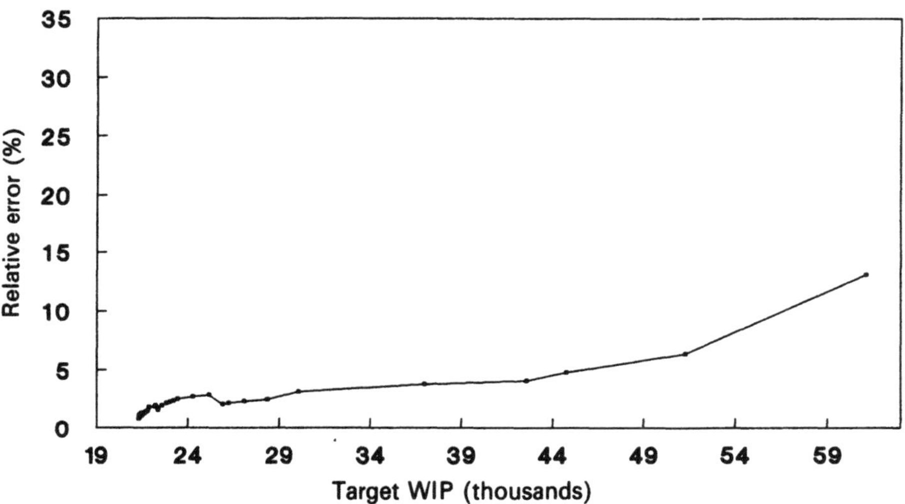

Fig. 2.4. System MA1 – relative error set of rules SA2. ·, Set of rules SA2.

30%), the improved set of rules are able to cut the relative errors substantially. Heuristic SA1 produces large relative errors when the last 'greedy' server added by the set of rules creates a large difference between the WIP of the final allocation and W_T. If this is the case, the improved set of rules have a large 'space' between W_T and the WIP produced by SA1 to improve upon. This is not the case when the relative errors produced by SA1 are small. In most cases where SA1 produced small relative errors (<2%), SA2 produced the same solution. Another improvement of SA2 over SA1 is the monotonic behavior of the relative errors. Although the overall behavior of (2.1) for SA1 is decreasing when W_T decreases (this is to be expected since the constraints get tighter because of the convex behavior of $L(m)$), the individual behavior of (2.1) for SA1 is very unpredictable. The relative errors produced by SA2, however, show an almost monotonic decreasing behavior. Hence, when W_T decreases, SA2 almost certainly gives a better solution. In Table 2.1 we show the average relative errors over all target WIP values and the corresponding standard deviations for the heuristics. We see that SA2 improves the quality of the solution by a considerable amount.

Table 2.1. System MA1

Heuristic	*Average relative error (%)*	*Standard deviation*
SA1	7.24	0.062
SA2	2.11	0.02

The results given in Table 2.2 and Figs 2.5 and 2.6 for system MA2 show a similar behavior of the different heuristics.

As for the (SR) problem, we show the trade-off curves we obtained with set of rules SA2 when applied to, respectively, system MA1 and system MA2. The trade-off curves act as a sort of efficient frontier for the machine allocations. This means that each allocation that lies to the left or below the trade-off curve cannot be achieved. The curves reveal important information about the manufacturing process. One can use the trade-off curve in the design phase of the manufacturing system. The curve shows what level of WIP performance can be achieved when a certain budget is available. Furthermore, given a certain target WIP the associated minimal costs can be found. It is interesting to see that on both curves there is a WIP level (Fig. 2.7: around 22 000; Fig. 2.8: around 2680) past which adding machines does not really improve the performance of the system anymore. From a manager's point of view this might be a good allocation to consider, since it represents an almost optimal performance level of the system with corresponding minimal costs.

The results show that the design of a manufacturing system can greatly benefit from applying the set of rules as described in this section. First of all, applying these rules gives insight in how to make optimal use, in terms of

Table 2.2. System MA2

Heuristic	*Average relative error (%)*	*Standard deviation*
SA1	4.71	0.028
SA2	1.62	0.006

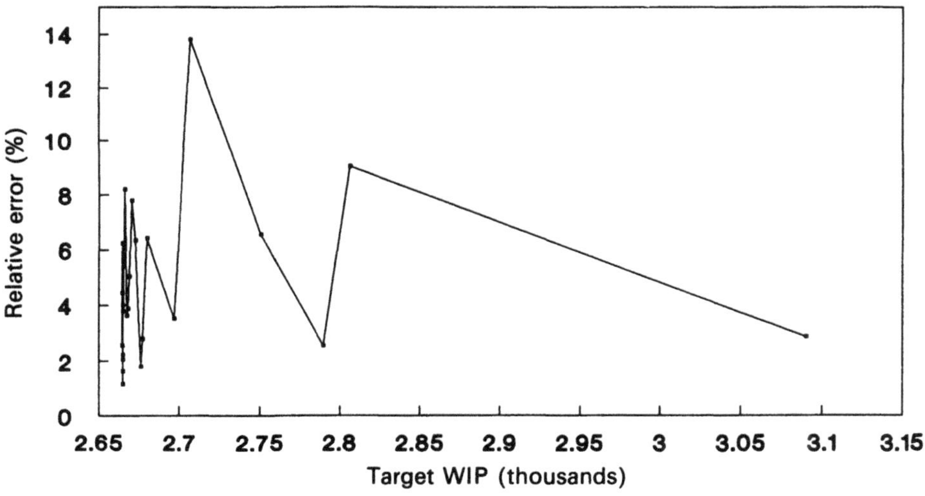

Fig. 2.5. System MA2 – relative error set of rules SA1. ·, Set of rules SA1.

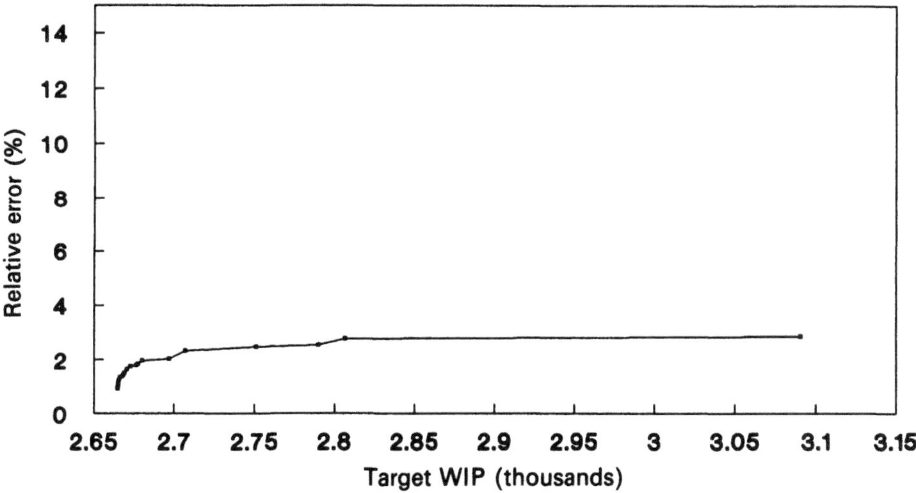

Fig. 2.6. System MA2 – relative error set of rules SA2. ·, Set of rules SA2.

logistical performance, of a given investment budget (Figs 2.7 and 2.8) or a set of machinery (Figs 2.1 and 2.2). Furthermore, the figures show that the rules can be used to give insight into the maximal budget one should invest is the logistical performance of the system should be optimized. In Fig. 2.7 one sees that past investment costs of 40 000 the logistical performance does not substantially improve anymore.

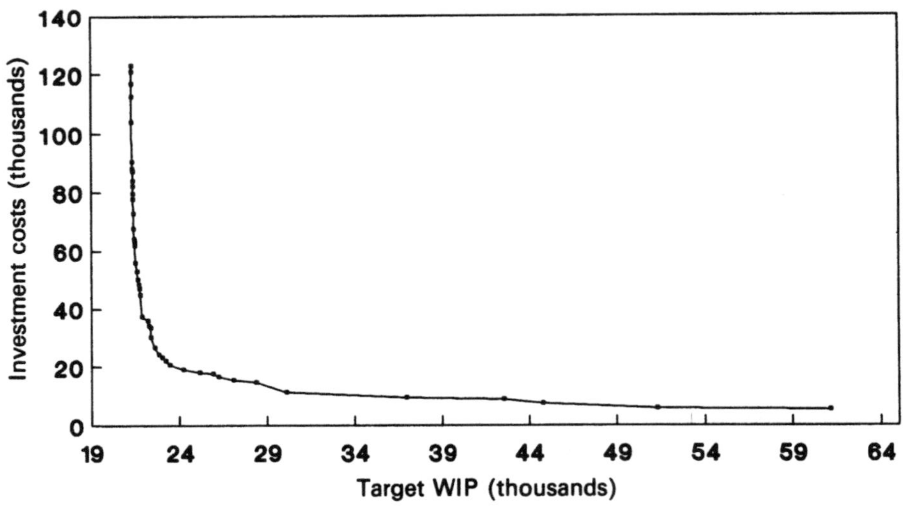

Fig. 2.7. System MA1 – trade-off curve server allocation. ·, Set of rules SA2.

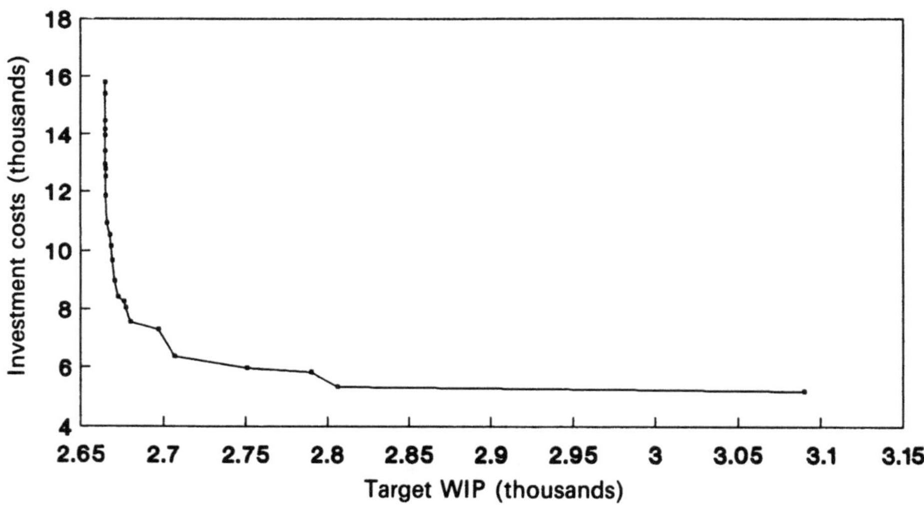

Fig. 2.8. System MA2 – trade-off curve server allocation. ·, Set of rules SA2.

2.3.3 Traffic allocation

(a) Introduction

In this section we consider a traffic allocation problem for a specific kind of manufacturing system. In traffic allocation one studies the optimal allocation of incoming traffic of products over a set of workstations. One can distinguish the notion of traffic allocation into the following two broad classes.

1. Static traffic allocation
2. Dynamic traffic allocation

(*b*) *Static traffic allocation*

In static traffic allocation one allocates the products according to allocation rules which are set *a priori*. This is important information when designing a system, since the infrastructure of the system should be adequate to handle the traffic flow within the system. Therefore, optimal static traffic allocation policies are important for manufacturing system design. However, one can also consider dynamic traffic allocation policies (or routing policies). When studying routing policies one considers the routing of individual products within the system. The routing policy is dynamic in the sense that the routing of the products may depend on the state of the system. Examples of such policies are policies in which products arriving into the system join the shortest queue or policies in which the routing of the products depends on the status of the workstations (in repair, in service, etc.). For a survey on dynamic routing policies we refer to Wang and Morris (1985). Since traffic allocation is considered within the context of the design of manufacturing systems, the focus in this paper is on static traffic allocation rather than on dynamic traffic allocation.

Consider a manufacturing system with J workstations in parallel. Each workstation consists of a single machine with an infinite capacity buffer. The system produces N product types. Each product visits one workstation to be processed and then leaves the system. The machines are flexible in the sense that every product can be processed by each workstation. Products of type i arrive into the system according to a general stochastic process characterized by an interarrival time distribution with mean arrival rate λ^i. A machine at workstation j has a general service time distribution with a mean processing rate μ_j.

Upon arrival in the system, a product of type i visits workstation j with probability p_{ij} ($\Sigma_{j=1}^{J} p_{ij} = 1, i \in \{1, \ldots, N\}$). The rate of the composite arrival stream at workstation j is denoted by λ_j. In Tang and Van Vliet (1993) two kinds of arrival processes are discussed; an arrival process behaving as a Poisson process and an arrival process with a general interarrival time distribution. For these two arrival processes, the composite arrival rate at workstation j is given by the following expression.

$$\lambda_j = \sum_{i=1}^{N} p_{ij} \lambda^i \quad (j \in \{1, \ldots, J\}) \tag{2.2}$$

In the sequel we focus upon a general interarrival time distribution.

(c) *The traffic allocation problem*

Given the heterogeneity of the workstations, it is necessary to allocate the traffic among the workstations in an effective manner. This observation motivates us to consider a traffic allocation problem. In this problem, we are interested in determining the type and the amount of work to be sent to each individual workstation (i.e. to determine the values of the p_{ij} variables). Our objective concerns the total number of products that are on average queueing at the workstations (or average queue length). The analysis also extends to other measures like average queueing time or average sojourn time (queueing + service time) or objectives in which queueing times or queue lengths are weighted with cost factors (Tang and Van Vliet, 1993). However, to simplify the exposition we use the above mentioned objective.

The performance analysis phase (Section 2.2.1) is used to come up with a functional relationship between the average queue length at workstation j (denoted, as in Section 2.3.2 by L_j) and the traffic offered to that workstation, i.e. the variables $p_{1j}, \ldots, p_{Nj}$. This functional relationship is denoted by

$$L_j(p_{1j}, \ldots, p_{Nj}). \tag{2.3}$$

In Tang and Van Vliet (1993) it is shown how one can come up with this functional relationship for several types of queueing systems. Here, however, we use this relationship, as before, as a 'black box', and focus upon the optimization phase to solve the traffic allocation problem.

The optimization problem that describes the traffic allocation problem is now as follows.

Problem (TA)

$$\min_{p_{ij}} \sum_{j=1}^{J} L_j(p_{1j}, \ldots, p_{Nj})$$

$$\text{s.t.} \sum_{j=1}^{J} p_{ij} = 1 \quad (i \in \{1, \ldots, N\}) \tag{2.4}$$

$$\sum_{i=1}^{N} p_{ij}\lambda^i < \mu_j \quad (j \in \{1, \ldots, J\}) \tag{2.5}$$

$$p_{ij} \geqslant 0 \quad (i \in \{1, \ldots, N\},\ j \in \{1, \ldots, J\})$$

In the above problem formulation, equation (2.4) ensures that all the traffic associated with product type i is assigned to that workstation. Equation (2.5) ensures the stability of the queueing network (all traffic intensities smaller than one).

The nice feature about the above problem formulation is that the constraint set consists of linear equations that constitute a polytope. Hence, the difficulty to solve problem (TA) depends upon the shape of the objective function. Tang and Van Vliet (1993) show that one cannot prove concavity or convexity of (2.3) under general conditions. This means that (TA) can have several local minima. To find a local minimum to problem (TA) we shall apply a method developed by Frank and Wolfe (1956). Although the method was originally developed for quadratic programming, it can also be used to find local minima of problems with a non-linear objective function restricted on a polytope (cf. Minoux, 1986, pp. 177–180).

The Frank–Wolfe algorithm is an iterative scheme. It starts with a feasible allocation. Given this allocation, it then calculates the gradient of the objective function. This can be compared with the local improvement criterion that we used in the set of rules for the machine allocation problems. The gradient is then used as the objective function in an LP that has the same set of constraints as problem (TA). The solution to this LP is another allocation. With a line search method, the allocation point in between the two above allocation points is found that minimizes (2.3). This allocation is then the next allocation in the Frank–Wolfe algorithm, and the above procedure is repeated. The algorithm stops if the difference in objective function value of two consecutive allocations is below a predefined threshold. Mathematically, the Frank–Wolfe algorithm is stated as follows.

Frank–Wolfe algorithm

Notation By p, p_k and $z_k \in \mathbf{R}^{N \times J}$ we denote feasible solutions to the constraint set as given by equations (2.4)–(2.5). By $L(p_k)$, $\nabla L(p_k)$ we denote, respectively, the objective function (2.3) and the gradient of (2.3) evaluated at p_k.

Step 1 (Initialization) Let p_0 be any feasible solution. Set $k:=0$. Choose termination parameter $\varepsilon > 0$.

Step 2 (Solve Subproblem) Let z_k be a solution to

$$\min_p \nabla L(p_k)p$$

$$\text{s.t. (2.4)–(2.5)}$$

Step 3 (Line Search) Choose p_{k+1} as to minimize (2.3) on the segment $[p_k, z_k]$.

Step 4 (Termination) If $|L(p_{k+1}) - L(p_k)| < \varepsilon$ then terminate, else set $k:=k+1$ and go to step 2.

As indicated before, the problem can have different local minima. We performed the Frank–Wolfe algorithm for several feasible start solutions. Results (Section 2.3.3.d) obtained show that the objective values evaluated at the different local minima are very close to each other. Tang and Van Vliet (1993) also give a theoretical justification to show that the different local minima produced by the Frank–Wolfe algorithm should be very close to each other. Hence, the local minimum as generated by the Frank–Wolfe algorithm with an arbitrary feasible start solution is likely to have an objective function value which is very close to the global minimum.

Note That the Frank–Wolfe algorithm is extremely simple to apply in practice. In fact, one only needs a good LP code, which is currently commonly available in any commercial environment. The other procedure used includes calculating a gradient and performing a simple line search. Our experience indicates that even when not 'state-of-the-art' techniques are used for these calculations, the algorithm is robust and produces very good results.

(*d*) *Numerical results*

We tested the Frank–Wolfe algorithm on three different manufacturing systems. We feel that the manufacturing systems are a representative set of systems to test the performance of the algorithm. The three systems range in size, measured by the number of product types and the number of workstations, from small ($N=4$, $M=5$) to large ($N=10$, $M=8$). Furthermore, systems in which the number of products is larger than the number of machines, as well as the reverse are represented. Since in manufacturing large variations in inter-arrival and service times are very scarce, the systems tested all have scv's that are smaller than one. The data on the three manufacturing systems are stated in Tang and Van Vliet (1993).

A common traffic allocation policy is to assign traffic to workstations in such a way that the traffic intensity ($\equiv \rho_j = \lambda_j/\mu_j$ for workstation j) is equal for all workstations. To study the performance of such a traffic allocation policy (which we refer to as the balanced allocation policy) for the manufacturing systems tested, we compared the Frank–Wolfe algorithm with the allocation policy based on balancing the traffic intensities.

To show how the Frank–Wolfe algorithm performs when different feasible solutions are used as start solutions, we give the results of the Frank–Wolfe algorithm for four different start solutions (Table 2.3). As seen from Table 2.3 the different local minima found have objective function values that are very close to each other. This indicates that the objective function value of the global minimum will be very close to the objective function values of the observed local minima (see also the discussion in Section 2.3.3.c).

To study the performance of the two traffic allocation policies, the Frank–Wolfe algorithm and the above mentioned balanced allocation, we

Table 2.3.

System	*No. of product types*	*No. of workstations*	*Local min. 1*	*Local min. 2*	*Local min. 3*	*Local min. 4*
1	4	5	27.59	27.59	27.27	27.72
2	8	3	3.59	3.36	3.32	3.39
3	10	8	9.25	9.27	9.48	9.47

compared the allocation policies for a wide range of states of the three manufacturing systems. We obtained the different system states by changing the average traffic intensity in the system. If the average traffic intensity increases, one could say that the manufacturing system gets more and more 'busy' with work.

In Figs 2.9–2.11 we show the effect of the average traffic intensity of the systems on the performance of the allocation policies. The traffic intensity as denoted in the figures is taken as an average over all the workstations. The allocation as given by the Frank–Wolfe algorithm clearly outperforms the balanced allocation. We see that the difference between the allocations as given by the Frank–Wolfe algorithm and the balanced policy increases as the average traffic intensity of the system increases. Apparently, in systems experiencing heavy workload, it pays off to use more sophisticated traffic allocation rules than often applied rules like balanced allocation.

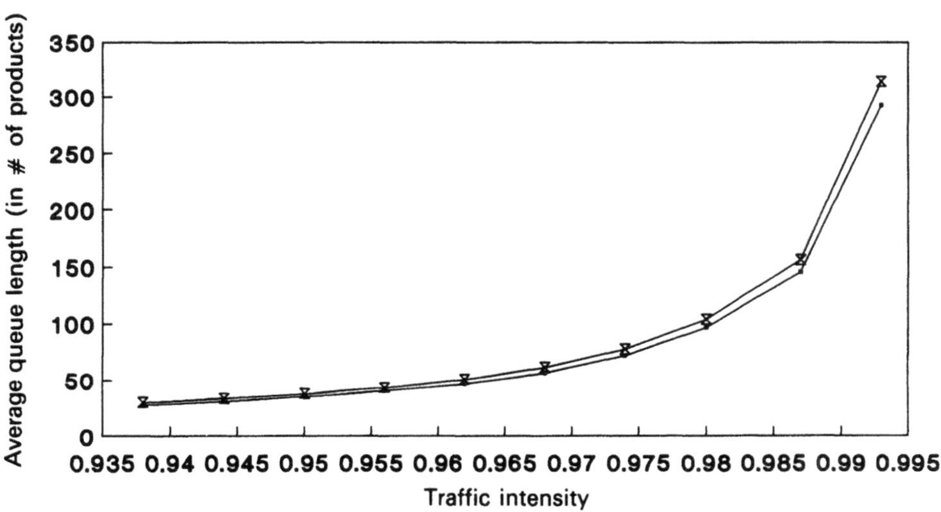

Fig. 2.9. System TA1 – changes in $\rho_j (j \in \{1, \ldots, J\})$. FW algorithm. X, Balanced allocation.

2.4 CONCLUDING REMARKS

We have presented rules for two allocation problems that occur within the area of manufacturing system design. The two allocation problems represent important issues when designing a manufacturing system: how to allocate capacity over workstations (machine allocation) and how to allocate incoming traffic of products over workstations (traffic allocation). We have justified the theoretical quality of the rules by means of error bound and worst case analysis. These theoretical results show that the allocations as

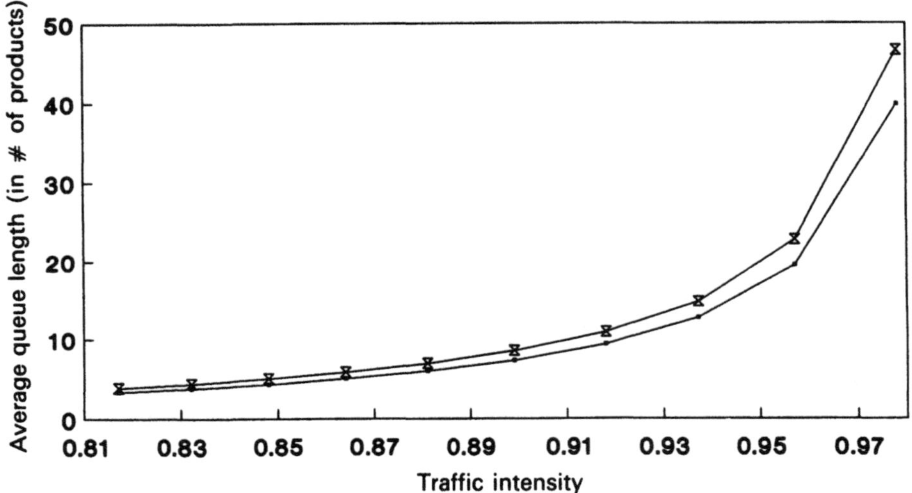

Fig. 2.10. System TA2 – changes in $\rho_j (j \in \{1, \ldots, J\})$. •, FW algorithm. X, Balanced allocation.

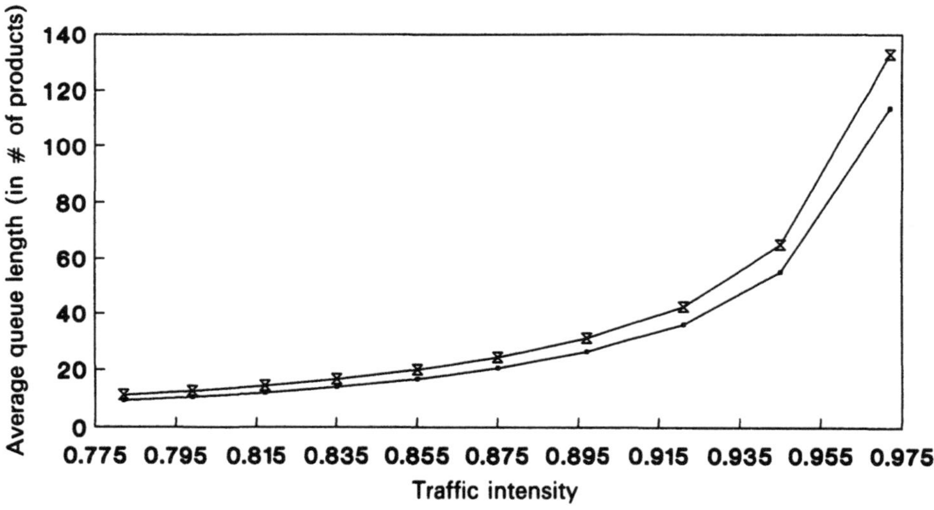

Fig. 2.11. System TA3 – changes in $\rho_j (j \in \{1, \ldots, J\})$. •, FW algorithm. X, Balanced allocation.

given by the rules are of a very high quality with regard to solving the relevant allocation problems.

The rules have been applied to a large variety of real life manufacturing systems. The results show that next to theoretical quality, the rules also give very good results when applied in practice. Furthermore, the results show

how a manufacturing system designer can get insight into the relevant options when designing a manufacturing system. For example, in the machine allocation problems, the results show that there is a logistic threshold in terms of minimal WIP. This means that no matter how much one invests in the manufacturing system in term of capacity, the level of the minimal WIP (and the related minimal product leadtime) cannot be surpassed. Furthermore, the results also show the designer how such this threshold can be achieved by using a minimal level of investment costs. The results in traffic allocation show that commonly believed, and applied, principles like traffic allocation based upon 'balancing' the manufacturing system, are outperformed by the traffic allocation rules as given in this paper. This means that if these rules are applied in practice, the logistic performance greatly improves.

The above mentioned rules can very easily be implemented in expert systems. The rules mainly use information that is available within the expert knowledge of the designer. Our experience indicates that, for example, the marginal change in queue lengths at workstation, due to changes in the workstation's capacity or the incoming level of traffic, is knowledge that is available within the company, either with the designer or with the people on the workfloor. In fact, such information is the main information that the above mentioned rules use. The expert system using the above rules is a system that is used by the designer of a manufacturing system as a reliable tool to do fast evaluation of different system designs. Therefore, the designer gets a much better insight into the relevant investment options and can compare these options in terms of quantitive figures of WIP, leadtime of products, utilization of machines, etc. Based on this information the designer can make a sound decision in terms of how to design a manufacturing system such that its logistic performance is optimal.

Most expert systems focus on management support with regard to short-term (operational) problems. The decision problems discussed here fall into the category of mid- to long-term decisions. We feel that the latter category has received far too little attention in the recent OR/AI literature. The results discussed in this paper show that expert systems using OR/AI rules be very useful in addressing these longer term decisions. It is our conviction that the future will hold many more beautiful developments in this area.

REFERENCES

Bitran, G.R. and Tirupati, D. (1989) Trade-off curves, targeting, and balancing in manufacturing networks. *Operations Research*, **37**(4), 547–64.

Boxma, O.J., Rinnooy Kan, A.H.G. and Van Vliet, M. (1990) Machine allocation problems in manufacturing networks. *European Journal of Operational Research*, **45**(1), 47–54.

Frank, M. and Wolfe, P. (1956) An algorithm for quadratic programming. *Naval Research Logistics Quarterly*, **3**, 95–110.

Frenk, J.B.G., Labbé, M., Van Vliet, M. and Zhang, S.Z. (1991) Improved algorithms for machine allocation in manufacturing systems. *Tech. Rep. 9132/A*, Econometric Institute, Erasmus University, Rotterdam. To appear in *Operations Research.*
Little, J.D.C. (1961) A proof for the queueing formula $L = \lambda W$. *Operations Research*, **9**, 383–9.
Minoux, M. (1986) *Mathematical Programming: Theory and Algorithms*, John Wiley & Sons, Chichester.
Tang, C.S. and Van Vliet, M. (1993) Traffic allocation for manufacturing systems. *European Journal of Operational Research*, **70**, 1–15.
Van Vliet, M. and Rinnooy Kan, A.H.G. (1991) Machine allocation algorithms for job shop manufacturing. *Journal of Intelligent Manufacturing*, **2**, 83–94.
Wang, Y.-T. and Morris, R.J. (1985) Load sharing in distributed systems. *IEEE Transactions on Computers*, **C-34**(3), 204–17.

CHAPTER 3

A common skeletal framework for knowledge-based solutions to a representative set of manufacturing problems

M. Marefat and P. Banerjee

3.1 INTRODUCTION

Intelligent systems have been increasingly studied and used in the fields of flexible design and manufacture. Some recent applications include knowledge-based systems for process planning, scheduling, facilities layout and intelligent design environments (Kumara *et al.*, 1986; Kusiak and Chen, 1988; Steffan, 1986). The evolution of intelligent systems can be thought of as having passed through several phases. While logic and knowledge representation were the themes during the 1970s, the early 1980s were when expert systems gained popularity and the late 1980s acknowledged the development of many knowledge-based application programs as well as software tools. For the knowledge-based methods to advance, we need a deeper investigation of the fundamentals behind successful knowledge-based applications. In order to demonstrate that knowledge-based methods are not *ad hoc*, we need to study the common underlying themes, generalize particular solutions into problem solving methodologies, study the properties of the underlying solving structures and investigate the impact of representation. Such an investigation provides us with a framework and a deeper generic understanding of the solution methodology. This framework and understanding can be used in the new generation of intelligent systems to:

- Evaluate the capabilities of certain classes of systems.
- Enhance and improve the capabilities of existing and new systems through incorporation of those elements of the framework which are lacking in these systems.

- Design new problem solving systems by applying the framework to particular problem domains and instantiating its key elements.

Expert systems or traditional knowledge-based approaches are the only general purpose approach to date. The utility of expert systems has been mostly judged by their opportunistic reasoning power, comparing their performance against human opportunistic performance. However, although opportunistic reasoning considers local optimization in decision making, it does not consider the longer range optimality impacts of decision making. In addition, the reasoning in expert systems is most often rule-based, but the magnitude and complexity of the manufacturing system design is most often best addressed by a system which can draw upon different types of reasoning. For example, in the later sections we will show an application of our framework to a system for process planning which utilizes both rule-based and uncertainty-based reasoning. The naive rule-based expert systems do not provide an effective generalization into a methodology. This is because the reason structure is encoded in the rules and most of the reason structure varies from problem to problem with very little separation of generic elements from problem specific elements.

Any framework which attempts to address broad classes of problems in manufacturing system design must evolve by taking into consideration the common characteristic of many manufacturing system design problems. Certain characteristics observed in many of the problems which we have researched are as follows:

1. The problems involve understanding based on pattern recognition in its broad sense, i.e. conclusions about certain features of the problem has to be drawn by detecting and analyzing certain generic patterns which are present in the problem. For example, the automated process planning one has to locate the machining primitive features such as slots and pockets, and use the machining knowledge associated with them to generate a plan for producing the part. In facilities layout, one has to locate certain anomalous zones in the layout which indicate a high probability of solution improvement. The points of interest in the solution domain are a minor fraction of the total number of possible solution points. The points of interest are often the non-linear transitions from one state to another, e.g. (a) existence and non-existence of anomaly instances (such as significant empty spaces) in the facilities layout, and (b) existence and non-existence of a cavity type (like an angular cavity) in a part. An aggregate mathematical modeling remains unsolvable for such non-linear cases of interest. The problem is often more efficient to represent qualitatively and perform non-linear reasoning to select the most promising prospective solution states at every stage in the iteration.
2. The problems involve a multitude of relationships, some of which are apparently in conflict.

3. Most of the essential deterministic relationships can be suitably represented by graphs. The wealth of the graph theoretic methods can be exploited to extract and manipulate the information captured in these graphs. We will show in later sections how graph-based methods provide invaluable cues to the type of shape features in the boundary of a part and to the presence or absence of anomalies in a facilities layout.
4. The problems have a progressive unraveling quality, i.e. the finer characteristics of the problem become clear as one decides upon the gross characteristics. It is either difficult to recognize all the characteristics in the beginning or the finer characteristics emerge only as a result of decisions on the gross characteristics. Because of the progressive unraveling quality the initial set of assumptions or constraints are too soft. A frequently adopted principle for solving such problems is that of 'least commitment' [e.g. MOLGEN planning system (Stefik, 1981)]. In least commitment, a decision is delayed until one reaches a dead end.
5. The progressive unraveling quality makes it impossible to formulate a systematic solution search strategy in its totality. A preferred alternative is to vary the order of exploration of different gross characteristics in isolated solution attempts, and follow up each solution attempt by analyzing the specific details.

Studying the above observations one arrives at the conclusion that a knowledge-based framework for manufacturing system design needs to:

- Be able to *iteratively improve* its solution. In such a framework there is a blend of opportunistic local reasoning and globally satisfying solution. Opportunistic reasoning determines the best path under the most immediate set of conditions, and globally satisfying solution determines the best path under a more enlarged set of conditions.
- Effectively utilize *different types of reasoning*. In this framework, successful solutions result from integrated modeling. Most manufacturing systems design problems usually involve symbolic manipulation as well as quantitative manipulation. Integrated modeling takes advantage of both symbolic and quantitative reasoning. Although symbolic manipulation has been emphasized in expert systems, the quantitative manipulation also has to be emphasized at places. The scope of quantitative manipulation needs to be expanded from its narrow use in evaluating thresholds to a much broader scope: that of studying the impact of local change on the global state in a combinatorial problem. These concepts can be implemented by a variety of reasoning methods, namely optimization, qualitative reasoning, uncertainty reasoning, constrained-based reasoning and opportunistic reasoning.
- To be capable of representing and reasoning with information at *different abstraction levels*. Such a framework can exploit different abstraction levels to avoid the large number of states which otherwise have to be searched [see, e.g. ABSTRIPS planning system (Sacerdoti, 1974) for an

initial development of this concept]. Abstraction levels also allow the framework to structure its knowledge such that the information can be easily manipulated and at the same time the necessary details are readily accessible for decision making. Ease of manipulation and accuracy of problem representation are often mutually conflicting goals. To achieve this capability, the framework uses a hierarchical scheme for representation and reasoning. Just as an expert process planner looks at a part at a higher abstraction level, and decides that to achieve the final desired shape of the part, a slot should be machined at a certain place, and then after considering the lower abstract information of the dimensions, tolerances, etc., of the particular region the expert decides which specific operations and tools to use. Similarly, a knowledge-based framework should structure and exploit its knowledge hierarchically.

3.2 A SIMPLIFIED FRAMEWORK

In the previous section, one of the conclusions we made was that a framework for manufacturing system design should be able to iteratively improve its solution. We observed earlier that the solutions to many problems which are the aim of such systems involve a multitude of conflicting relationships. Some mathematical techniques such as dynamic programming (Denardo, 1986) have been suggested for finding an optimal solution to the combination of the inter-related decisions which must be made, but the pior information needed for the articulation of the dynamic programming problem is not always available. However, opportunistic reasoning provides an alternative solution methodology for these problems.

Design is inherently an iterative process. The skeletal framework for a simple system exploiting iterative improvement is shown in Fig. 3.1. Such a framework yields a state space solution structure. In this structure, the solution at each stage is represented by a state. The underlying space of all the possible states in an implicit graph which is made explicit only if all states are enumerated. The final solution to the problem can also be represented by a state. The information in each state is related to the actual proposed solution by a representation scheme, which determines the structure of the information, as well as the semantics of the elements in the structure. To obtain a solution to the problem, an initial solution is formulated and is represented by the system as the initial state. This initial solution state may be no more than the representation of the given problem in the representation scheme used. The problem solving then progresses by iteratively modifying the present solution state to obtain new solution states which are closer to an acceptable final solution state. Problem solving stops when an acceptable final solution state is found. Figure 3.2 shows the knowledge-based problem solving by a typical application program which uses this simplified framework.

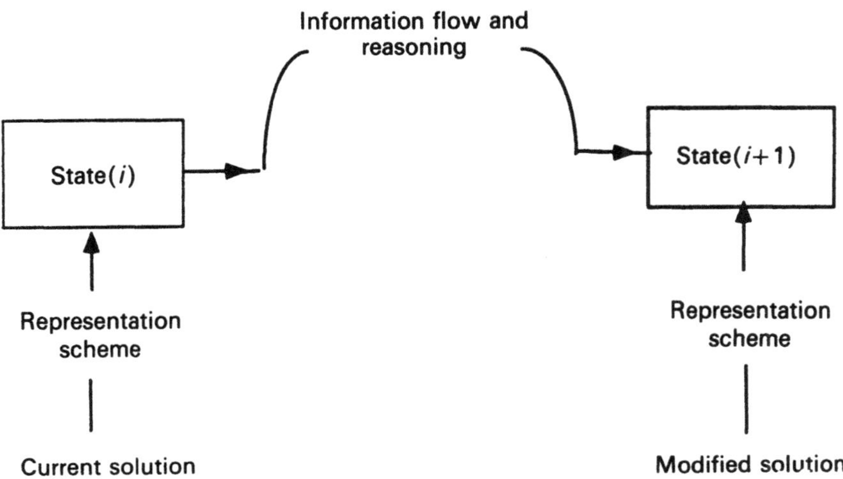

Fig. 3.1 The common skeleton for a simplified iterative framework. In this framework, the proposed solution at each stage is represented by a state. In each cycle, the information flow and reasoning operates on the current solution to produce the next modified solution. This cycle is where the system applies its knowledge to solve the problem at hand. The representation scheme must be complete and precise and yet powerful for efficient reasoning. Such a framework yields a state space solution structure.

In the above scheme, the control is cyclic for every iteration. A cycle is completed when one returns to the next iteration. In every cycle, the information flow and reasoning operate on the current solution state to produce the next modified solution state. This cycle is where the system employs its knowledge to solve the problem. Currently most of the knowledge-based systems in manufacturing are rule oriented [e.g. expert systems in the areas of process planning, scheduling, facilities layout and manufacturability evaluation, to name a few (for reviews, refer to Heragu and Kusiak, 1988; Hummel, 1989; Montreuil, Venkatadri and Ratliff, 1989).

The skeletal components of the above framework which characterize its application to solution of different problems are: the knowledge representation scheme, the preliminary design/solution, the iterative improvement method and the validation scheme. Because visiting solution states is costly, a goal in the design of knowledge-based systems is to minimize the number of states visited in the path to finding an acceptable solution state. To achieve this goal, the systems based on the simplified framework build a variety of strategies which employ backtracking into the information flow and reasoning cycle. The effect of these strategies is that the solution state space is visited in a best-first, depth-first, breadth-first or a combination order; with the net outcome being the need to explore a large number of solution states. Such a strategy by itself has had a very limited success. The reasons for the shortcoming are many-fold.

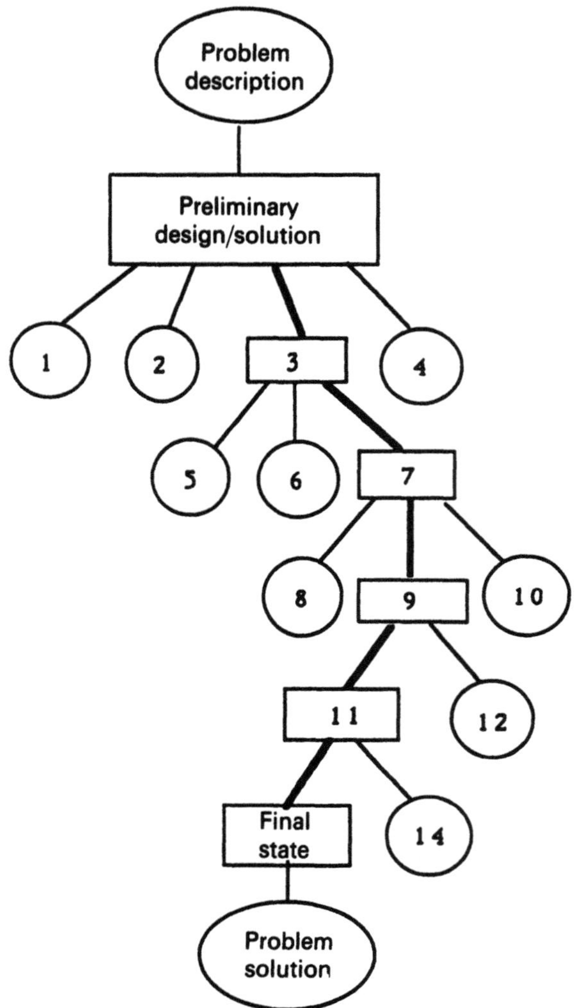

Fig. 3.2. Schematic diagram of knowledge-based problem solving by an application which uses the framework in Fig. 3.1. A preliminary design/solution is obtained based on the initial problem description. Acceptable final solution(s) can also be represented by a state. In each iteration many different next-states are possible, but only the most promising one (shown by rectangles) is selected and the others ignored or kept for back tracking. The path from the preliminary design/solution to final state (shown by thick lines) may be interpreted as the reasoning performed.

Firstly, the strategies are mostly based on simple heuristic needing a very exhaustive evaluation of the solution states to decide how the problem solving should be pursued further. However, to develop systems which are useful for solving real world problems the knowledge-based approaches should rely on a rich representational structure; simple heuristics (alone) would not be effective. A rich representational structure has many implicit constraints by which it is often possible to discriminate the most promising solution states. The use of a richer problem representational structure leads to a reduction in the search space because of elimination of inferior alternatives by a more precise and easily comprehensible problem representation in the form of a more accurate, interactively tractable and easily manipulatable problem representational structure. The investment made in a rich representational structure of the solution state at each iteration enables one to decipher the relevant situation fragments (primitives) and quickly focus on the most promising solution states. Once the representational structure becomes rich and loaded with domain specific information, much more computation is needed to apply a problem solving structure based on systematic blind (uninformed) search techniques (such as depth-first, breadth-first and best-first). Hence a relatively weak (less exhaustive) solution search strategy (mainly generate and test, and hill climbing) is applied. This seems to be a better alternative than a highly abstract (simple) representational structure coupled with a more exhaustive solution search strategy. The rich representational structure can serve as the equivalent of an informative evaluation function, for evaluation of prospective alternative solution states. This serves as a knowledge-based alternative for a systematic uninformed search technique.

Secondly, the above framework is not developed to accommodate structuring information and reasoning with it in a combination of different abstraction levels. Many of the developed systems are either developed with very abstract and oversimplified information so that they can only handle toy or very simple problems (therefore, they are not appropriate in real world applications), or the representation adapted by the system contains so much detail (not necessary in every solution step) that, except for a handful of problems, the problem solving process gets so bogged down with processing irrelevant information that it will not find an appropriate solution or it becomes computationally inefficient. To overcome these problems the knowledge-based systems must employ a hierarchical approach to representation and use of their knowledge. This approach allows them to use the higher level abstract concepts to process information efficiently and apply the detailed information as it becomes needed.

Thirdly, in the above framework, one kind of information flow and reasoning is exploited in every cycle. This aspect of the framework is not helpful for exploiting different types of reasoning such as constraint-based and uncertainty-based reasoning. In the above framework, different types of reasoning at best can be employed by encoding them in various rules which would get fired as the rule matches a condition in the solution state. However, this leads to much *ad hoc*ness, because we do not always know what other rules get fired in the same cycle that a particular rule is being fired, because we do not know in what order the rules get fired, and because we do not always know whether a rule will not fire in situations in which we want it not to fire. [This difficulty becomes particularly nagging when an existing system gets extended; see Forbus (1990) for an interesting example of a robot performing household tasks which explains the reasons for inadequacy of existing expert systems). A knowledge-based framework which supports different paths for information flow and reasoning enables a more systematic exploitation of various types of reasoning.

3.3 A HIERARCHICAL PROBLEM-SOLVING FRAMEWORK

In order to overcome the above mentioned shortcomings we have investigated a new framework. Figure 3.3 shows the skeletal representation of this framework. This framework uses a hierarchical approach to structure and reason with its knowledge, therefore it can exploit a rich representational scheme. It has the power to reason with higher level abstract concepts, and has the power to use the abstract concepts to structure and draw inferences based on the lower level details. The framework also allows

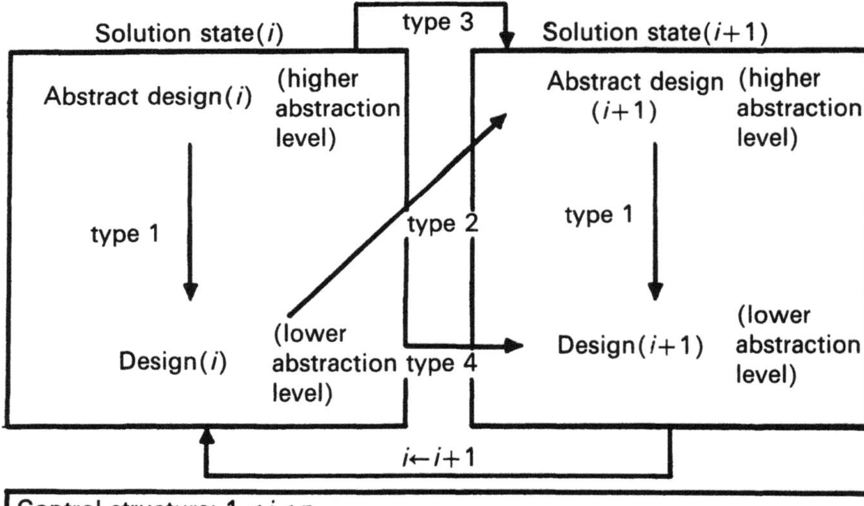

Fig. 3.3. A hierarchical skeletal framework for iterative movement of solution states. This iterative problem solving framework supports structuring and reasoning with abstractions. It can therefore, focus its reasoning and exploit a rich representation efficiently using abstractions. There are four information flow and reasoning paths indicated by the arrows. Information flow *e* is between the abstractions and the low level detailed information at each state. Information flow type 2 draws inferences about the gross characteristics of solution at the next state from the lower level information at current state. Information flow type 3 performs reasoning with the higher level abstractions. Different information flows may be based on different reasoning schemes.

taking advantage of different kinds of reasoning in an iterative underlying structure in a more systematic way. Multiple information flow types in the hierarchical framework support this capability in a natural and modular manner. The characteristics of this framework which form the backbone of its problem solving approach are as follows:

- Hierarchical knowledge representation.
- Preliminary design or solution.
- Iterative modification.
- Four information flow and reasoning paths.
- Solution validation.

In the following subsections we discuss these characteristics in more detail, and in the following section we will show application of the framework to design of two knowledge-based systems for solving problems in facilities layout and understanding shape from solid model representations for process planning. We will also discuss the important components of this framework which should be identified in applying the framework to the design of new systems or evaluation of existing systems.

3.3.1 Hierarchical knowledge representation

A decision is made about a balance of information content that is easily manipulated and at the same time that too much accuracy is not lost. Ease of manipulation and accuracy of problem representation are often mutually conflicting goals. To achieve the balance, the information is abstracted into more than one level of detail: the abstract information being used for ease of manipulation and the necessary details being used for prevention of too much loss in accuracy in representing the problem. The abstraction may be conceived with a graph-based form of knowledge representation. In many manufacturing system design problems the information modeling benefit to effort ratio is large when two levels of information abstraction are employed (the graph being the higher level of information abstraction). If there is less than two levels then one is dealing with a highly monolithic structure, which does not separate out the details, is difficult to manipulate and is often very inefficient. If there are more than two levels, the problem tends to be overabstracted and one may spend too much time in traversing between the levels of abstraction which also make the solution procedure inefficient. Figure 3.3 illustrates the rich problem representational structure which is characterized by two levels of abstraction. The graph-based knowledge represents the higher abstraction level (indicated as abstract design) and this is for ease of manipulation, while the information more specific to the problem under consideration indicated as design at the lower abstraction level and its purpose is to prevent too much loss in accuracy. Both these abstraction levels get updated as the solution moves from state(i) to state($i+1$).

Based on observation 3 in the Introduction, a lot of the essential relationship knowledge in manufacturing system design can be suitably represented by graph-based methods. In order to do this, the essential knowledge is first filtered out by a process of information abstraction. The graph-based information is then easily and efficiently represented using objects in an object-oriented language or using a special data structure in a different language.

3.3.2 Preliminary design or solution

Utilizing the abstract information in the initial problem, a preliminary design or a preliminary solution is constituted. This solution is based on decisions on a set of gross characteristics of the problem. The main goal of the preliminary design or solution is to characterize a representation that is well suited for the abstract as well as detailed representation of the problem.

3.3.3 Iterative modification

An important design decision is to formulate a process for transformation from one solution state to another. An iterative approach is used for defining this process of transformation. Generate-and-test and hill climbing are the two general purpose problem solving structures which are primarily employed. These two problem solving structures are fit for the progressive unraveling quality (observation 4 in the Introduction) of the manufacturing systems design problem space. Generate-and-test has the least amount of information on the prospective solution states and is employed to gather sensitivity information of the local neighborhoods of the current subgoals. Generate and test is often combined with some type of backtracking strategy (e.g. best-first). Hill climbing has no provision for retracting to discarded solution states from previous iteration. The rich representational structure (graph-based knowledge representation as well as the detailed problem representation already committed during the initial design or solution) is designed to aid this process. The transition from state(i) to the next state($i+1$) represents the iterative problem solving structure. The problem solving control structure involves sequential subgoaling. The role of each subgoal is to attempt to bring the present solution state [i.e. state(i)] closer to an acceptable solution state [i.e. state(n)]. The sequential subgoaling transforms the problem through a series of solution states, until a satisfactory solution state(n) or a non-determinable state(n) is reached. The exact nature of state(n) is often not known in advance. The subgoal structure is adopted because it decomposes the problem into subproblems. Problem decomposition reduces search because each decomposition tends to divide the exponent in exponential search spaces (Minsky, 1963), the idea has been further explained in Korf (1987).

3.3.4 Four information flow and reasoning paths

The representational and problem solving structures involve flow of information which has been classified under information flow types 1 through 4, the exact nature of which is determined in the respective domain instantiation. The information flow between the states(i) and ($i+1$) in Fig. 3.3 indicates this process. A non-linear situation-based information processing (picking out the situation pattern primitives which are relevant and discarding the ones which are not relevant) is employed for sequential interstate transition.

The higher abstraction level is designed to capture the globality of the solution and is termed abstract design. The lower abstraction level is designed to explore the nature of violation of local constraints which were suppressed at the higher abstraction level. Information flow type 1 relates the higher abstraction level to the lower level by imposing additional constraints to the ones already existing at the higher level. The information flow type 1 is used to assimilate a combination or reorganization of the total set of constraints. For example, in the problem of understanding the shape from solid models for process planning (Marefat and Kashyap, 1990; Marefat, Feghhi and Kashyap, 1990; Shah and Rogers, 1988), the description of the part in the higher abstraction level may contain its interpretation in terms of the machinable features such as pockets and holes, and information flow type 1 imposes additional constraining information (in the form of constraints relating pockets and holes to neighboring faces and edges) and then reorganizes these constraints (eliminating the references to pockets and holes) to describe the part in terms of its faces and edges. In some applications such as image understanding and inspection (Marefat, Timke and Kashyap, 1990; Knoll and Jain, 1986), the higher level constraints are combined by information flow type 1 with lower level constraints which are provided by the sensor reading to overcome the inadequacy, incompleteness and noise in the sensor input. Thus the higher level constraints are assimilated with lower level sensor input constraints to construct the total constraining information.

Information flow type 2 is used to screen out the detailing constraints in the lower abstraction level to retain the constraints which are involved in capturing the globality of the solution. The solution is driven to the next solution state by information flow type 2. The detailing local constraints in the lower abstraction level of the previous solution state are utilized to improve upon the global solution state which is represented by the higher abstraction level, either in terms of quality (for design representation-type problems) or in terms of accuracy (for diagnostic-type problems). Usually some form of generate-and-test-based sensitivity analysis is carried out based on the total set of constraints at the lower level. For example, in the facilities layout design problem, the lower abstraction level involves a layout of cells representing the local constraints along with the global ones and the higher abstraction level involves a graphical representation of the global constraints. Information flow type 2 generates constraints from anomalous regions in the layout (e.g. an empty space between two cells which have material flow between them) and translates these constraints into the higher abstraction level by preparing specifications for organization of the graphical constraints at the higher abstraction level such that the spotted anomalies in the layout are attempted to be rectified without significantly deteriorating a global quantitative objective function. Each step in generate-and-test may involve some form of local optimization, e.g. linear objective function optimization in facilities layout and 0–1 matching criterion function optimization in part feature identification for process planning.

Information flow type 3 is used to directly reorganize and/or modify the constraints which determine the globality of the solution from one solution state to the next without making use of the detailing constraints. For example, in the problem of understanding the shape of a part for developing process plans to manufacture it, a current abstract design may interpret the part in terms of pocket and slot machinable feature entities. However, if a particular identified slot is blocked at one end (this situation can be recognized without incorporating the detailing constraints in the form of faces and edges representation) and therefore inaccessible then the abstract part interpretation is exchanged by information flow type 3 to modify the higher abstraction level constraint 'slot' into another higher abstraction level constraint 'blind-slot' without having to proceed through the lower abstraction level detailing constraints.

Information flow type 4 transforms the total set of constraints at the lower abstraction level from one solution state to the next solution state without going through the higher abstraction level. Even though this feature is not used frequently in the studied manufacturing problem domains, it provides a useful tool for user interaction during the lower abstraction level design process.

3.3.5 Solution validation

The solutions in this framework combine rich representational structure with general purpose problem solving methods such as generate-and-test or hill climbing. In order to strengthen the combination, a reasoning validation mechanism is conceived. The validation is based on a combination of opportunistic reasoning and globally satisfying solution. Because the state space may not be searched in a systematic manner by the problem solving structure, the validation mechanism is used to ensure correctness while optimizing the search.

3.4 APPLYING THE FRAMEWORK

There are several important components which should be identified when applying the framework to the evaluation of existing systems or the design of new systems. These components include the representation of each state, the instantiation of the information flow and reasoning types, construction of a preliminary design/solution state, and evaluation of solution states. In this section we will briefly present applications of the developed hierarchical problem solving framework to two problems. The first application is concerned with integration of design and process planning. It discusses methods for machine understanding of the shape of parts designed with a solid modeler and effectively using this information for constructing process plans. In the second application the framework is applied to design a facilities layout with the objective of minimizing the material flow travel score along a tree-shaped flow network structure subject to a variety of linear cell dimension constraints. For each application, an introductory section introduces an overall overview of the problem, its significance and the solution. Then, the following sections discuss the components of the framework in the particular domain and how they work together to solve the problem.

3.4.1 Integration of design and process planning

Integration of design and process planning attempts to generate a process plan for producing a part from stock material using a solid model representation of the part. Automatic construction of process plans from solid model representations can be split up into two problems. The first problem, which is an interpretation problem, consists of the automatic interpretation of the geometry and topology associated with a part (Choi, Barash and Anderson, 1984; Liu and Srinirasan, 1984; Prinilla, Finger and Prinz, 1989; Wang and Chang, 1987). Current commercial computer-aided design (CAD) systems produce geometric description of parts in terms of the low level entities like faces and edges, but to manufacture the parts semantic information associated with entities like slots and pockets, and the relations between them are needed (Luby, Dixons and Simmons, 1986; Jakiela, 1989; Requicha, 1986), and these are not available by simple graph-based template matching of faces and edges. More detailed geometric reasoning methods are needed. The second problem, which is a planning problem, is the automatic generation of a plan to achieve the desired geometry and topology. While the interpretation constructs meaningful descriptions for the parts, the process planning (Chang and Wsyk, 1985; Descotte and Letombe, 1986) is the activity which produces a sequence of operations including machining and tool specifications to transform an initial stock into final product. The part is assumed to be represented by a boundary representation scheme [a form of solid modeling (Mortenson, 1985)]. Semantic descriptions are built from this representation by use of geometric reasoning based on a hypothesis generate–eliminate scheme. This scheme, which uses cavity graphs to represent depressions, iteratively generates hypotheses by decomposing the cavity graphs into maximal constituents (Marefat and Kashyap, 1990). Also, since the interaction between primitives may remove the adjacency between the faces of a primitive, the concept of virtual links and a method based on uncertainty reasoning to find virtual links to be augmented to the cavity graphs is used. To generate a plan to produce the part, a case-based planning method to map the constructed descriptions into a set of ordered manufacturing operations is used. The case-based planner keeps a memory of the previous cases, and automatically selects, abstracts and tailors an appropriate old plan to fit the part at hand.

(*a*) *Automatic generation of process plans from solid model representations*

As mentioned earlier, integration of design and process planning is broken up into two problems: (1) computer understanding (or interpreting) of the physical shape of a part and (2) constructing a locally optimal plan to produce the desired shape. A solution to the above problem for general three-dimensional polyhedral parts is discussed and its natural evolution from the proposed common skeletal framework is shown.

Given a solid model representation for the example part shown in Fig. 3.4(a), for instance, an integrated system generating high level plans interprets the shape of the part and produces a sequence of operations, for transforming the stock, such as the one shown in Fig. 3.4(b) without human intervention.

(*b*) *Problem representation*

Boundary Representation (BRep), which is the most commonly used form of solid modeling, describes a part by vertices, edges and faces which form its boundary (Requicha, 1980) (this description is not necessarily unique). Although this representation describes the part completely, it is too detailed to be used for direct reasoning and it does not contain the information needed to machine the part.

In order to reason about the part, the information at two levels of abstraction in the common skeletal framework is exploited. At the design state (the less abstract level) the knowledge about the part is represented as the depressions in the boundary of the part. (Note that this representation associates semantic meaning to a region of the part, because a slot region, for example, can be produced by one of the known machining procedures, where as a collection of faces and edges by itself does not have this information). At the final solution state, the design level contains a correct description of all the depressions of the part in terms of known primitive machining features (such as slots, pockets, etc.) and the relations between these primitives. A correct description of depressions of the part in terms of primitives and relations between them is called one interpretation for the part. At the abstract design state (the more abstract level), the cavities of the part are represented by a set of cavity graphs [refer to G(i) in Fig. 3.7 for an example of cavity graphs of the part shown in Fig. 3.4(a)]. The cavity graphs are labeled graphs which capture the gross topology and geometry of the depressions. In the cavity graph(s) for a depression: (1) each face of the part, which is part of the depression, is represented by a node of the graph, (2) for each pair of faces of the depression which are concavely adjacent, there is a link in the cavity graph which connects the corresponding nodes of the graph, and (3) each node of the graph is labeled by a symbol from the set $\{B, -B, +X, -X, +Y, -Y\}$ which expresses the dominant direction of the outward normal to the corresponding face.

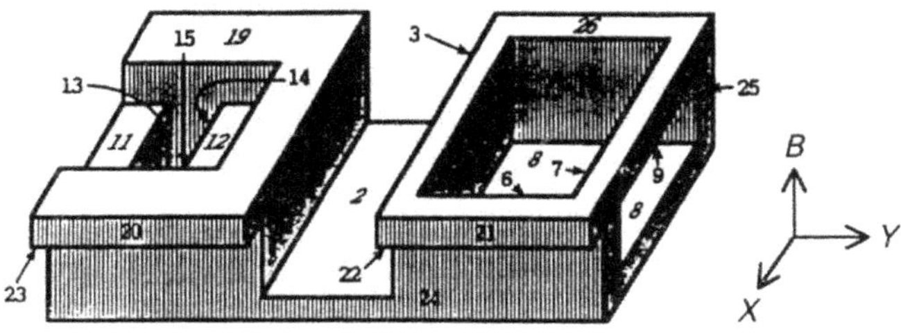

(a)

Primitive 1
(pocket Base: 15; Sides:10,14,17,13)
[1] SELECT FOLLOWING PLAN (certainty 1.0)
1. END-MILLING
2. FINE-END-MILLING
3. JIG-GRINDING
[2] LEVEL 1 OPERATION: 'END MILLING'
[3] TOOL--> 'END-MILL (certainty 1.0)
[4] LEVEL 1 OPERATION: 'FINE END MILLING'
[5] TOOL--> 'END-MILL (certainty 1.0)
[6] LEVEL 1 OPERATION: 'JIG GRINDING'
[7] TOOL--> 'MOUNTED POINT (certainty 1.0)

Primitive 2
(pocket Base: 8; Sides: 4,5,6,7)
[1] SELECT FOLLOWING PLAN (certainty 0.8)
1. END-MILLING
2. FINE-END-MILLING
3. JIG-GRINDING
[2] LEVEL 1 OPERATION: 'END MILLING'
[3] TOOL--> 'END-MILL (certainty 1.0)
[4] LEVEL 1 OPERATION: 'FINE END MILLING'
[5] TOOL--> 'END-MILL (certainty 1.0)
[6] LEVEL 1 OPERATION: 'JIG GRINDING'
[7] TOOL--> 'MOUNTED POINT (certainty 1.0)

Primitive 3
(pocket Base: 4; Sides: 9,5,8,6)
[1] SELECT FOLLOWING PLAN (certainty 0.8)
1. END-MILLING
2. FINE-END-MILLING
3. JIG-GRINDING
[2] LEVEL 1 OPERATION: 'END MILLING'
[3] TOOL--> 'END-MILL (certainty 1.0)
[4] LEVEL 1 OPERATION: 'FINE-END-MILLING'
[5] TOOL--> 'END-MILL (certainty 1.0)
[6] LEVEL 1 OPERATION: 'JIG GRINDING'
[7] TOOL--> 'MOUNTED POINT (certainty 1.0)

Primitive 4
(step Base: 24; Sides: (22, 23)
[1] SELECT FOLLOWING PLAN (certainty 1.0)
1. FACE MILLING
2. FINE-END MILLING
3. SURF-LAPPING
[2] LEVEL 1 OPERATION: 'FACE MILLING'
[3] TOOL--> 'FACE-MILL (certainty 1.0)
[4] LEVEL 1 OPERATION: 'FINE-END-MILLING'
[5] TOOL--> 'END-MILL (certainty 1.0)
[6] LEVEL 1 OPERATION: 'SURF LAPPING'
[7] TOOL--> 'SURF-LAP (certainty 1.0)

Primitive 5
(slot Base: 2; Sides: 1, 3)
[1] SELECT FOLLOWING PLAN (certainty 1.0)
1. END-MILLING
[2] LEVEL 1 OPERATION: 'END MILLING'
[3] TOOL--> 'END-MILL (certainty 1.0)

Primitive 6
(blind_slot Base: (11,12); Sides: 10,17,18)
[1] SELECT FOLLOWING PLAN (certainty 1.0)
1. END MILLING
2. FINE END MILLING
[2] LEVEL 1 OPERATION: 'END MILLING'
[3] TOOL--> 'END MILL (certainty 1.0)
[4] LEVEL 1 OPERATION: 'FINE END MILLING'
[5] TOOL--> 'END MILL (certainty 1.0)

(b)

Fig. 3.4. An example part and a high level process plan to machine it. (a) An example part; (b) a high level process plan for the example part.

(c) *Problem solving structure*

The geometric reasoning has a hypothesis generate–eliminate method at its base. The cavity graph(s) representing a depression are used to generate hypotheses about the primitive features present in the depression. There are unique graph representations for non-interacting primitive machining features. (Figure 3.5 shows the representations for an isolated pocket and an isolated thru-slot). Hypothesis generation is accomplished by decomposing the cavity graph(s) into subgraph constituents which represent the primitive features. Hypothesis elimination is accomplished by examining a set of generated hypotheses in detail and eliminating the incorrect hypotheses. This task is performed by firing rules from independent experts on the proposed hypotheses. Clearly, it is possible to decompose a cavity graph into different groups of constituting subgraphs and consequently produce different sets of hypotheses.

The problem solving process progresses by generating hypotheses based on given cavity graphs, examining the proposed hypotheses, modifying the cavity graphs into new cavity graphs by augmenting them with virtual links or unifying some of their nodes, generating a new set of hypotheses based on

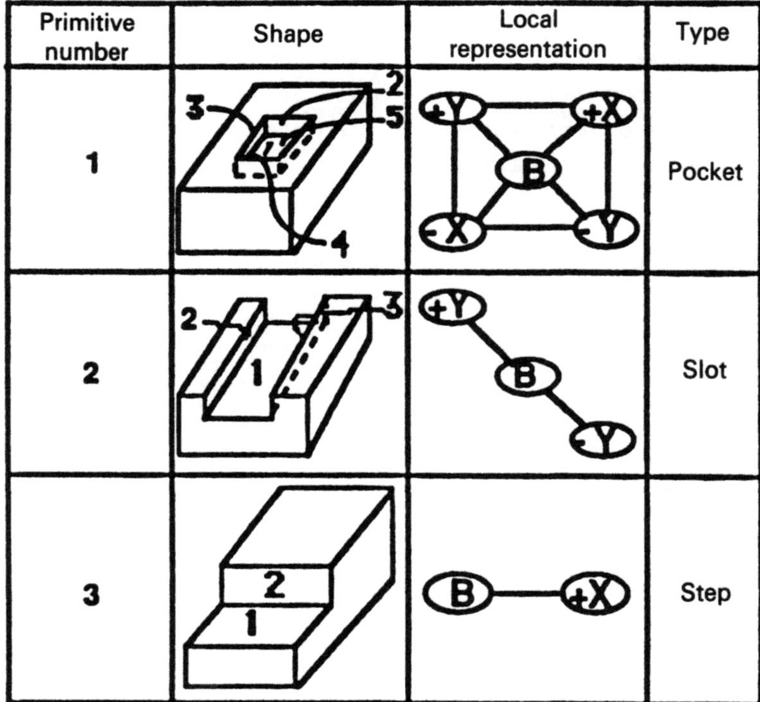

Fig. 3.5. Some primitive machining features and their labeled graph representations.

the new graphs, and continue refining the graphs and the hypotheses by exploiting the information in the graphs, hypotheses and failures of previous stages until an acceptable solution is found. Because the interaction between primitive features may cause one or more faces of a primitive to become divided into disconnected components or it may cause one or more faces of a primitive (which are adjacent when not interacting) to become disconnected, we should be able to modify the cavity graphs. For example, if two faces of a primitive are separated in the depression as a result of the interaction of primitives, there will be no link between the nodes of the corresponding faces of the cavity graph. Since hypotheses are generated by decomposing the cavity graphs into graph representations for primitives, the hypothesis space will not contain the correct hypothesis for the particular primitive. By modifying the cavity graphs we mold them into graphs which contain the graph representations of the primitives (correctly forming the depression) as subgraphs. Modification of the cavity graphs by adding virtual links is accomplished by a method of uncertainty reasoning which utilizes the Dempster–Shafer decision theory (Kusiak and Chen, 1988) as a tool.

(d) The reasoning process and the generic framework

Figure 3.6 summarizes the above discussed overview of solution approach and shows how the method evolves from the generic proposed framework for construction of intelligent solutions. The abstract design state (*i*) consists of the cavity-graphs G(*i*) and a decomposition of these cavity graphs into a set of constituent subgraphs, H(*i*), which represent the hypotheses. The union of the subgraphs in H(*i*) produces G(*i*).

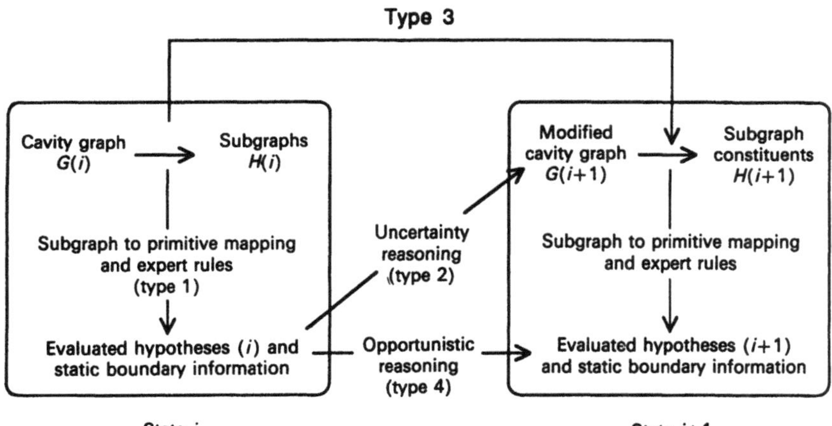

Fig. 3.6. Instantiation of common skeletal framework for integration of design and process planning. Information flow type 3 is based on a maximal decomposition strategy and uses the history of previous decompositions to avoid regenerating a previously unsuccessful set of hypotheses.

The information flow type 1 maps the constituent subgraphs at state i into hypotheses stating the existence of various primitive features and also determines which hypotheses are correct. The representations for primitives are used to map the subgraphs into hypotheses and, as mentioned earlier, rules fired by rule-based experts determine the correctness of each hypothesis. The information flow type 2 generates a modified cavity graph for state $i+1$ by utilizing the design information consisting of the depressions at state i and the lower level part information (edges and faces). The lower level part information is static and is carried through every design state. The modification augments the cavity graph with virtual links. To decide which virtual links (if any) should be added to a cavity graph uncertainty reasoning is employed. Information flow type 3 has two functions. The primary function is to decide how the cavity graph(s) at state $i+1$ should be decomposed into subgraphs which represent the primitive features. To do this, we use the decomposition executed at state i which includes the history of previous decompositions in conjunction with a maximal decomposition strategy. While the history helps us not to repeat a previously generated set of hypotheses, the maximal decomposition gives precedence to decomposition of the cavity graph into maximal subgraphs over its decomposition into smaller subgraph patterns. Information flow type 4 uses the failed hypotheses of state i to directly refine those hypotheses and propose new hypotheses for state $i+1$. This inference path does not concern the cavity graphs. It is in a form of opportunistic reasoning that considers which precondition of a fired rule failed and alters the involved hypothesis accordingly. The methods by which the above information flows achieve their tasks are explained by means of an example.

(*e*) *An expository example*

The concepts and reasoning methods which are briefly mentioned above in describing different information flows are illustrated in this section. The part shown in Fig. 3.5(a) is used throughout this section. Most of the previously proposed systems have difficulty in identifying and classifying the primitive machining features of this part, because of the intricacy of the interaction between the interacting primitives. The example part has four depressions. The middle depression is simply a thru-slot (faces 1, 2 and 3) and the depression on the front is a step (faces 23, 22 and 24). However, the other two depressions are combinations of primitives. The depression on the left consists of a blind-slot (faces 11, 12, 13 and 14) and a vertical pocket opening into the blind-slot. The depression on the right consists of two pockets opening into each other, one of which is vertical (base-face 8, and side-faces 4, 5, 6, 7) and the other one is horizontal (base-face 4, and side-faces 5, 6, 8, 9). Figure 3.7 shows two instantiated states, which our implemented system [we have implemented integrated design and planning (IDP) (Marefat, Feghhi and Kashyap, 1990) in Franz Lisp on a Sun 3/60 workstation] actually followed in interpreting the part and creating a process plan for it. The top part of state *i* shows the four

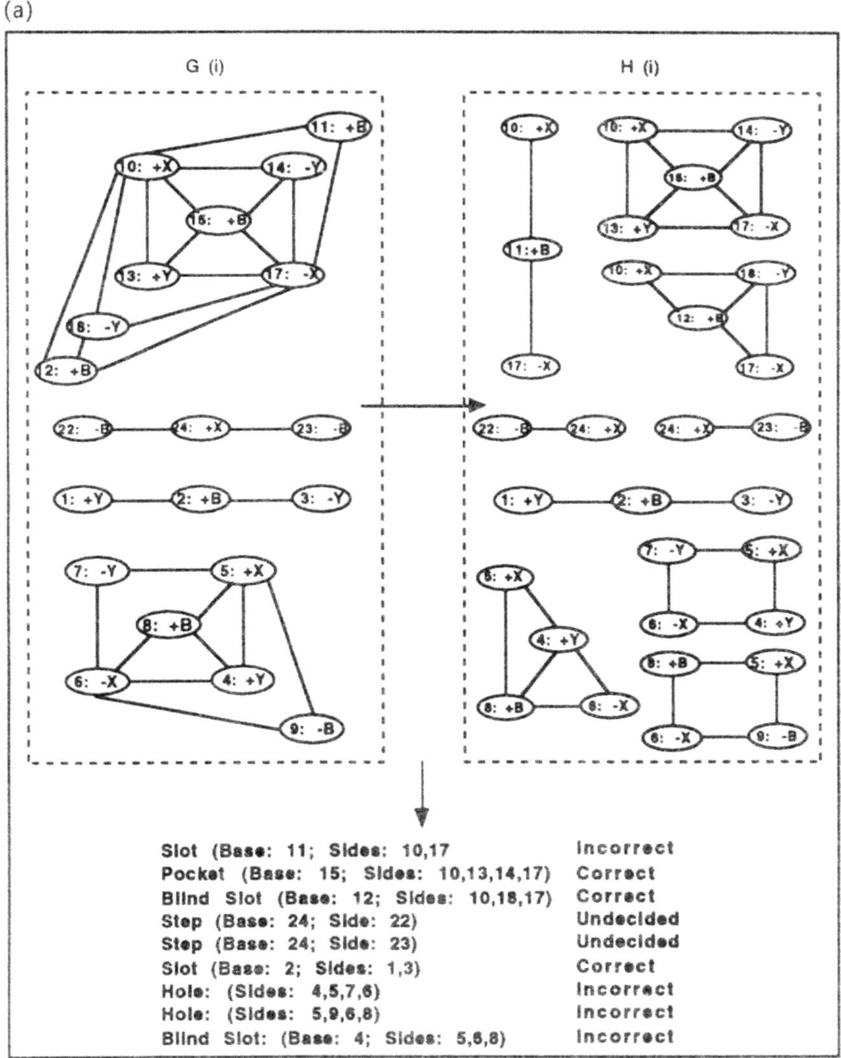

Fig. 3.7. (a) Example of instatiation of the generic framework for state *ti* in finding a correct semantic interpretation for the part shown in Fig. 3.4(a); (b) example of instantiation of the generic framework for state $t+1i$ in finding a correct semantic interpretation for the part shown in Fig. 3.4(a); (c) example of instantiation of the generic framework for two states in finding a correct semantic interpretation for the part shown in Fig. 3.4(a).

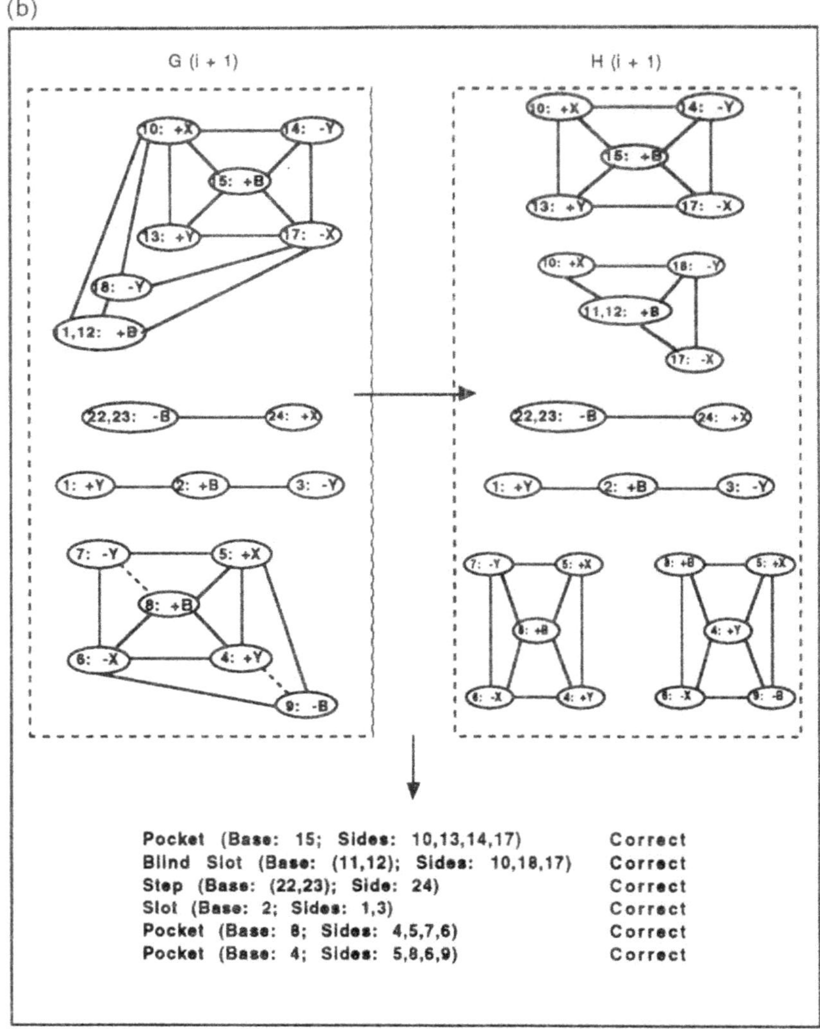

Fig. 3.7. *Contd.*

cavity graphs, G(*i*), that initially correspond to the depressions of the example part. By comparing the part and these cavity graphs with the representations shown in Fig. 3.5 we notice the following:

1. The interaction of the middle slot and the step has resulted one of the faces of the step to become divided into two disconnected faces 22 and 23. Therefore, the initial cavity graph of the step has three nodes instead of two.
2. The interaction of the two pockets of the depression on the right has caused the side-face 7 of the vertical pocket to become disconnected from

(c)

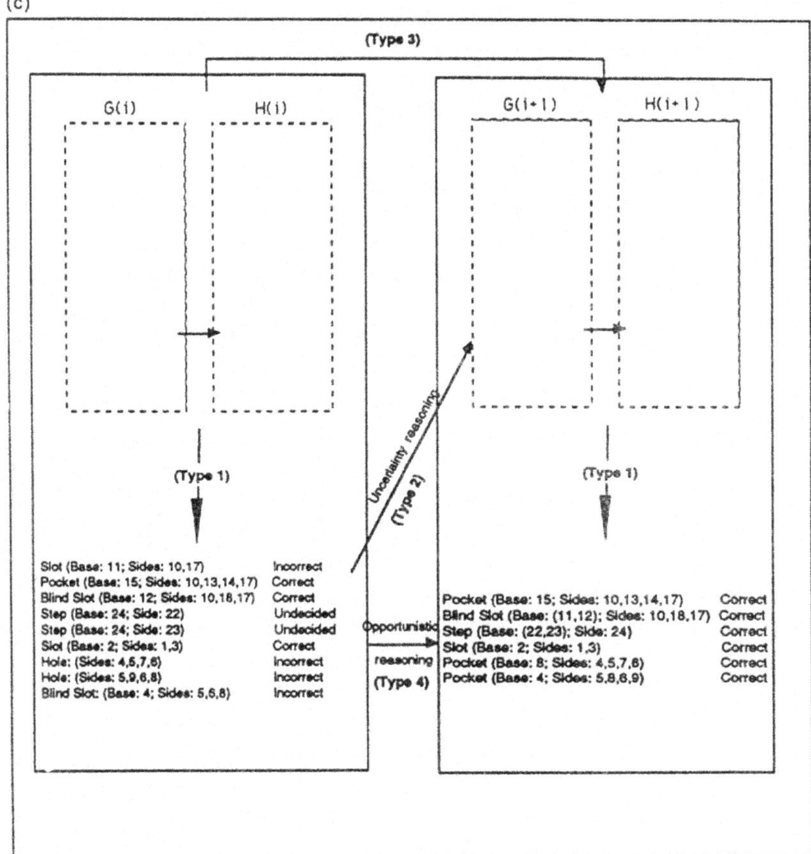

Fig. 3.7. *Contd.*

its base (face 8) and the side-face 9 of the horizontal pocket to become disconnected from the pocket's base (face 4). Therefore, in the initial cavity graph there are no links between the nodes of the corresponding faces and the nodes corresponding to the bases.

It is apparent that a unique graph representation corresponds to each primitive, but a unique primitive does not correspond to each graph representation. Hence, one cannot simply decompose the initial cavity graphs into representations for primitive features, and expect to find a correct interpretation for the depressions of a part at hand. The top part of state *i* in Fig. 3.7 shows the maximal subgraph constituents, $H(i)$, of the four cavity graphs of state *i*.

- Information flow type 1 maps the constituent subgraphs, $H(i)$, into hypotheses for primitives and determines their correctness. The hypotheses for state *i* are shown on the lower part of state *i* of Fig. 3.7. To generate the hypotheses the representations for primitives such as those shown in Fig. 3.5 are used. Correctness of the hypotheses is determined by expert rules modeling possible configurations for a primitive. The following is how one example rule for a thru-slot looks like:

(FEATURE thru-slot: base(b1) sides:(side-list))
IF
(sides are concavely adjacent to base) and
(sides are opposite to one-another) and
(sides are-not concavely adjacent to another face) and
(base has no concave edge between intersection-edges of sides).

As shown in the lower part of state *i*, several of the generated hypotheses in that state (one for a thru-slot, two for polyhedral holes and one for a blind-slot) are determined to be incorrect.

- Information flow type 2 builds a modified cavity graph for state $i+1$ from the depression information and the static boundary information of state *i*. The static boundary information is not shown in Fig. 3.7. After comparing the cavity graphs shown on the top part of state $i+1$ with those on the top part of state *i* we observe that two new virtual links (shown by dotted lines between nodes 7 and 8, and between nodes 4 and 9) are added to the previous cavity graphs. We also observe that some nodes (nodes representing face groups [22, 23] and [11, 12]) represent more than one face. These new nodes are formed by unifying (Marefat, Feghhi and Kashyap, 1990) the corresponding previous nodes together. To determine the virtual links to be augmented to the cavity graph an uncertainty-reasoning technique, which uses the Dempster–Shafer method (Shafer, 1976) for combination of evidence, is developed. In this technique, every link which does not already exist in the cavity graph is a potential virtual link for the cavity-graph. Once the set of all such links is identified, the method combines the geometric evidence in favor and against every link in the set, and picks the most promising links as the new links to be attached to the cavity graph. The geometric evidences are basic probability assignments (bpa) which stem from pieces of geometric information, such as whether two faces are almost parallel or not. The combination rule is Dempster's rule, which assigns the product of the bpas for two sets to their intersection. A clustering technique ranks and clusters the links based on their combined assignments to select the most promising links.
- Information flow type 3 decides how the cavity graphs at state $i+1$ should be decomposed to form hypotheses based on the previous decomposition history and a maximal decomposition strategy. The maximal decomposition strategy prevents the generation of many useless hypothesis sets and hence visiting many unnecessary states. The labeled graph representation for many machining primitives is such that the representation for a primitive is the supergraph of the representations for many others. In Fig. 3.5, the representations for a thru-slot and a step are subgraphs of the representation for a pocket. However, if a pocket is present in a region of the part, we do not want to interpret the region as a

combination of thru-slots or steps. Therefore, the maximal decomposition strategy forms the hypothesis for a pocket first, before looking for the other primitives. However, if the hypothesis for a pocket is found to be incorrect in the earlier states, then the strategy decomposes the cavity graph into other subgraphs in the later states. It also keeps track of the previously proposed hypotheses, so that it will not resubmit an incorrect hypothesis for evaluation. In state $i+1$ of the example, a maximal decomposition of the cavity graphs produces the set of hypotheses correctly modeling the depressions of the part and, therefore, there is no need to attempt a decomposition into smaller constituents.

- Information flow type 4 refines the incorrect hypothesis of solution state (i) and proposes new improved hypotheses for state ($i+1$) without going through cavity graphs. Our system did not use this inference path for the discussed states of the example part, but we demonstrate how the method works. The method considers which antecedent of a rule has failed and refines the hypothesis (to which the rule was applied) accordingly to reflect the next closest correct alternative. Let us consider an expert rule shown for a thru-slot in describing information flow type 1. Now, let us assume that for a given hypothesis the third antecedent in the IF part failed due to one of the side-faces being concavely adjacent to some other face besides the base. The knowledge that a thru-slot with a side concave to a face besides its base is probably a blind-slot together with the identity of other concavely adjacent face (let us call it New-Face) is then used to modify the old hypothesis to form a new hypothesis for a blind-slot. In the new hypothesis, the blind-slot will have the same base as the old thru-slot and the side-list will have the New-Face appended to the old list.

The stopping criteria for the above method for understanding a part's shape is reached when a correct interpretation for all depressions of the part is found. A computer understanding of the part as a whole is reached only when the correct subset of hypotheses in a state properly assigns all the geometric entities (faces and edges) of the depressions to one or more primitive machining features. There should not be any dangling nodes or links in the cavity graph(s) which are not part of a correct hypothesis. Extensive experimentation with our system empirically established that the system rarely visited more than five states before the final solution state was found. Figure 3.8 shows part of the output of our system, which shows the primitive features found for the discussed example part. As shown in Fig. 3.8, once the correct primitives are found, a complete description, including axis, length, etc., is extracted for each primitive from the boundary information. Finally, this expository section is summarized by making the following observations:

1. It is possible to have multiple correct interpretations, and therefore final solution states, for many parts with interacting primitives.

Primitive 1
(pocket Base: 15; Sides: 10,14,17,13)
opening face (11, 12)
position ((2.0) (1.5) (2.0))
length 4.0 ((1.0) (0.0) (0.0))
width 1.5 ((0.0) (1.0) (0.0))
depth 8.5 ((0.0) (0.0) (1.0))
approach ((0.0) (0.0) (-1.0))
feed ((0.0) (0.0) (1.0))

Primitive 2
(pocket Base: 8; Sides: 4,5,6,7)
opening face 26
position ((1.0) (14.0) (1.0))
length 5.0 ((1.0) (0.0) (0.0))
width 4.5 ((0.0) (1.0) (0.0))
depth 11.0 ((0.0) (0.0) (1.0))
approach ((0.0) (0.0) (-1.0))
feed ((0.0) (0.0) (1.0))

Primitive 3
(pocket Base: 4; Sides: 9,5,8,6)
opening face 25
position ((6.0) (14.0) (1.0))
length 10.0 ((0.0) (0.0) (1.0))
width 5.0 ((0.0) (1.0) (0.0))
depth 5.5 ((0.0) (1.0) (0.0))
approach ((0.0) (1.0) (0.0))
feed ((0.0) (-1.0) (0.0))

Primitive 4
(step Base: 24; Sides: 22,23)
opening face ((20, 21), 29)
position ((7.0) (0.0) (11.0))
length 19.5 ((0.0) (1.0) (0.0))
width 11.0 ((0.0) (0.0) (-1.0))
depth 1.0 ((1.0) (0.0) (0.0))
approach (((-1.0) (0.0) (0.0)) ((0.0) (0.0) (-1.0)))
feed (((0.0) (1.0) (0.0)) ((0.0) (-1.0) (0.0)))

Primitive 5
(slot Base: 2; Sides: 1,3)
opening face (19, 26)
position ((0.0) (6.5) (2.0))
length 7.0 ((1.0) (0.0) (0.0))
width 6.5 ((0.0) (1.0) (0.0))
depth 10.0 ((0.0) (0.0) (1.0))
approach ((0.0) (0.0) (-1.0))
feed (((0.0) (1.0) (0.0)) ((0.0) (-1.0) (0.0)))

Primitive 6
(blind_slot Base: (11,12); Sides: 10,17,18)
opening face (27, 19)
position ((2.0) (0.0) (10.5))
length 4.5 ((0.0) (1.0) (0.0))
width 4.0 ((1.0) (0.0) (0.0))
depth 1.5 ((0.0) (0.0) (1.0))
approach ((0.0) (0.0) (1.0))
feed ((0.0) (1.0) (0.0))

Fig. 3.8. Descriptions of the primitives found by the system in the depressions of the example part of Fig. 3.4(a).

2. Although it is possible to reencounter one or more primitive features again in a solution path, the same interpretation for the part was never revisited.

(*e*) *Generating the process plan*

The descriptions produced for the part by the above scheme are used to construct a high-level machining process plan for the part. In planning to machine the part, alternative courses of action are possible, because different valid interpretations for a part may exist. A modified version of case-based planner, TOLTEC (Tsatsoulis and Kashyap, 1988a,b), developed at the knowledge-based systems laboratory at Purdue University is used for this purpose. TOLTEC uses its memory of previous cases (Schank, 1982) to choose a most appropriate old plan, and automatically modifies the old plan to fit the new part. To create a process plan, first, the description of the part is mapped to the highest abstract space by discarding the less relevant attributes. An abstract plan for the part is selected based on this information. The planner then iteratively details the selected abstract plan until all the operations are elementary and cannot be detailed further. Figure 3.4(b) shows the high-level plan that the planner constructed for the example part of Fig. 3.4(a).

3.4.2 Facilities layout

The facilities layout problem is a multicriteria and non-linear problem which cannot be completely captured by a rigorous optimization model. Although a number of heuristics and a few optimization models have been proposed [a few recent references are Foulds, Gibbons and Griffin (1985), Golany and Rosenblatt (1989), Hassan and Hogg (1989), Picone and Willhelm (1984) and Scriabin and Vergin (1985)], many of the decisions are taken by the designer on an *ad hoc* basis. The problem serves as an excellent domain for the application of the common framework for knowledge-based solutions in manufacturing. Existing knowledge-based approaches (Fisher, 1986; Hreagu and Kusiak, 1988; Kumara, Kashyap and Moodie, 1988; Malakooti and Tsurushima, 1989) present difficulties in exhaustively prespecifying a majority of knowledge input requirements. The application of the common framework emphasizes the construction of such knowledge as has been previously addressed in Banerjee *et al.* (1990) and Flemming *et al.* (1988).

The problem presented here is the design of layout around a prespecified tree-shaped flow network. The flow network is usually obtained after clustering the intercell flow data. The network consists of a set of interstation flow values. Each manufacturing cell is assumed to contain one station, the station being a node in the flow network. The layout is evaluated by an objective function involving flow and distance terms. The objective function score quantitatively expresses the quality of the layout. Representative local qualitative (non-linear) situational patterns, which the designer dislikes and would like to rectify, are represented as constraints and these are analyzed by the described hierarchical problem solving framework.

(a) Local non-linear reasoning based facilities layout design

Layout design involves manipulation of directional information. Traditionally the directional information is addressed by dividing the continuous space into grid blocks so that the relative location of a grid block can be easily identified with respect to others. A richer alternative suggested in layout literature is to use linear programming (LP) approximation with a continuous space representation (Montreuil, Venkatadri and Ratliff, 1989). By using such a representation, the relative directions among cells in the layout get represented by binary integer variables and not by linear continuous variables. One such representation is shown by constraints (3.1)–(3.4) below:

$$\text{minimize} \sum_{\forall i,j \in F} f_{ij} * (|x_i - x_j| + |y_i - y_j|)$$

subject to cell length, width, perimeter bounds and station location within cell constraints [all these can be expressed as linear constraints with linear continuous variables (Montreuil, Venkatadri and Ratliff, 1989)]

$$R_{i|j}+R_{j|i}+R_{i/j}+R_{j/i}\leqslant 3 \quad \forall(i<j, i, j\in C) \tag{3.1}$$

$$R_{i|j}=0 \text{ or } 1,\ R_{i/j}=0 \text{ or } 1 \tag{3.2}$$

$$\bar{x}_i\leqslant \mathbf{x}_j+MR_{i|j} \quad \forall(i\neq j, i, j\in C) \tag{3.3}$$

$$\bar{y}_i\leqslant \mathbf{y}_j+MR_{j/i} \quad \forall(i\neq j, i, j\in C) \tag{3.4}$$

where f_{ij} represents flow from cell i to cell j; x_i and y_i represent the station location of cell i; F is the set of flows; C is the set of cells; $R_{i|j}=0$ if i is imposed to be to the west of (left of) cell j, 1 if the above imposition is relaxed; $R_{i/j}=0$ if cell i is imposed to be to the north of (above) cell j, 1 if the above imposition is relaxed; M is a very large positive number; $\bar{x}_i$ and $\bar{y}_i$ represent upper length and width coordinates of cell i; and $\mathbf{x}_i$ and $\mathbf{y}_i$ represent lower length and width coordinates of cell i.

The binary integer variables lead to a mixed integer programming (MIP) problem. The number of constraints from equation (3.1) is $8C4-1$ (i.e. 69) for each pair of cells and there are $nC2$ pairs for n cells, i.e. the number of pairs is of the order n^2. It is difficult to quickly optimize such a problem without reducing the number of constraints and is also difficult to optimize in polynomial time because of the binary integer variables. The existing solution methodologies for such MIP rely on interactive layout design based on *ad hoc* decisions about layout manipulations by the designer (Montreuil and Ratliff 1989) to eliminate a set of interior constraints (i.e. unimportant constraints or constraints with slack) and to heuristically eliminate the binary integer variables in order to convert the MIP into an LP problem with reduced set of constraints. Such a problem is then optimized in polynomial time by an LP solver (Montreuil, Venkatadri and Ratliff, 1989).

Instead of *ad hoc* layout manipulations described above, an alternative approach is described using the framework outlined in Fig. 3.9. The globality of the layout problem is first represented by a weighted graph, which signifies the higher level of abstraction. This graph represents the flow values between cells after clustering them in the form of a tree. In general, the graph may indicate one or more of the following parameters: (1) adjacency weights between cells, (2) relationship weights between cells and (3) flow values between cells, with or without clustering. The only constraints imposed on the flow graph is edges proportional to the prospective inter-centroid distance among neighboring cells and connections only between a certain set of nodes (determined by the flow clustering criteria) to form a tree structure. Qualitative constraints for relative intercell directions, namely: east, west, north and south are derived from the relative inter-node locations in the graph.

Information flow type 1 builds a more detailed representation of the problem in the form of block layout or gross layout (a gross layout is a

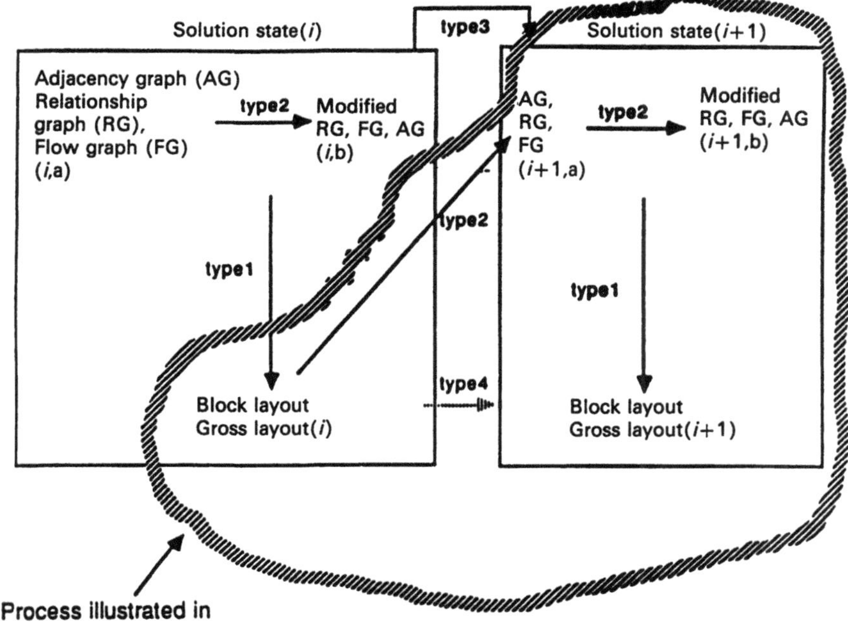

Fig. 3.9. Instantiation of common skeletal framework for facilities layout. The constraints in the higher abstraction level (graphical representation) capture the globality of the solution. The lower abstraction level (layout representation) is obtained after adding additional constraints and reorganizing the entire set of constraints. Information type 1 usually represent some form of optimization subject to an objective function and the constraints indicated by the graph. Information flow type 2 represents identification of focal points for solution improvement by making use of detailed local constraints in the layout and translating these constraints to the graphical level. Qualitative reasoning is one of the methods of achieving this effect. Information flow types 3 and 4 are interface tools which allow the user to intuitively improve the solution based on certain problem specific information which could not be generalized as part of a standard solution methodology.

block layout with a flow path structure embedded in it) by incorporating additional constraints, reorganizing them and optimizing the total set of constraints with respect to the objective function shown above. Making use of the qualitative constraints for relative intercell directions in the abstract design, information flow type 1 heuristically adds more detailed numerical constraints based on neighborhood relative directions among nodes to reduce the number of constraints of type (3.3) and (3.4) shown above, and also implicitly incorporate the conditions imposed by constraints (3.1) and (3.2).

Information flow type 2 is designed to manipulate the above set of relative intercell direction constraints for improving the layout flow travel score (objective shown above). This is done by qualitative reasoning. A set of qualitative layout anomalies (QLAs) is identified in the gross layout (see Fig. 3.10). Rectification of some of these anomalies are considered to be the probable local causes for layout score improvement. Thus the QLAs identify certain boundary values in the solution space for the system formulated above which are prime sources for potential solution improvement. One type of QLA is the empty spaces in the layout (abbreviated ESHZs in Fig. 3.10). Omitting the details and exceptions, it is stated that the empty spaces can be identified between east–west neighboring cells i and j which have

$$\bar{x}_i \underset{t}{<} \mathbf{x}_j \tag{3.5}$$

or between north–south neighboring cells which have

$$\bar{y}_i \underset{t}{<} \mathbf{y}_j \tag{3.6}$$

where $\underset{t}{<}$ indicates lower than by an amount above a certain heuristic threshold, the variables are the same as shown previously. The actual identification algorithm of the empty spaces uses a number of steps which are elaborated in Banerjee (1990). Such empty spaces are attractive candidates for determining the tight or important constraints (i.e. corner constraints in the language of the simplex linear programming algorithm) and leaving out the interior constraints from the entire set of constraints of types (3.1)–(3.5) of the MIP problem stated earlier and reducing it to an LP.

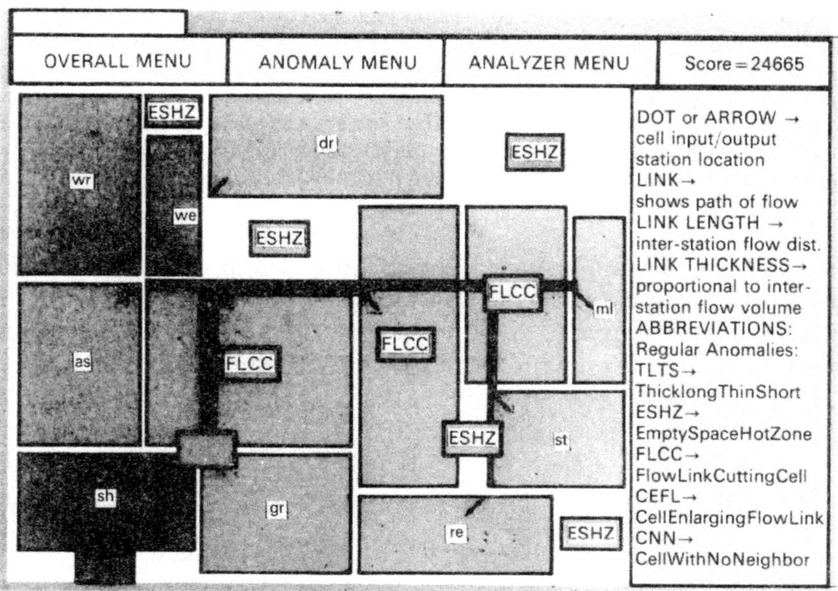

Fig. 3.10. Initial layout state showing score and qualitative anomalies.

The qualitative layout manipulations are derived from such empty spaces. The manipulations involve discrete relocations of cells (or nodes in the flow graph, which abstract the relative locations of cells) to induce a change in the relative intercell direction constraints among the cells. An altered set of relative intercell direction constraints in the LP, in the local neighborhood of the empty space and aimed at filling up the empty space, is thus obtained. The LP-solver is then used to optimize the LP with the altered set of constraints to judge the impact of the local layout manipulation on the objective function score.

A similar strategy is employed for rectification attempts on other types of QLAs, i.e. flow link cutting cell (FLCC), and thick (high volume) and long flow link and thin (low volume) and short flow link emanating from a common station node (TLTS) (for details, see Banerjee, 1990).

It is interesting to realize that although the LP formulation of the layout problem involves a single objective as shown previously, the rectification of each of the qualitative anomaly is considered as additional objective. Instead of directly incorporating the anomaly rectification as additional objective functions, they are implicitly incorporated in the form of alteration of a set of relative intercell direction constraints resulting from qualitative layout manipulations based on anomalies. Hence the layout problem is considered as an implicit multiobjective optimization problem instead of a direct multiobjective optimization problem.

Information flow type 3 consists of predicting local modifications in the graph structure and performing them without judging the global implications of these local manipulation on the actual layout. Information flow type 4 involves direct layout refinements without going through the graph structure. Information flow types 3 and 4 are good tools for interactive designer manipulation of the problem to fit the needs of a specific problem in case the general reasoning strategy (information flows 1 and 2 is either conflicting or is inefficient in meeting certain specific needs of a problem.

(*b*) *An expository example*

As an illustration, Fig. 3.10 shows a layout with a set of QLAs. The layout also shows material flow links. Note that there is one-to-one connection between material flow links in the lower abstraction level and the edges in the flow graph constituting the higher abstraction level. The thickness of the links is drawn proportional to the material flow volume. The arrows indicate input/output (I/O) or pickup/delivery (P/D) stations in the cells. There is also a one-to-one correspondence between the nodes in the flow graph and these stations in the gross layout.

The data for the case shown here are taken from (Montreuil, Venkatadri and Ratliff, 1989). The layout(i), graph($i+1$, a), graph($i+1$, b) and layout($i+1$) solution states are illustrated in Fig. 3.11. The graph ($i+1$, a) is

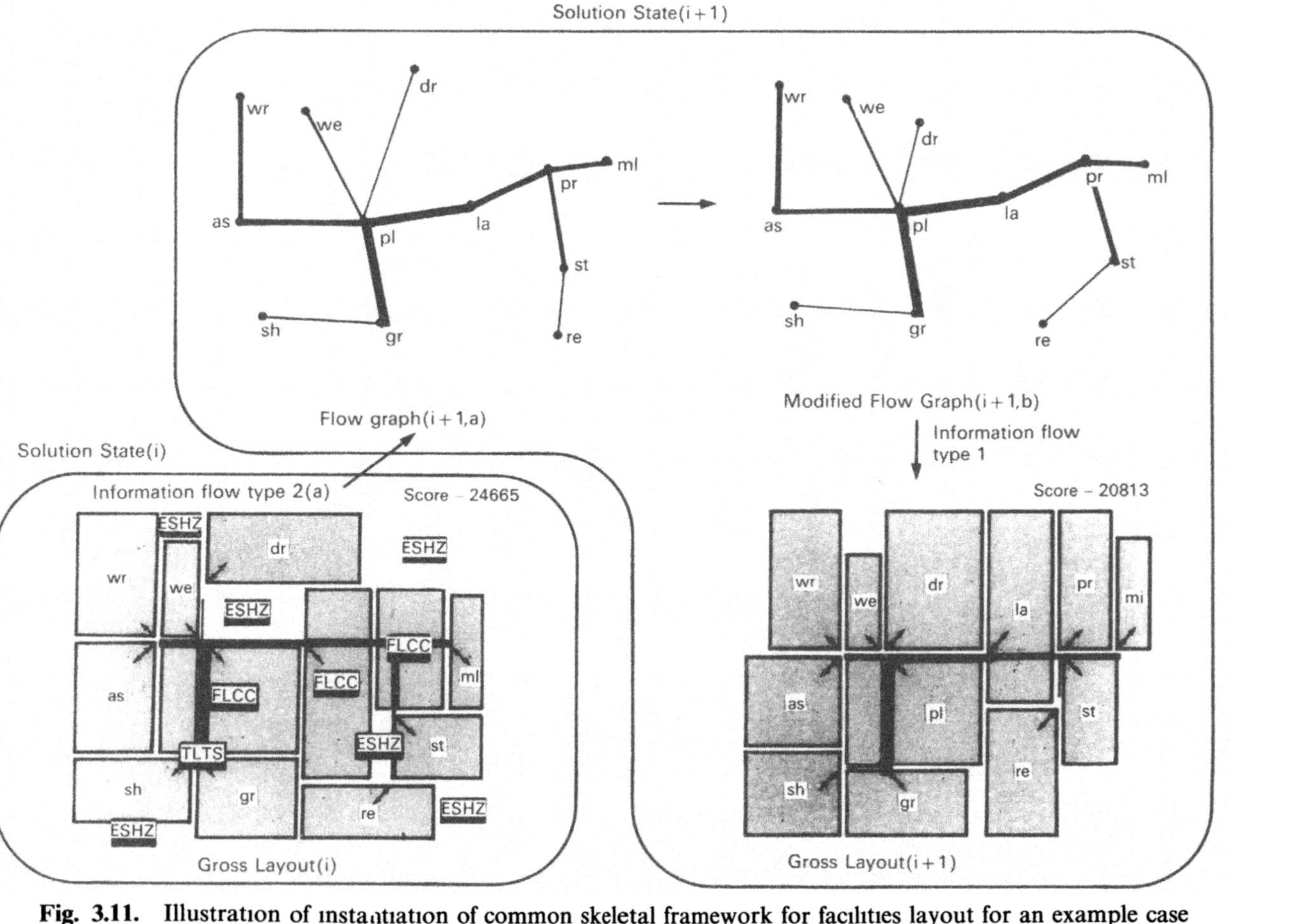

Fig. 3.11. Illustration of instantiation of common skeletal framework for facilities layout for an example case

generated by taking the cell centroid locations from the layout (i). This represents information flow type 2(a). Gross Layout (i) is the same as the one shown in Fig. 3.10. It shows the identified QLAs. Rectification attempts on these QLAs are performed by manipulations on graph ($i+1$). This represents the qualitative reasoning and is indicated by information flow type 2(b). The exact nature of this reasoning is explained with reference to the particular example in Fig. 3.11. A promising anomaly instance [shown by an empty space (ESHZ) instance marked * in Fig. 3.11] is selected and a sensitivity analysis is performed as follows: a set of different edge movements on the flow graph($i+1$, a) is performed around the local neighborhood of a superimposition of the empty space region in the layout on the flow graph. A linear programming solver is used to see the impact of the constraint alterations after each manipulation on the flow graph on the resultant layout objective function. Each such impact is collected as sensitivity information and the graph corresponding to the best acceptable alternative is shown as state($i+1$, b). The layout already generated as part of the sensitivity analysis is shown as layout($i+1$). This actual incorporation of layout($i+1$) from graph($i+1$, b) represents information flow type 1 for solution state $i+1$.

The process continues and the layout obtained at state (n) which signifies the final state as shown in Fig. 3.12. Note that the layout shown in Fig. 3.12

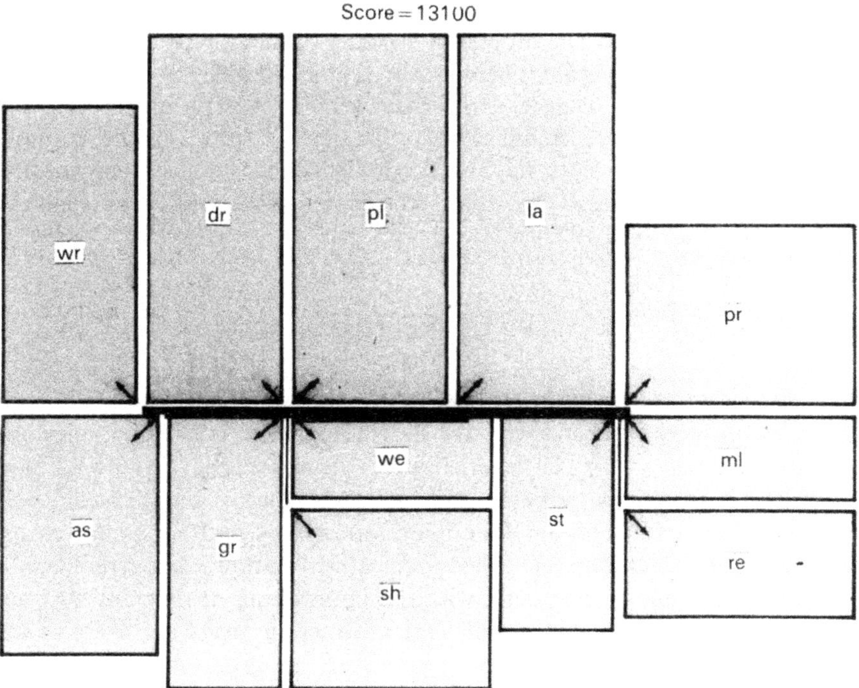

Fig. 3.12. Layout obtained at solution state (n) after removal of all targeted anomalies.

is not the best in terms of score (as better scores have been obtained) but it represents the best layout after removal of all targeted anomalies from a chosen starting solution state.

(c) *Discussion of problem representational and solving structures*

The rich problem representational structure is shown by:

1. Ability to automatically detect local layout anomaly instances in layout(i).
2. Ability to perform local manipulation on graph($i+1$, a) based on a selected anomaly instance on layout(i).
3. Ability to automatically generate an optimal layout($i+1$) based on the constraints generated from graph($i+1$, b).

The adopted problem solving structure is represented by: (1) The process of selecting an anomaly instance from among the anomalies in layout(i). This is presently done by a simple heuristic rule. Because of the heuristic nature of this anomaly selection procedure often multiple combinations of heuristic rules are used for experimental purposes. (2) The process of preventing selection of anomaly instances in layout(i) which have already been selected within the last n solution states. Presently a value of $n=2$–5 is being experimented. For the cases studied, it has been observed that a value of $n=2$–5 prevents solution states from getting trapped in cycles, to a considerable extent. (3) The process of deciding on a stopping criteria for the solution process. This again is heuristic in nature and different combinations of heuristic rules are being experimented. (4) The process of selecting the next best solution state. A hill climbing strategy is shown in the example case. The above concepts have been implemented in a system entitled Qualitative Layout Analysis using Automated Recognition of Patterns (QLAARP) (Banerjee, 1990).

3.5 CONCLUDING DISCUSSION

We have presented a skeletal framework for knowledge-based solutions in manufacturing systems and we have shown applications of this framework to two domains. The characteristics of the developed framework are hierarchical knowledge representation scheme, iterative modification, preliminary design or solution, four information flow and reasoning paths, and solution validation. The hierarchical utilization and structure of the knowledge allows reasoning with the information at different abstraction levels in the framework. Different abstraction levels are effective for determining the focus of reasoning based on the gross characteristics of the problem and for the efficient manipulation a needed rich representation structure. The framework is also capable of effectively exploiting both symbolic and quantitative reasoning. The iterative problem solving nature is suitable for the idea of starting from where the shortcoming exists in a proposed solution instead of beginning from the raw data.

The problem solving structure in each of the domains selectively explores a few solution states by making use of a rich representational structure and selects the most promising ones. The use of a richer problem representational structure leads to a reduction in the search space because of the elimination of inferior alternatives by more accurate, interactively tractable and easily manipulatable problem representational structure. A richer and more accurate problem representation, in contrast to a relatively simple problem representation, takes the advantage of the ability to extract situation specific quick problem solving pattern primitives. Once the representation structure becomes loaded with domain specific information, much more information is needed to apply a problem solving structure based on systematic blind search techniques (like depth-first, breadth-first and best-first). Hence a different problem solving structure (mainly generate-and-test and hill climbing) is employed. Thus a trade-off has been made: a rich problem representational structure is treated as more relevant for getting quick acceptable results compared with uninformed or blind (but rigorous) problem solving search technique.

The empirical results show that a combination of generate-and-test and hill climbing problem solving structures making use of a rich problem representational structure leads to good solutions quite rapidly for the domains studied. The generate-and-test procedure is used to search the local neighborhood by local heuristics like macro operators and production rules at each solution state when too little information about prospective solution states are available (which is frequently the case). The hill climbing strategy is more global and often involves some form of global evaluation (like LP optimization in facilities layout). Integration of design and process planning is more of a local problem compared to the other domain. This indicates that potentially knowledge-based approaches can serve as a good tool for analyzing manufacturing problem domains which are non-linear and NP-hard.

We are currently investigating applications of the developed framework to a system for evaluation of the castability of casting design (Requicha, 1980). In this system the castings are represented as a combination of a set of sweep components. Each sweep component is described symbolically in terms of its cross section and its axis by a set of generic pattern primitives. In the system being developed, knowledge-based geometric analysis of this symbolic information is used to provide rapid castability assessment and feedback before a time consuming and complete final element analysis set-up is attempted.

REFERENCES

Banerjee, P. (1990) An automated reasoning architecture for a representative set of human designer tasks in manufacturing systems layout organization by designing default knowledge combining linear objective optimization and non-linear qualitative analysis, *PhD Dissertation*, Industrial Engineering, Purdue University.

Banerjee, P., Montreuil, B., Moodie, C.L. and Kashyap, R.L. (1990a) A qualitative reasoning-based interactive optimization methodology for layout design. *IIE Conference Proceedings*, 230–5.

Chang, T.-C. and Wysk, R.A. (1985) *An Introduction to Automated Process Planning Systems*, Prentice-Hall, Englewood Cliffs, NJ.

Choi, B.K., Barash, M.M. and Anderson, D.C. (1984) Automatic recognition of machined surfaces from a 3D solid modelor. *Computer-Aided Design* **16**(2), 81–6.

Davies, B.J. and Darbyshire, I.L. (1984) The use of expert systems in process planning. *Annals of the CIRP*, **33**(1), 303–6.

Denardo, E.V. (1986) *Dynamic Programming Theory and Applications*, Prentice-Hall, Englewood Cliffs, NJ.

Descotte, Y. and Latombe, J.C. (1981) *GARI: A Problem Solver that Plans how to Machine Mechanical Parts.* IJCAI-7, pp. 766–72.

Fisher, E.L. (1986) An AI-based methodology for factory design. *AI Magazine*, **7**(4), 72–85.

Flemming, U., Coyne, R., Glavin, T. and Rychener, M. (1988) A generative expert system for the design of building layouts—version 2, in *Artificial Intelligence in Engineering*: *Design*, (ed. J.S. Gero), Elsevier, Amsterdam, pp. 445–64.

Forbus, K.D. (1990) Qualitative physics: past, present, and future, in *Qualitative Reasoning about Physical Systems* (eds D.S. Weld and J. de Kleer), Morgan Kaufmann, San Francisco.

Foulds, L.R., Gibbons, P.B. and Giffin, J.W. (1985) Facilities layout adjacency determination: an experimental comparison of three graph theoretic heuristics. *Operations Research*, **33**(5), 1091–106.

Golany, B. and Rosenblatt, M.J. (1989) A heuristic algorithm for the quadratic assignment formulation to the plant layout problem. *International Journal of Production Research*, **27**(2), 293–308.

Gossard, D.C., Zuffante, R.P. and Hiroshi, S. (1988) Representing dimensions, tolerances, and features in MCAE systems. *IEEE Computer Graphics and Applications*, pp. 51–9.

Hassan, M.M.D. and Hogg, G.L. (1989) On converting a dual graph into a block layout. *International Journal of Production Research*, **27**(7), 1149–60.

Heragu, S. and Kusiak, A. (1988) Knowledge based system for machine layout (KBML). *IIE Conference Proceedings*, pp. 159–64.

Hummel, K.E. (1989) Coupling rule-based and object-oriented programming for the classification of Mach. *Computers in Engineering*, **1**, 409–18.

Ikeuchi, K. and Takeo, K. (1988) Automatic generation of object recognition programs. *Proceedings of the IEEE*, **76**(8), 1016–35.

Jakiela, M.J. (1989) Design and implementation of a prototype 'intelligent' CAD system. *Journal of Mechanics, Transfer and Automation in Design*, **3**, 252–8.

Karinthi, R.R. and Nau, D.S. (1989) *Geometric Reasoning as a Guide to Process Planning.* Proceedings of the ASME International Computers in Engineering Conference, July 30–August 3, pp. 609–16.

Knoll, T.F. and Jain, R.C. (1986) Recognizing partially visible objects using feature indexed hypotheses. *IEEE Journal of Robotics and Automation*, **RA-2**(1), 3–13.

Korf, R.E. (1987) Planning as search: a quantitative approach. *Artificial Intelligence*, **33**, 65–88.

Kumara, S.R.T., Joshi, S., Kashyap, R.L. *et al.* (1986) Expert systems in industrial engineering. *International Journal Production Research*, **24**(5), 1107–25.

Kumara, S.R.T., Kashyap, R.L. and Moodie, C.L. (1988) Application of expert systems and pattern recognition methodologies to facilities layout planning. *International Journal of Production Research*, **26**(5), 905–30.

Kusiak, A. and Chen, M. (1988) Expert systems for planning and scheduling manufacturing systems. *European Journal Operations Research*, **34**, 113–30.

Liu, C.R. and Srinivasan, R. (1984) Generative process planning using syntactic pattern recognition. *Computers in Mechanical Engineering*, 63–6.

Luby, S.C., Dixon, J.R. and Simmons, M.K. (1986) Design with features: creating and using a feature data base for evaluation of manufacturability of castings. *Computers in Mechanical Engineering*, **5**(3), 25–33.

Malakooti, B. and Tsurushima, A. (1989) An expert system using priorities for solving multiple-criteria facility layout problems. *International Journal of Production Research*, **27**(5), 793–808.

Marefat, M. and Kashyap, R.L. (1990) Geometric reasoning for recognition of three dimensional object features. *IEEE Transactions on Pattern Analysis and Machine Intelligence* **TPAMI-12**(10),

Marefat, M., Feghhi, S.J. and Kashyap, R.L. (1990) *IDP: Automating the CAD/CAM Link by Reasoning about Shape.* The Sixth Conference on Artificial Intelligence Applications, Santa Barbara, California.

Marefat, M., Timke, M. and Kashyap, R.L. (1990) *A Framework for Image Interpretation in manufacturing applications.* Proceedings of IEEE International Conference on Systems, Man, and Cybernetics (SMC), Los Angeles, California.

Minsky, M. (1963) Steps toward artificial intelligence, in *Computer and Thought*, (ed. Feigenbaum and Feldman), McGraw-Hill, New York, pp. 441–3.

Montreuil, B. and Ratliff, H.D. (1989) Utilizing cut trees as design skeletons for facility layout. *IIE Transactions*, **21**(2), 136–143.

Montreuil, B., Venkatadri, U. and Ratliff, H.D. (1989) Generating a layout from a design skeleton. *Document 89-01*, Department of Operations & Decision Systems, Laval University, Quebec, Canada (to appear in *Management Science*).

Mortenson, M.E. (1985) *Geometric Modeling*, John Wiley & Sons, New York.

Newell, A. and Simon, H.A. (1972) *Human Problem Solving*, Prentice-Hall, Englewood Cliffs, NJ.

Picone, C.J. and Wilhelm, W.E. (1984) A perturbation scheme to improve Hillier's solution to the facilities layout problem. *Management Science*, **30**(10), 1238–49.

Pinilla, J.M., Finger, S. and Prinz, F.B. (1989) Shape feature description and recognition using an augmented topology graph grammar. NSF Engineering Design Reserach Conference, June 11–14, Amherst.

Requicha, A.A.G. (1980) Representations for rigid solids: theory, methods, and systems. *IEEE Computer Graphics and Applications*, **12**(4), 45–60.

Requicha, A.A.G. and Chan, S. (1986) Representation of geometric features, tolerances, and attributes in solid modelers based on constructive geometry. *IEEE Journal of Robotics and Automation*, **RA-2**(3), 156–66.

Sacerdoti, E.D. (1974) Planning in a hierarchy of abstraction spaces. *Artificial Intelligence*, **5**, 115–35.

Schank, R.C. (1982) *Dynamic Memory: A Theory of Reminding and Learning in Computers and People*, Cambridge University Press, Cambridge.

Scriabin, M. and Vergin, R.C. (1985) A cluster-analytic approach to facility layout. *Management Science*, **31**(1), 33–49.

Shafer, G.A. (1976) *A Mathematical Theory of Evidence*, Princeton University Press, Princeton, NJ.

Shah, J.J. and Rogers, M.T. (1988) Expert form feature modelling shell. *Computer-Aided Design*, **20**(9), 515–24.

Steffan, M.S. (1986) A survey of artificial intelligence-based scheduling systems. *IIE Conference Proceedings*, 395–405.

Stefik, M. (1981) Planning with constraints (MOLGEN: Part 1). *Artificial Intelligence*, **16**, 111–40.

Tsatsoulis, C. and Kashyap, R.L. (1988a) A case-based system for process planning. *International Journal of Robotics and Computer-Integrated Manufacturing*, **4**, 557–570.

Tsatsoulis, C. and Kashyap, R.L. (1988a) A system for knowledge-based process planning. *Artificial Intelligence in Engineering*, **3**(2), 66–75.

Wang, H.-P. and Chang, H. (1987) Automated classification and coding based on extracted surface features in a CAD data base. *International Journal of Advanced Manufacturing Technology*, **2**(1), 25–38.

Woo, T.C. (1982) *Feature Extraction by Volume Decomposition*. Proceedings of the Conference on CAD/CAM Technology in Mechanical Engineering, MIT, Massachusetts, pp. 76–94.

You, I.C., Chu, C.N. and Kashyap, R.L. (1989) Expert system for castability evaluation using a fixed-features based approach. *Robotics and Computer-Integrated Manufacturing*, **6**(3),

Zeigler, B.P. (1990) *Object-Oriented Simulation with Hierarchical Modular Models*, Academic Press, New York.

CHAPTER 4

A general purpose knowledge-based system and its application to design problems

Setsuo Ohsuga and Jiebo Guan

4.1 INTRODUCTION

In this paper we discuss a general purpose knowledge-based system that can solve problems mostly autonomously. We need to know a general method of problem solving for the purpose which is free from any specific condition coming from each specific domain. Most problems are dependent on each problem domain and therefore specific knowledge must be used to solve the problems. Thus, the common part of problem solving should be separated from the domain specific part and a framework should be created that is common to all problem domains but which works specifically by the given domain specific knowledge. It is generally known that this is the purpose of ordinary expert systems; however, the authors assert that these systems are not powerful enough to solve any problem given to the systems unintentionally. In particular, these systems are not suited for supporting creative works such as design.

One of the reasons lies in the very fundamentals of system design, i.e. ordinary expert systems lack the idea of model building. Model building is a key concept not only in design problem but in any problem solving by human beings. In many scientific and engineering domains, knowledge has been created through scientific research based on the scientific method which is formalized as to make a model of an object being studied and analyze it. This knowledge must be used effectively in the design system in combination with the other expertise to produce results. This prompted us to develop a language with which we can represent any object model for a given problem and also knowledge on the way of solving the problem in terms of the model. Here an object model must be represented as a combination of a model structure and a set of its functionalities.

A mechanism only to collect knowledge and apply it to the model is not sufficient to develop a good design system because, if the amount of knowledge increases (as in the ordinary case of problems in the real world) and knowledge is selected without any guiding principles, an explosion of combinations can occur which can limit ordinary expert systems. When human beings solve such problems, they build a problem solving process as a guide to the application of knowledge to the model. In order to adopt this method, it is necessary to classify knowledge to various classes (but allowing some part of knowledge to belong to different classes) and represent a problem solving process by assigning an order of applications of the classes as the guide. An explicit reprsentation of information for controlling the process becomes necessary. It needs the representation of the process. These representations must be in a higher level than the knowledge to be applied directly to the model (object knowledge) because the process representation and its control knowledge describe the object knowledge. This process representation may not always be the optimal but must be modified so that a better one is obtained. This means that the problem solving process itself becomes the design object. These conditions require an expert system to be provided with a language and/or logical architecture suited for representing any object model, model transformation rules as problem solving knowledge, problem solving process as well as its control knowledge and its modification.

The authors developed such a system that meets these conditions. This system has been applied to a number of design problems including mechanical design, chemical compound design, feedback control system design, etc. Among these applications, the feedback control system design system has been made as an automatic design system.

In this chapter, after discussing the idea and approach to the general purpose system we present first the outline of this knowledge-based system. Then we present the feedback control system design as an application of the system.

4.2 MODEL BUILDING

4.2.1 Modeling as a basis of problem solving

A problem is created by a human being and a model is built to represent the problem. Modeling is the basis of problem solving irrespective of the problem solver being a human being or a computer. The model may be made implicit and be invisible from outside in the case of problem solving by human beings; however, in the case of problem solving by computers, the model must be represented explicitly. This requires a new software technology which is referred here as a 'model-based method of problem solving'.

Modeling is based on an assumption that every problem is concerned with some entity, either real or abstract, in the real world. To represent a problem is to represent an (or more) object in relation with which the problem is created. Every object has a body as a real existence or an abstract concept which has various facets or aspects showing different functionalities. The term functionality is used here to mean property, function, behavior in a certain condition, relation with other objects, etc. In general, an object can have an indefinite set of facets. Therefore only facets with which the current problem is concerned are represented explicitly. This is the object model. Information modeling is a technology to represent a model explicitly in computer.

Model functionalities are dependent on the model body. Therefore these representations cannot be made separate. The model body and its functionalities must be the basic unit of model representation. If either the model body or some of its functionalities are unknown and are required to be made known then a problem arises. This is the model-based representation of a problem.

Thus, problem solving is formalized as being composed of model building and the following operations to derive the required information from the model. In the first half, descriptions of a problem are given and a model is built based on them. In some cases, some functionalities are given as the requirements to be satisfied by the model and model building is used to find a model body to satisfy these requirements. Very often this is a non-deterministic process involving trial-and-error operations and, in many real cases, human–computer interactions are necessary.

On the other hand, once a model body is fixed, the dependent functionalities of the model are obtainable by the deterministic operations represented in terms of model representations. Thus the latter half of ordinary problem solving is the deterministic process.

Model representaton is different for every problem; however, the basic technique of modeling is common to a wide range of probems. Therefore it is possible to provide computer systems with this model building method in advance so that these systems can support a large part of human model building. Once a model is built, problem solving is to specify a set of operations to derive some information from this model. In this model-based method of problem solving, every operation must be represented in terms of the model representation. What is required here is a common modeling scheme.

The model-based method of problem solving has many advantages. Ohsuga (1990) discussed that it is desirable to separate the representation of a problem from that of the problem-solving process so that different experts can be responsible to each of them. In ordinary problem solving by human beings; an end user can focus attention on specifying his/her problems before giving attention to the method to solve it. When there is an expert, the user can ask him/her to solve this problem. Computer systems provided with the model-based method of problem representation and problem solving can replace this human expert. In this case the model plays another important role to aid communication between the human being and the computer because the model is an embodiment of the human intention.

To specify a model, however, is not an easy task because (1) model building is a non-deterministic process and (2) human beings cannot always express their ideas correctly. A human being expresses his/her idea in words which he/she thinks appropriate; however, these may be incorrect and can be changed afterward. The systems must be designed to allow a user to take this process. A trial-and-error operation is necessary here. It must be composed of tentative model building, model analysis and its evaluation, and model modification. Easy addition, deletion and updating of information to the computer systems are required for the purpose.

Knowledge processing systems are suited for realizing such requirements. One of the characteristics of knowledge processing technology that enables this is the modularity of knowledge representation which is realized by the use of declarative language. It brings flexibility and expendability to these systems.

4.2.2 Conditions of model building support

In order for model building methods to be provided to computer systems, the model building process must be formalized and represented explicitly. The ordinary design process is referred to here as a typical example of model building because design is an activity to make up a model that has the functionalities given as the requirements. This is a creative work because the method of producing such an object cannot be foreseen in advance.

Design as a functional reasoning was first discussed in Freeman and Newell (1977). However, there has been no theory which can describe total design as the model building process. Any creative work needs trial-and-error model building as shown in Fig. 4.1. This is the method to approach the goal, i.e. the model representation that satisfies all the requirements, starting from the tentative model and repeating the model analysis/evaluation and model modification.

Thus, computerization of the model-based method requires us to represent explicitly (1) a model for a given problem, (2) model analysis/evaluation and modification rules defined in terms of the model and (3) the model building process as shown in Fig. 4.1 (Ohsuga, 1989). It is desirable that all the representations are made in a declarative way to assure system flexibility and evolutionality.

(*a*) *Model representation*

The body of a complex object is composed of constituents which are either real or abstract entities. Every real object also has a specific shape. These are represented by a data structure. At the same time the complete object or its

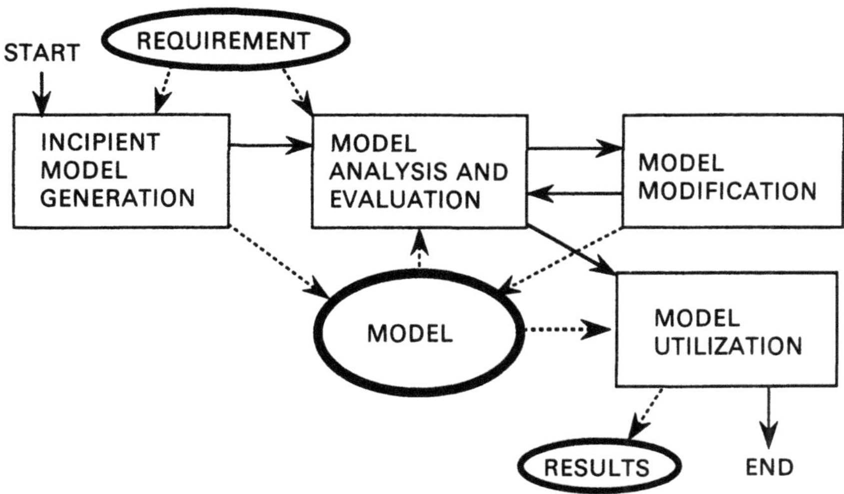

Fig. 4.1. Model building process.

components have functionalities. Let each functionality be represented by a formalized declarative language such as predicate logic. Then an object is represented as a composite of data structure to represent the object body and a set of predicates to represent functionalities. As an object structure and its functionalities are dependent on each other and inseparable, a compound of these must be defined as the unit of model representation. In order to meet this condition, predicate logic to represent functionality must be extended to include any data structure representing object structure as a term.

(*b*) *Model transformation rules*

Model manipulation (analysis/evaluation and modification) is a transformation of the model and is represented in the form of rules in the knowledge processing system. This represents a way of modifying an object structure and/or its functionalities according to changes in each. For example, it must be of the form, 'If Object-model-1 and (Object-structure1 → Object-structure2) then Object-model2 with Object Functionalities2'.

This means that if there is an object model 1 (with the Object-structurel and the Functionalitiesl) and Object-structure1 is modified to Object-structure2, then Object-Functionalities1 change to Object-Functionalities2 resulting in the other object model representation Object-Model2. This includes the structural modification of the model as well as the accompanying modification of the functionalities.

Model transformation is necessary not only for model modification but also for model analysis/evaluation. In many cases, existing computer programs may be used effectively for model analysis. These programs have been developed by human beings based on specific model representations. Hence, in order to use a program in the model-based method of problem solving, this implicit model may have to be created by transforming the object model representation.

(c) *Model building process*

When a design object gets large, it tends to be difficult to reach the goal in a reasonable time by a random application of model manipulation rules. It is necessary to represent explicitly the model building process and control it based on this process representation. This process representation must be independent of both representations of the object model and model transformation rules to ensure the generality of representation. Let it be called a process model. As a process model is a description of the transformation rules, the former must be arranged in a higher representation level than the latter. Let the representation level in which an object model and its transformation rules are located be the object level. Then, the process model must be in the meta-level.

Referring to design again, a design activity and the consequent design product are affected by the design process taken by a designer. For example, if the design process is specified strictly to the detail, then the freedom of any model building activity is restricted and there is no room for innovation. If the process is not so restrictive, however, there is more room for innovation but with less efficiency. Hence, there must be some trade-off between the freedom and the efficiency of design. There is a need for designing and controlling the design process. As the model building process will be represented in the meta-level, this is the design problem in the meta-level. In this case a design object is the process model composed of a structure (a control structure for the object level activity) and its functionalities (control strategies). In order to represent this problem, a level much higher than the meta-level, i.e. a meta-meta-level, is necessary, as the object level and the meta-level were needed to represent the ordinary object model design problem. In order to ensure the optimal solution, the process model must also be represented completely free from the method to obtain the result and a model-based method as shown in Fig. 4.1 must also be used here. There must be some rules to evaluate and modify the process model. Let us note that this design problem in the meta-level is a type of scheduling problem to decide the operations in the object level.

Thus the representation of the processes extends at least up to the meta-meta-levels, but it is not limited there. For example, it is possible and sometimes necessary to introduce a learning capability to modify a design strategy in the meta-meta-level. Hence, the design process model in the meta-meta-level cannot be fixed but must be modified by a still higher level strategy.

Of course, the number of levels used for problem representation must be finite. It depends on each problem. The more sophisticated the functions are required, such as problem solving combined with learning, the higher the level necessary. It is desirable from the viewpoint of these representations that a knowedge-based system be designed in such a way that allows users to define any finite number of levels. Ohsuga (1992) showed that ordinary large-scale problem solving requires at least three levels for representation.

It may be desirable to permit free communication between the levels from the viewpoint of giving the systems high expressibility. However, this induces a difficult problem, as discussed in Ohsuga (1990). It is necessary therefore to restrict this communication but to perform it under the control of the system. Thus, there must be a manager that manages the operation extending over the levels and controls the communication between levels. We call it a level manager.

4.2.3 Built-in knowledge structure and dynamic structuring

If such a process model as shown in Fig. 4.1 can be represented and managed in the system as the standard process of problem solving, then an end user can ignore the problem solving process and focus attention to specifying an object model. The process model is represented in the meta-level. Ordinarily, both model analysis/evaluation and model modification are model transformations in the broad sense. Some operations in this process are complex and need to be expanded to a set of primitive operations which are also represented in the form of transformation rules. An operation is performed by applying these rules in a knowledge base to the result of the last transformation, i.e. a knowledge base is used to carry out a transformation. In order to represent a process model the knowledge base is classified into separate knowledge sources corresponding to the different operations and these knowledge sources are arranged in the order of application.

In many cases some existing analysis methods are used for these transformations. Then each transformation may require a model representation specified to the method. For example, aerodynamic analysis programs, the structural analysis programs, the electronic circuit analysis programs, etc., have been developed in the engineering field. Each analysis problem belongs to the different problem domain and a specific model representation is used, which has been determined in the most convenient way for solving this problem in each domain. Thus, in order to use these existing methods, the system must be able to create the specific analysis models from the general object model in this system.

4.2.4 Formalizing model building

In order to assure the generality of the system, a built-in method of building a process model must be general enough, on the one hand, and at the same time it must be realistic, on the other hand. In this sense, Fig. 4.1 is too simple to represent a real process. Usually more than one requirements are given to a model building process. Even requirement is analyzed and evaluated. The set of the functionalities obtained as the results of these analyses corresponding to the given requirements represents the state of the model or, in the other words, the distance between the current model and the goal. Thus such a model modification rule that reduces this distance must be selected based on this state representation.

In order to represent this state, a set of pairs of the obtained and the required functionality is made first. It is referred as ModelStatus in the sequel, based on which a model modification strategy selects one or more model modification rules. Let all the obtained and required functionalities be put in the object level. Then it is convenient to represent the Model Status by a structure in the meta-level. Its manipulation rules are also put in the meta-level.

Assuming we make this ModelStatus as an intermediate of the model analysis/evaluation and model modification, the model building process of Fig. 4.1 should be rewritten as shown in Fig. 4.2. Figure 4.2 also shows that, if the current model is very far from the goal, a new model should be reconstructed rather than be obtained by modifying the old model. Figure 4.2(a) shows a process flow while Fig. 4.2(b) shows the knowledge sources to represent this process in the real system.

There are a number of model modification rules (in the object level). Each modification rule acts on the model and modifies some aspects (functionalities). One or a set of modification rules which is considered to lead the model to the goal effectively must be selected referring to the ModelStatus. Let these model modification rules be called the ModifRuleSet. Thus the objective of the model modification strategy is to map the ModelStatus into a ModifRuleSet. In general, this transformation requires very sophisticated decision making by looking through the effect of the modification. Since it is difficult to look through the effect correctly, to select the best rule is often difficult. Instead, a finite set of rules may be selected out of the large number of possible modification rules. This set should be small enough to reach the goal as early as possible but be large enough to assure that the best rule is included. Since the set of rules in the object level can be specified by a structure in the meta-level, to decide the set of rules is to construct a new structure in the meta-level. This is a very dynamic operation.

Formalization of this mapping from ModelStatus to ModifRuleSet may not, however, be possible in every case. Therefore part of this decision making may be left to human beings. In this case a process must include a man–machine interaction.

When more than one modification rule is selected, model building becomes a search problem to select one out of a number of possible candidates produced by these modifications. This search algorithm is not simple for model building in the real world. It cannot be categorized simply

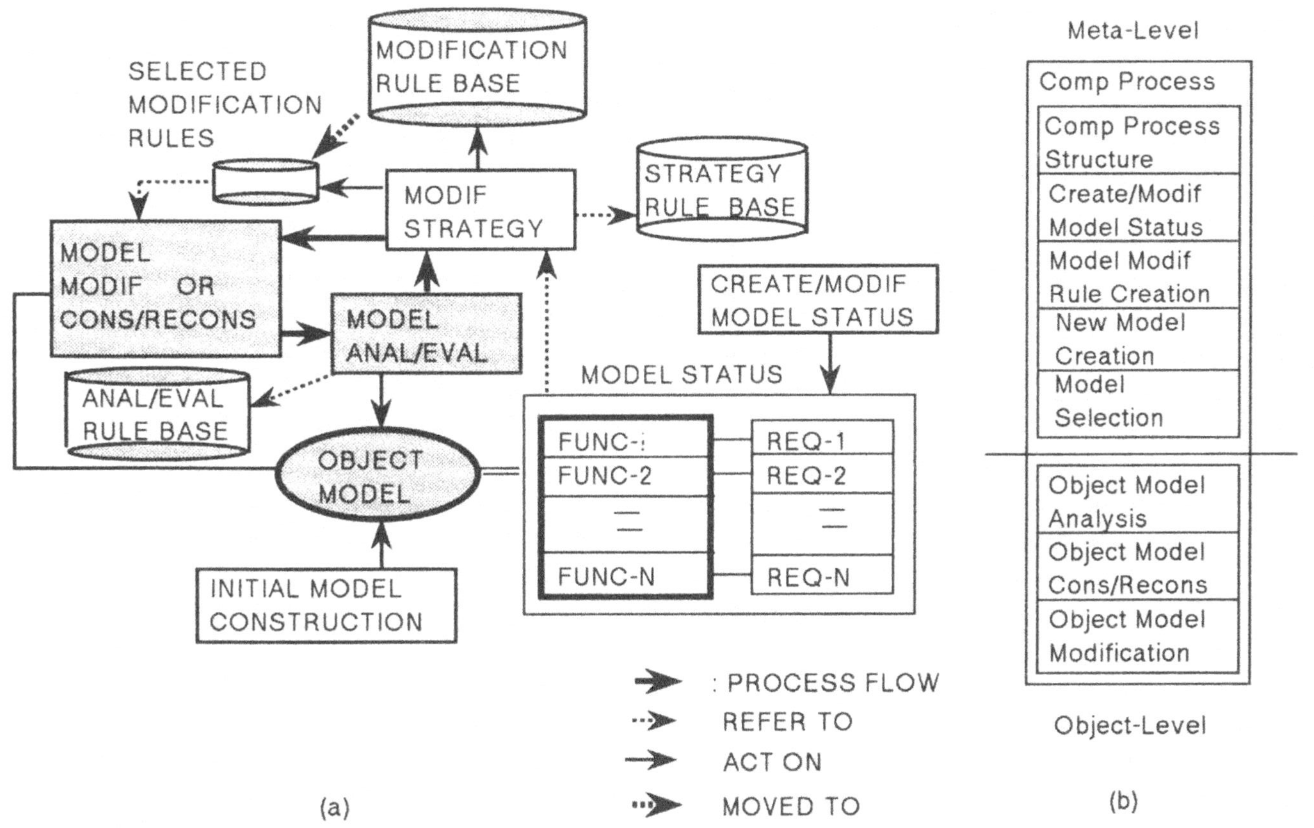

Fig. 4.2. Object model building process.

as being either depth-first or breadth-first but must be combined by the cases. For example, a set of selected modification rules are applied to the current model to create a set of candidate models and these created models are displayed. Let the set of models be denoted ModelSet. If there is a method of evaluating these candidates, the best one may be selected, otherwise a human being selects one out of them. This selected model is analyzed to create a new ModelStatus for the model. This new ModelStatus is compared with that of the old model. If this new one is evaluated as being better than the old one, this model becomes the new model for the next cycle. Otherwise, this one is abandoned and another one is selected from ModelSet. There must be a method to evaluate the new model by comparing its ModelStatus with that of the old model. If there is no such method, a human being must do it. This process is repeated until the desirable model is found. The remaining candidates in ModelSet may be saved if necessary for latter use.

It is possible to change the order of the model selection and its evaluation so that, after the candidate models have been generated, ModelStatus is created for every alternate candidate immediately and a model is selected looking over these ModelStatuses. It seems better than the method mentioned above in selecting better candidates; however, when the set of candidates gets large, this requires a long time and a lage memory. In the following therefore the former method is considered; however it is possible to switch between them dynamically.

If there is no model better than the old one in ModelSet, then the model building process ends in failure. In this case, the control must be sent back to a human being to modify the requirements. When some requirements are changed or added by a human being, some items in ModelStatus are changed or renewed. Then a new set of modification rules must be selected according to the modification of ModelStatus and model modification is executed.

Here, at least three very high level decision-making processes are included in this process which may often exceed the capability of current knowldge processing technology by computers; selection of the modification rules as the mapping from ModelStatus, selection of a new model from ModelSet and evaluation of the new model. Therefore, a knowledge-based method and human decisions must be combined for executing these. The more knowledge that is accumulated on these operations and the more sophisticated reasoning that is developed, then the larger the degree of automation that can be achieved.

4.2.5 Examples of model building

This model-based problem solving system has been applied to some problems by using a knowledge-based system which the authors and their colleagues have developed. The authors and their colleagues have already covered a number of applications such as aircraft wing design (Takasu *et al.*, 1989), a feedback control system design (Guan and Ohsuga, 1988) and chemical compound structure design (Akutsu and Ohsuga, 1988). Two applications among them are picked up here to show that a human being can decide his/her role in problem solving himself/herself with this system. One is to design a feedback control system and the other a chemical compound design. In the first example, a system is designed to create a feedback control system for a given control object which should satisfy a number of (up to seven) requirements represented in the form of performance indices such as rise-time, delay-time, peak-time, etc. This example is presented in the next section. The authors give the system a set of model transformation rules and a rule selection strategy which selects strictly one model transformation rule out of many possible rules in a knowledge base by analyzing the model status. Thus, no selection problems occur and consequently an automatic design system can be obtained.

In the second example, the same system was used to create new chemical compounds. Model transformation rules were collected from previous instances of chemical reactions. These instances were analyzed in advance and only the changed part (substructure) of a chemical structure before and after a reaction was extracted with an accompanying change in the characteristics (functionalities). This is the source of the transformation rules. The design process starts from a known chemical compound as the starting model and applies the transformation rules in the knowledge base to produce new compounds. In this case, chemists as the end users did not want computers to do everything. Instead, they wanted to keep the control of the design process in their hands by making decision themselves. Figure 4.3 shows an example of a demonstration of the operation of this system. Starting from the top chemical compound, the new chemical structures are created successively. These created structures are sent to a data-base of existing chemical compounds and checked there unless the created ones already exist. Then, the compounds are analyzed to obtain their functionalities. Some functionalities, e.g. toxity, are obtained analytically or by using a database in a computer. If these models are not satisfactory, this process is repeated. In this process, transformation rules are classified by the effects which are expected to be caused by applying the rules. For example, rules are classified by whether their applications are expected to cause to decrease in the toxicity of a chemical compound or not. This classification of the rules is represented in the meta-level by a structure and the characteristic (functionalities) of the rules in each group is represented and linked with it. A list of these characteristics of the transformation groups is shown on the display screen together with the current model. The chemist, looking at the display screen, decides in what way he/she should proceed. He/she can indicate his/her decision by selecting one or more items in the list of characteristics. The computer system creates a set of modification rules accordingly and applies them to the current model to

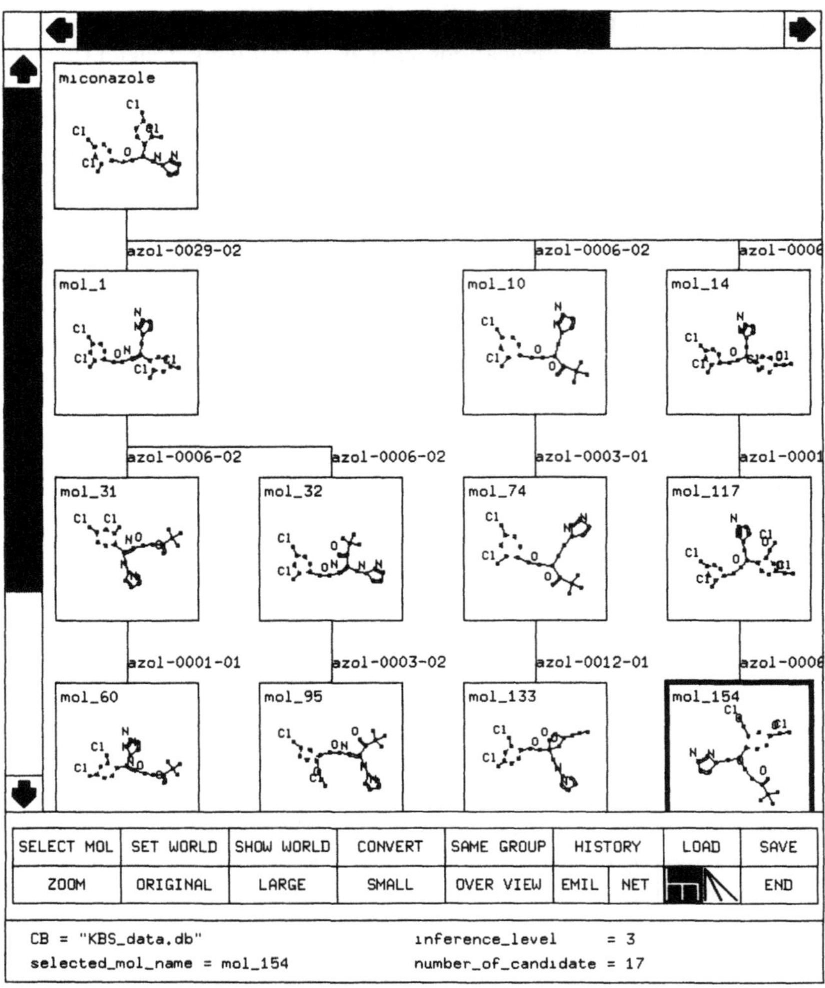

Fig. 4.3. An application of the model-based method to chemical compound design.

produce new models. Usually more than one new model is created and shown. Then the chemist again selects the one which he/she thinks the best to reach his/her goal as fast as possible. This is the opposite case from the first example because automatic design is avoided. This depends on the way of making the strategy as shown in Fig. 4.2.

4.3. KNOWLEDGE BASES FOR FEEDBACK CONTROL SYSTEMS DESIGN

To demonstrate the model building process described in Section 4.2, the authors have developed a knowledge base for automatically designing a feedback control system for a given object. This knowledge base is called MIDUS and is based on a knowledge-based system named KAUS (Knowledge Acquisition and Utilization System). In this section, we describe in detail how the model building process of design is carried out.

4.3.1 Knowledge base configuration

From the viewpoint of inference, the knowledge base is divided into two levels, i.e. object level and meta-level. In the object level, there are data structures and predicates for model representation, and rules for model evaluation and model modification. Model evaluation is an inference process to calculate performance indices, i.e. the design requirements. They are transient response indicated with overshoot, rise-time, delay-time, peak-time and settling-time, and steady response indicated with steady state error coefficient. On the other hand, model transformation is another inference process to transform the structure or to modify the attributes of the model in order to satisfy the design requirements. Structure is a data structure giving the information about the signal flow between the controlled object and the controlling elements. In our system, structure is in the form of a signal graph, which is an equivalent of the well-known block diagram. Attributes are represented as a chunk of information such as transfer functions of blocks, substitutions of values for parameters in the transfer functions.

As described in Section 4.2, the model building process is carried out by repeating the model evaluation and modification circle. On each step, how the model is transformed depends on which trasformation rule is selected. Each time the model is transformed, the new model, which is also called the current model, is analyzed and evaluated. The results are performance indices and are checked with the design requirements. If all the design requirements are satisfied, the current model is the solution. If not, the difference between the performance indices and the design requirements, described as model status in Section 4.2, is calculated in order to select a model transformation rule out of the many possible rules. By applying the selected rule to the current model, a new model is generated and the operation continues. In the case that no more transformation rules are applicable, the design process ends in a failure.

Selecting one transformation rule out of many possible rules is an inference process other than that of model evaluation or model modification. Rules for this inference process should be put in a level higher than the object level, i.e. the meta-level. Terms of meta-level rules can be rules in the object level. In other words, meta-level rules can be knowledge about object level rules. For instance, supposing that there is an object level rule named lead_compensator_1, which is a model transformation rule and whose purpose is to insert a cascade lead compensator into the model, there may be a meta level rule as following.

Meta_rule_1:
(effect lead_compensator_1 ⟨"rise_time", "settling_time"⟩, 95).

This rule is read as: applying lead_compensator_1 to the model may improve the performance indices, which are rise-time and settling-time, with effect of score 95.

When performing inference, more precisely, SLD resolution, in the meta-level, object level rules should not be applied because SLD resolution is only proven complete and sound in a closed set of first-order predicate logic formulas (Akutsu and Ohsuga, 1988).

Thus, rules giving a description of effects of model transformation rules, rules to calculate model status, rules to select model transformation rules, etc., and put in the meta-level. Using these rules, the design process is defined as a search process in the meta-level. Before describing this search process, we give the syntax of the KAUS language in the next section.

4.3.2 KAUS language

In this section, we describe briefly MLL (multi-layer logic), knowledge base system KAUS and the language provided by KAUS. We call this language KAUS language hereafter. In the later sections, we will describe a searching algorithm for design problems using this language.

MLL is a first-order predicate logic extended from many sorted logic (see Hostetter *et al.*, 1982; Yamauchi and Ohsuga, 1985) and KAUS is a knowledge base management system intended to be a general purpose tool for various knowledge-based systems, especially for design aid systems. KAUS provides a powerful knowledge representation language basing on MLL. KAUS is characteristic in that terms can be structures and a set of built-in predicates which simulate the inference mechanism have been provided for meta-level inference. Structures are not only useful in representing models but also useful in representing knowledge structures in the meta-level. Thus, although KAUS uses a fixed computation rule which always selects the left-most atom as the selected atom and a depth-first search strategy the same as that of Prolog, i.e. rules are applied in the order they have been declared, it is possible to obtain stronger completeness in object level.

There is a set of primitive relations involved in MLL as shown in the following.

1. 'Element-of' relation. This is denoted as $x \in X$ meaning an entity x is an element of a set X.
2. 'Power-set of' relation. This is denoted as $Y = *X$ meaning Y is a set obtained by excluding the empty set from the power set of a set X.

3. 'Product-set-of' relation. This is denoted as $Y = \Pi_{i=1}^{m} X_i$ meaning Y is the Cartesian product of $X_1, X_2, \ldots, X_m$.
4. 'Component-of' relation. This is denoted as $X \lhd R$ meaning X is a component of Y.

Various relations can be derived from these primitive relations as compound relations. For example, 'subset-of' relation denoted as $X \subset Y$ meaning X is a subset of Y can be composed of $Z = *Y$ and $X \in Z$. Among others, the 'component-of' relation is the most important one in MLL. It provides a method to represent object models. Unlike 'subset-of' or 'element-of' relations, which represent a mathematical relation between entities, the 'component-of' relation can be declared between any entities to represent any structures in favor of individual interpretation. For example, in this paper, 'component-of' relation is used to represent knowledge structures in the meta-level. A knowledge structure, say K_s, in the meta-level has several sets of object level knowledge as its components. Let one of them be K_{s1}. Then it is clear that $K_{s1} \lhd K_s$ holds. Between K_s and elements of K_{s1} there exists a so-called 'part-of' relation. The former and the latter are said to form two different layers and the former is one layer higher relative to the latter. Also, any element of K_{s1} can be a structure like K_s such that multi-layers can be formed to represent complicated structures.

The syntax of formulas in MLL is similar to that of many sorted logic (Enderton, 1972) in the sense that quantified variables can have a domain name in their prefixes. The differences are that quantified variables can be used as domain names for the other variables and that not only 'element-of' relations but also 'part-of' relations can appear in the prefixes of formulas. Thus KAUS consists of two parts, i.e. commands to define structures and MLL formulas.

There are three primitive commands for user to declare structures:

1. The command of the form
   !sk_p $*n_1 X_1 * n_2 X_2 \ldots *n_k X_k$;
   declares concept nodes where n_i is a positive integer and X_i is a node name.
2. The command of the form
   !sk_e *set* $e_1 e_2 \ldots e_n$;
   declares 'element-of' relation.
3. The command of the form
   !sk_c *object* $p_1 p_2 \ldots p_n$;
   declares 'part-of' relation.

Any structure defined using these commands can be involved in MLL formulas. An MLL formula can be any AND–OR structure of predicates composed of three logical connectives: $\sim$ (NOT) & (AND) and | (OR). The $\rightarrow$ (IMPLY) is represented by $\sim$ and | connectives because $A \rightarrow B$ is logically equivalent to $\sim A | B$. A well-known example is given as follows where 'A' is the universal quantifier. The AND–OR expression in the matrix part is represented as a prefix form. By the way, the existential quantifier is '*E*'.

[A*X*/man] (mortal *X*).

KAUS is implemented with C language on the UNIX operating system. The user can write built-in predicates such as predicates to perform mathematical calculations using C language. Such predicates are called procedural type atom (PTAs). For example, the user can call a procedure with a predefined name as follows:

[EY?/float]($sin *Y* 1.0)?

where *Y* is bound to sin(1.0) as if a ground atom of the form:

(sin 0.841471 1.0).

has been declared. The dollar sign is used to distinguish PTAs from ordinary atoms, which are called non-procedural type atoms (NTAs).

What concerns us most here is a set of PTAs provided by KAUS for meta-level inference. Only those used in this paper are described.

First, it is necessary to make it clear how rules are identified in KAUS. Rules are identified with rule identifiers or their predicate names. When defining a rule, the user can give an identifier to it as follows:

\\rule_1: [AX/*int] [EY#/*X*] [AMax#/*X*] (| (max *X* Max) ~($ge Max *Y*)).

After the rule has been defined, the predicate name 'max' can be used in other rules, which are often meta-rules, as well as the identifier '\\rule_1'. For example, suppose 'insert_lead_compensator' is a model transformation rule in object level, a meta-level rule can be written as

(| (design Model Goal)
~ ($eq Goal "settling time")
~(apply_rule [insert_lead_compensator Model])
).

This rule can be read as: if the settling time of 'Model' is to be improved then apply 'insert_lead_compensator' in object level.

Now, two PTAs which are most important for meta level inference are given as follows.

1. 'scopeKUs', PTA to make scope of applicable rules. As indicated in the previous sections, it is necessary to specify the object level rules and in what order should they be applied. In fact, that is the control information for object level inference. This information is given as an ordered set that is called a local world or world. The most general world is the whole knowledge base represented by a reserved symbol 'univ'. 'scopeKUs' is a means to make use of such information. After 'scopeKUs' has been

applied, the search of applicable rules for successive subgoals is restricted to the world. For example, the query

(& ($scopeKUs '*World*1) (*solve X*))?

resolves the subgoal (*solve X*) only with rules belonging to the set "*World*1'. If the current knowledge base is

*r*1: (*solve* 1).
*r*2: (*solve* 2).
*r*3: (*solve* 3).

and "*World*1' is $\{r2, r1\}$, the answer will be

$X = 2$
$X = 1$

while the answer will be

$X = 1$
$X = 2$
X = 3

when 'scopeKUs' not being applied. Note that not only the number of the answers but also the order of them is different. By the way, atom names such as "*World*1' are preceded by a "' symbol and variables all begin with an upper case character in KAUS language.

2. 'resolve', PTA to resolve the given query in the given world. When control information for the object level has been obtained in the form of a world, the next thing to do is to perform object level inference within the world. 'resolve' is the PTA for the purpose. The resolution is based on the fundamental SLD-refutation procedure with depth-first search strategy. For example,

   ($resolve *Query*1 '*World*1 *Truth Result*)?

   will resolve '*Query*1' in the world "*World1*' and the variable '*Truth*' is 1 if the empty clause was derivated and 0 if it was not. '*Result*' is bound to a list holding the result of unification, i.e. the correct answer substitutions.

4.3.3 Object model representation

When a human designer designs a feedback control system basing on classical control theory, he/she is used to the block diagram model as shown in Fig. 4.4. The system consists of elements indicated with blocks and the function of each block is given with its transfer function. Signal flow between blocks is given by signal lines with an arrow on the head showing the signal's transfer direction. Usually, there is at least one input signal and at least one output signal. In Fig. 4.4 they are R and Y, respectively.

In MIDUS, signal flow graphs are used instead of block diagrams. These two are equivalent representations. The corresponding signal flow graph of

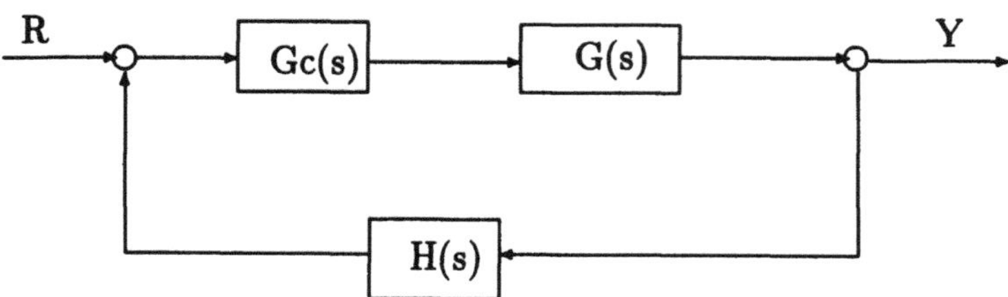

Fig. 4.4. Example of block diagram.

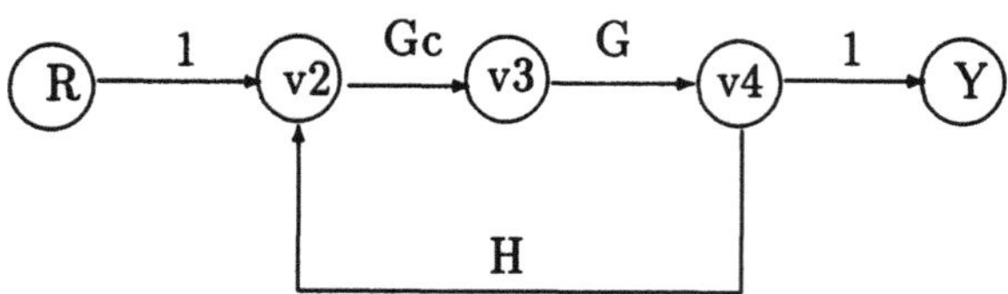

Fig. 4.5. Signal flow graph of Fig. 4.4.

the block diagram shown in Fig. 4.4 is given in Fig. 4.5. By using the signal flow graph, it is convenient to generate the closed loop and open loop transfer function from the input signal to output signal based on Mason's gain rule (Hostetter *et al.*, 1982). Basically, these two transfer functions are necessary for calculating performance indices. On the other hand, to a human being, block diagrams are more comprehensible than signal flow diagrams. MIDUS has a man-machine interface which allow the user to enter block diagrams into the system. A knowledge base was developed to transform block diagrams into signal graphs. Discussion of this diagram cognition knowledge base is out of the scope of this paper.

As shown in Fig. 4.6, a graph structure is represented with two sets, i.e. a vertex set and an arc set. The elements of the vertex set are vertexes of the graph, while the elements of the arc set are arcs of the graph. Each arc consists of three components. The first and the second one are elements from the vertex set that are connected by the arc. The third one is a string which is an expression giving the transfer function. Transfer functions are given in the syntax of REDUCE, an expression processing system.

4.3.4 Model evaluation

Model evaluation is an inference process to calculate performance indices of the model. Performance indices are checked with the design requirements. There are two kinds of performance indices, i.e. transient response and steady state response. Deriving the closed loop and open loop transfer functions from the signal flow graph is the first step of model evaluation.

```
!sk_c graph vertex arc;
!sk_e vertex #R #v2 #v3 #v4 #Y;
!sk_e arc #GO #Gc #G #G1 #H;
!sk_c #GO #R #v2 '1';
!sk_c #Gc #v2 #v3 '5*(s+1)/(s+2)';
!sk_c #G #v3 #v4 'k*s/(s**2+5*s+7)';
!sk_c #G1 #v4 #Y '1';
!sk_c #H #v4 #v2 '-1';
```

Fig. 4.6. Figure 4.5 represented in KAUS language.

4.3.5 Transfer function synthesis

Based on Mason's gain rule, given an input and an output signal, the corresponding closed loop transfer function can be synthesized in a systematic way. Basically, two steps are needed, where the first step is to retrieve information from the graph structure and the second step is to synthesize the transfer functions based on the information.

Firstly, it is necessary to generate all paths from the input signal to the output signal. All paths can be represented as a set of nodes on it or a set of arcs constituting the path. Paths set with each element as a node set and set with each element as an arc set are both necessary.

Secondly, it is necessary to generate the loop set from the input signal to the output signal. The same as a path, a loop can be represented as a set of nodes on it or a set of arcs constituting it. Two loops are said to be non-touching if there exists no node which belongs to both of the loops. Similarly, a path and a loop are said to be non-touching if there exists no node which belongs to both of them.

Thirdly, it is necessary to generate the power set of the loop set, and exclude from the power set the empty set and such sets that have at least two loop sets touching to each other. Clearly, sets with only one loop set as their elements are never excluded. The resultant set is divided into subsets C_l such that the number of loops of each loop set in C_i is i. All these operations can be performed with rules written in KAUS. In fact, like programs written in Prolog, these rules can be implemented in a method using recursion. However, it is more efficient to use procedural graph manipulation algorithms. In MIDUS, graph structure represented in KAUS language is transformed into data structures suitable for such algorithms and passed to the procedures implementing the algorithms, using the mechanism of PTAs described in the previous section. The results are passed back as the variable binding of resolution.

Thus we are ready to describe the concepts of path gain, determinant of a signal flow graph and cofactor of a path, which are used in the step of transfer function synthesis. The path gain of a given path is the product of all the transfer functions to arcs that constitute the path. In the same way, the loop gain of a given loop is the product of all the transfer functions to arcs that constitute the loop. The determinant of a signal graph is defined as:

$$\Delta = 1 + \sum_{i=1}^{N} (-1)^i \sum_{S_j \in C_i} \Pi_{k \in S_j} L_k,$$

where L_k is the loop gain of loop k. The cofactor of a path is the determinant of the graph generated by eliminating all loops touching the path.

The closed loop transfer function $T(s)$ of a signal graph is generated according to:

$$T(s) = \frac{\Sigma_{i=1}^{N} P_i \Delta_i}{\Delta}$$

where P_i and Δ_i are the path gain and the corresponding cofactor, respectively, and Δ is the determinant. The open loop transfer function is given as:

$$G(s) = \frac{T(s)}{1 - T(s)}$$

Obviously, transfer functions synthesized in this way are redundant ones. In MIDUS, such expressions are sent to REDUCE via the network and the reduced expressions are passed back. This is why expressions have to be input in the syntax of REDUCE. Once the transfer functions are generated, it is easy to obtain the performance indices.

4.3.6 Performance indices calculation

When the closed loop transfer function is given as

$$T(s) = \frac{b_1 s^{n-1} + b_2 s^{n-2} \cdots + b_{n-1} s + b_n}{s^n + a_1 s^{n-1} + \cdots + a_{n-1} s + a_n}$$

the corresponding state equation is

$$\frac{\mathrm{d}X(t)}{\mathrm{d}t} = \mathbf{A}X(t) + \mathbf{b}^T u(t) \; y(t) = \mathbf{c}^T X(t)$$

where

$$\mathbf{A} = \begin{bmatrix} 0 & 0 & \ldots & 0 & -a_n \\ 1 & 0 & \ldots & 0 & -a_{n-1} \\ 0 & 1 & \ldots & 0 & -a_{n-2} \\ \ldots & \ldots & \ldots & \ldots & \ldots \\ 0 & 0 & \ldots & 1 & -a_1 \end{bmatrix}$$

$$\mathbf{b}=[b_n \quad b_{n-1} \quad b_{n-2} \quad \dots \quad b_1]$$

$$\mathbf{c}^T=[0 \quad 0 \quad 0 \quad \dots \quad 1]$$

and $u(t)$ is the input signal that is a function of time.

Transient response is calculated by solving the state equation with $u(t)$ always having a value of 1. In MIDUS, there is a rule which generates a file containing data structure declarations and initializations for matrices **A**, **b** and $\mathbf{c}^T$, and, in addition to those, a function named $u(t)$ and returning 1.0 as its value. The generated data structures and function are written in C language. The generated file is compiled and linked with a subroutine which numerically solves the state equation and an executable module is obtained. This compilation and linkage are performed by executing the C compiler using a built-in predicate exec. In the same way, the resultant program is also executed with the exec predicate. Results of the calculation are asserted as facts so they can be referred to during model transformation.

Since the purpose of this section is to give an example of model building in the meta-level, we are not going to described the model evaluation process in more detail here.

4.3.7 Model transformation

In the case that not all the design requirements are satisfied, it is necessary to modify the model, including changing values of adjustable parameters of transfer functions and manipulating its graph structure. Manipulating the graph structure means inserting a block called a compensator in the forward or feedback path. The former is called cascade compensation while the latter is called feedback compensation. Model transformation can be performed by man–machine interaction but, more importantly, it is possible to be performed automatically using meta-level inference. The purpose of meta-level inference is to select one model transformation rule out of many such rules and apply it to the current model to generate a new model. Rules or knowledge for model transformation must be constructed so that the process of model evaluation and transformation can be carried out systematically.

4.3.8 The process of automatic design

Starting from the initial model, an object model that is assumed to satisfy all the design requirements is generated step by step. From another point of view, this is a search process with a space constructed with the initial model, tentative models and the final model.

In classic control theory, there is no deterministic method to synthesize a cascade or feedback compensator which can make a given model satisfy all the design requirements. Instead, only several fundamental compensators can be used. To make it simple, we only take cascade lead and cascade lag compensators into account in this paper. Cascade lead compensators are used to improve transient response, while cascade lag compensators are used to improve steady state response. Obviously, when using these fundamental compensators only part of the design requirements can be satisfied each time a new model is generated. Moreover, because satisfying some requirements may cause deterioration of other performance indices, it is possible for the search process to diverge. For instance, enhancing rise-time always deteriorates overshoot, and enhancing steady state response deteriorates transient response. Performance deterioration makes the design process both time and memory consuming. Thus when applying cascade compensators it is important to prevent such deterioration as far as possible. The process of automatic design can be given as below.

1. Evaluate the initial model to get performance indices. This is carried out with an inference process in the object level.
2. Check the performance indices with the design requirements. A subset of design requirements is obtained whose elements are ones that are not yet satisfied. If this set is empty, then the design succeeds. Otherwise continue.
3. From the set made at step (2), derive a set of compensators which have the possibility to improve performance indices corresponding to the design requirement set. Because compensators are really rules to apply them to a model, this process must be performed in the meta-level.
4. Get the first compensator from the set generated at step (3) and perform a precondition check with the current model. If the current model does satisfy the precondition, then apply the compensator to generate a new model. Next evaluate the new model to see if it is more improved than the current model. If so, recursively go back to step (2) with the new model as the current model. If it is not improved or if the current model does not satisfy the precondition, go on with the next compensator until there is no more compensator in the set.
5. In the case that no more compensator is left in the set, backtracking to upper recursive level occurs. If this backteracking is impossible, the design process fails.

4.3.9 Knowledge structure

As described above, it is necessary to associate object level knowledge with other knowledge. For example, a rule to determine whether the open loop dominant poles of a given model are real numbers is associated with a rule which implements the insertion of a cascade lead compensator into the model. The former is the precondition of the latter. In MIDUS, such an association is represented with a data structure called knowledge structure.

A knowledge structure consists of two components. The first one is a set giving preconditions. Each precondition is itself a set of goals, i.e. if each goal is inferred to be true then the precondition holds. The second one is a set containing a rule to insert a cascade compensator into a given model. This set is called action. Whichever precondition in the precondition set holds, the action can be performed to generate a new model. An example is given as below:

```
!sk_e theorySet leadCompensator_1;
!sk_c leadCompensator_1 leadCompensator_1_preconditions;
!sk_c leadCompensator_1 leadCompensator_1_action;
!sk_e leadCompensator_1_preconditions
{openLoopDominantPoleIsReal,systemTypeisMoreThanOne};
!sk_e leadCompensator_1_action insert_lead_compensator_1;
```

A knowldge structure named leadCompensator_1, which is an element of theorySet, is defined. Its preconditions and action components are also defined. The precondition component is a set with elements as a set of goals. The action component is a set of rules for cascade lead compensator insertion. The knowldge structure leadCompensator_1 gives information as, if a model's open loop dominant pole is a real number and its system type is greater than 1, then the rule insert_lead_compensator_1 can be applied to it to improve the transient response.

4.3.10 Meta-knowledge for compensator selection

Now we are ready to describe the knowledge base for compensator selection. Basically, this knowledge base consists of two parts, i.e. the static part and the dynamic part. The static part is as below;

[AGoal/{"velocity_constant","steady_error"}]
(effect lagCompensator Goal 100). (1)

[AGoal/{"rise_time","delay_time","peak_time","overshoot"}]
(effect lagCompensator Goal −20). (2)
(effect lagCompensator "settling_time" −100). (3)

[AGoal/{"rise_time","delay_time","peak_time","settling_time"}]
(effect leadCompensator_1 Goal 95). (4)
(effect leadCompensator_1 "overshoot" 10). (5)

The static part is a fact base that describes general knowledge about the effects of the compensators. Item (1) stands for that a cascade lag compensator represented by knowledge structure lagCompensator has the effect of improving velocity constant and steady error to an extent of score 100. Item (2) means that a cascade lag compensator will cause a deterioration in the performance indices of rise-time, delay-time, peak-time and overshoot to an extent of score 20. Item (3) means that a cascade lag compensator will cause a deterioration in settling time to a large extent. In the same way, items (3), (4) and (5) mean that the cascade compensator leadCompensator_1 has the effect of improving rise-time, peak-time, delay-time and settling-time to a large extent, and improving overshoot to a moderate extent. In MIDUS, six cascade lead compenators, one cascade lag compensator and one lead-lag compensator are implemented. Lead compensators differ in the way that they are constructed and they have different scores on performance indices.

Given a set of performance indices which needs to be improved, an ordered set that contains candidate compensators can be obtained from the fact base. This is an inference process in the meta-level. The order is determined in such a way that compensators that can possibly satisfy or improve more performance indices come nearer to the head of the set.

The first arguments of the effect predicates are knowledge structures described above. They consist of preconditions and actions. Checking the preconditions with the current model is the dynamic part of compensator selection. If the preconditions do not hold, the compensator is eliminated from the candidate set even it has a high score. Precondition checking is an inference process in the object level. For example, the precondition Open-LoopDominantPoleIsReal is an object level rule which is true only if the open loop dominant pole of the model is a real number. Thus a mechanism for initiating object level inference and getting the result of the inference is necessary. In fact, this is carried out with KAUS' level manager. KAUS' level manager provides a built-in rule called resolve to initiate an inference process in a level which is an object level relative to the current level. So, for example, even if a cascade lead compensator has the highest score, in the case that its preconditions do not hold, another cascade lead compensator with a lower score will be applied.

4.3.11 Rules for automatic design

In this section, we describe a rule called design which implements the automatic design process. The definition of the rule is given as below.

```
[AG/sgraph][AIV,TV/_][AGoals/*str](
|(design G IV TV)
 ~(eval_model G IV TV)
 ~(pickupGoals G IV TV Goals)
 ~(aux_design G IV TV Goals)
[AG/sgraph][AIV,TV/_][AGoals/*str](
|(aux_design G IV TV Goals)
 ~(isNullSet Goals)
 ~(printf2 "\nDesired model is %n\n" G)
).
```

```
[AG,NewGraph/sgraph][AIV,TV/_][AGoals/*str][AWorld/*int](
|(aux_design G IV TV Goals)
 ~(notNullSet Goals)
 ~(plan G IV TV Goals World)
 ~(apply_rule World G IV TV NewGraph Goals)
).

[AWorld/_][AG,NewGraph/sgraph][AIV,TV,Iv,Tv/_]
[AGoals,NewGoals/*str][AN,Truth/int][ATheory,Action,Rule/_](
|(apply_rule World G IV TV NewGraph Goals)
~(get_candidate World Theory)
~(check_preconditions Theory G IV TV)
~(getAction Action theory) ~(get_e Rule Action 1)
~(resolve [Rule G IV TV] univ Truth)
~(eq Truth 1)
~(getNewGraph G IV TV NewGraph Iv Tv)
~(eval_model NewGrah Iv Tv)
~(pickupGoals NewGraph Iv Tv NewGoals)
~(judge_effect Theory NewGoals NewGraph Iv Tv Goals G IV TV)
~(aux_design NewGraph Iv Tv NewGoals)
).
```

Given a signal flow graph G, which the current model, and its input signal IV and output signal TV, the design process begins by firstly evaluating the performance indices of the current model. The rule eval_model synthesizes transfer functions, generates an executable program and executes it to calculate all performance indices. Next, the rule pickGoals compares the performance indices with the design requirements and sets up a set containing performance indices, i.e. goals which are not satisfied yet. Then the rule aux_design is fired.

aux_design first checks if Goals is a null set. If so, then the design process ends in success and the current model is the desired one. If not, the rule plan selects candidate compensators out of the fact base described above and the result is the set World. Then the selected compensators are applied one by one by the rule apply_rule. apply_rule first gets out a knowledge structure from the set World and then checks its preconditions with the current model. If the preconditions do not hold, the next knowledge structure is tied by backtracking to the rule get_candidate. If the preconditions hold, it is applied to the current model and a new model is generated. Applying the rule contained in the knowledge structure is an inference process in the object level and is initiated with the resolve rule, which is a built-in rule provided by KAUS' level manager. The result is returned in a variable Truth and a value of 1 means that the inference in the object level succeeded. Next, the new model is evaluated and its performance indices are compared with the design requirements and NewGoals is obtained. The rule judge_effect tests to see if the new model has better performance indices than the current model. If so, the design process proceeds with the new model set to the current model. If not, backtracking to get_candidate occurs and the next compensators will be tried.

4.4 CONCLUSION

A knowledge-based system that can solve design problems was discussed. It is provided with a new method of model building. A model-based approach of problem solving is a general method which is used ordinarily by human beings. With this method, representations of the problem and of the problem-solving process can be separated. Thus, computers provided with this method allow an end user to focus his/her attention to the description of a problem without worrying about the description of the method to solve this problem by a computer. This method of problem solving using computers requires a representation scheme extending over the different levels of description. Such computers therefore must be provided with the multiple meta-level architecture including an object level, a meta-level, a meta-meta-level, etc.

This method is quite general and is applicable to problem solving in the different problem domains. The only parts that are different in each specific problem in this representation are the knowledge in the object-level to manipulate a model directly and a part of the control strategy in the meta-level to decide the way of doing and the extent to which a human being is concerned with the problem-solving process. This strategy can be decided by some expert of each problem domain in advance. Then end user can do everything without knowing about details of the problem-solving method and the computer or a few different strategies may be provided so that each user can select one out of them.

The authors and their colleagues have developed a prototype of a system which could implement this method and have proved its effectiveness.

REFERENCES

Akutsu, T. and Ohsuga, S. (1988) *CHEMILOG – A Logic Programming Language/System for Chemical Information Processing.* Proceeding of the 3rd International Conference on Fifth Generation Computing Systems, pp. 1176–83.

Enderton, H.B. (1972) *Mathematical Introduction to Logic*, Academic Press, New York.

Freeman, P. and Newell, A. (1977) *A Model for Functional Reasoning in Design.* Proceeding of the 2nd IJCAI.

Guan, J. and Ohsuga, S. (1988) An intelligent man–machine system based on KAUS for designing feedback control systems, in *Artificial Intelligence in Engineering: Design*, Elsevier Computational Mechanics Publications, Amsterdam, pp. 241–64.

Hostetter, G.H. *et al.* (1982) *Design of Feedback Control Systems*, Holt-Saunders Japan, Tokyo.

Ohsuga, S. and Yamauchi, H. (1985) *Multi-Layer Logic – A Predicate Logic Including Data Structure As Knowledge Representation Language, New Generation Computing, Vol. 4,* (*Special Issue on Knowledge Representation*), Ohmsha Springer-Verlag, Berlin.

Ohsuga, S. (1989) Toward intelligent CAD systems. *Computer Aided Design*, **21**(5), 315–37.

Ohsuga, S. (1990) Framework of knowledge based systems – Multiple meta-level architecture for representing problems and problem-solving processes. *Knowledge Based Systems*, **3**(4)

Ohsuga, S. (1992) How can knowledge based system solve large problems – model-based decomposition of problem solving. *Knowledge-Based Systems*, **5**(3)

Takasu, A. *et al.* (1989) *Intelligent Wing Design Support System.* Proceedings of the 2nd Scandinavian Conf. on AI

Yamauchi, H. and Ohsuga, S. (1985) *KAUS as a Tool for Model Building and Evaluation.* Proceedings of 5th International Workshop on Expert Systems and Their Applications, Avignon, France.

CHAPTER 5

A knowledge-based system for selection of resource allocation rules and algorithms

Gürsel A. Süer and Cihan H. Dagli

5.1 INTRODUCTION

Assembly lines have been used in industry since the early 1920s. Ford was the first to establish assembly lines in the automative industry. The productivity of the factories improved considerably when compared with previous production methodologies that were utilized (Wild, 1972).

The assembly line concept found wide acceptance and was later extended to many other industries.

The task of designing assembly lines became more complex in time due to increased product variety and options available to customers. Hence, the multi-model and mixed-model approaches emerged as ways of handling this problem. In single-model assembly lines, a line is dedicated to a single model (Fig. 5.1a) whereas in multi-model assembly lines the models are grouped and processed in batches (Fig. 5.1b). In mixed-model lines, the batch size makes the sequencing of models an important task (Fig. 5.1c). Mixed-model assembly lines are widely used in manufacturing systems that adopt a just-in-time philosophy.

In some industries, there may be limitations for the assignment of different models to assembly lines due to investment required for machinery and equipment. In others, the task of assigning models to lines may be highly flexible if operations are labor-intensive that require light weight tools and equipment or if robots are used to perform a variety of different operations in an automated area. Under these circumstances, no limitation on the number of lines nor the number of workstations in a line are imposed by the system. The latter case will be discussed in this chapter.

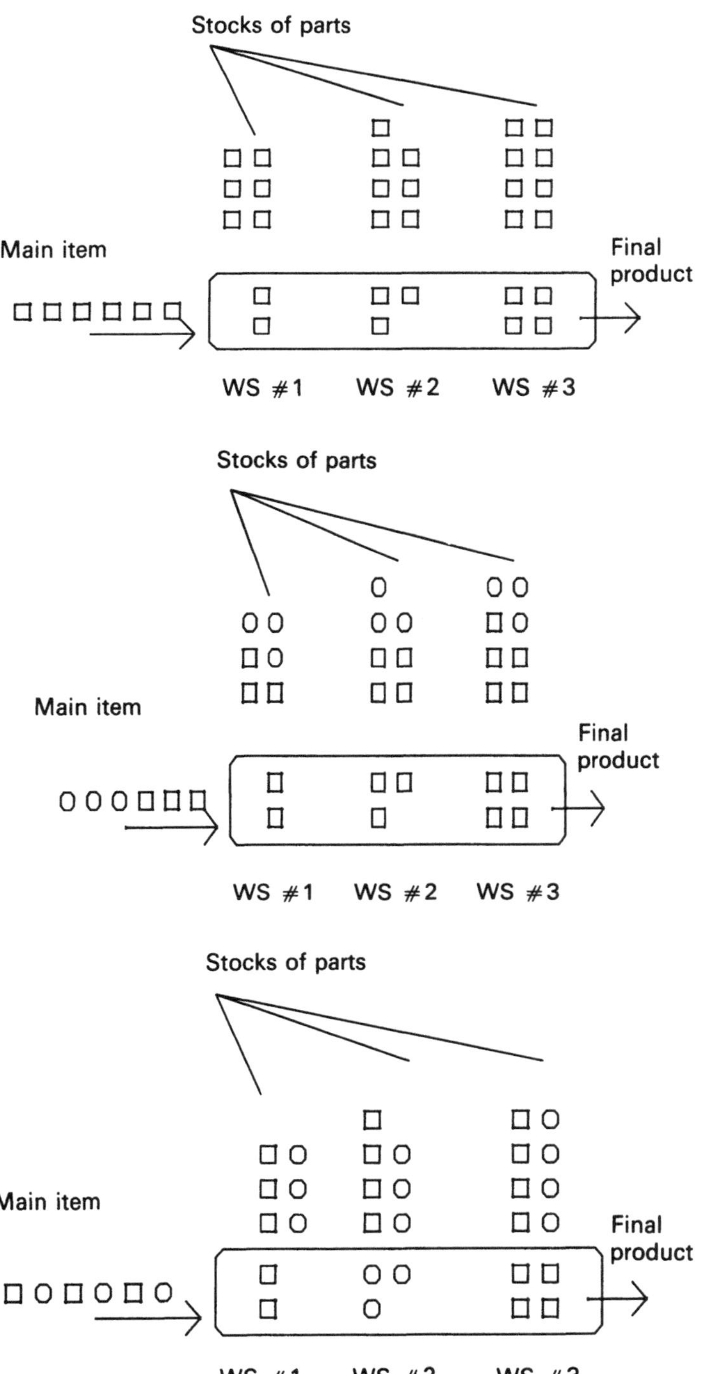

Fig. 5.1. Operation of assembly lines. (a) Single-model assembly lines. (b) Multi-model assembly lines. (c) Mixed-model assembly lines.

5.2 KNOWLEDGE-BASED SCHEDULING SYSTEMS

Knowledge-based scheduling is one of the new application areas for expert systems. There are only a few completed systems available. However, the interest in the area is growing among researchers. Most of the knowledge-based scheduling systems are developed as a combination of various subsystems such as simulation modules, database modules, operations research techniques and scheduling algorithms.

Miller (1984) developed a heuristic-based system for automatic generation of schedules in a semi-automated manufacturing environment. The approach used is the generation of a tree of possible schedules. The search space is limited to only feasible and promising schedules. Bruno, Elia and Laface (1986) developed an expert scheduling system for scheduling parts in a flexible environment with simulation subsystems for schedule evaluation. Ben-Arieh (1986) suggested a knowledge-based system for automated manufacturing with a simulation subsystem as well. Shen and Chang (1986) developed two real time scheduling algorithms for a flexible manufacturing environment. They suggested the integration of planning and scheduling functions, and the use of knowledge to improve the efficiency of an algorithm. Shaw (1986) described an artificial intelligence approach for solving a planning problem that can be decomposed into a number of subproblems. A combination of resource constraints and heuristic rules is used for flexible manufacturing system (FMS) scheduling. Morton and Smunt (1986) developed an expert system that has four levels of hierarchical structure for solving FMS problems including project planning and scheduling. Dagli and Cihangirli (1988) developed a prototype expert system for scheduling in a flexible manufacturing environment where the complexity of scheduling problems is simplified by breaking a problem into manageable subproblems.

Chang (1985) proposed a system that acts as a decision support system which allows for interactive uses for job shop scheduling. Erchler and Esquirol (1986) presented a job shop scheduling system. Bensama, Bel and Dubois (1988) developed OPAL which integrates a constraint-based analysis module with a rule-based decision support module for job shop scheduling.

O'Connor (1984) developed ISA and IMACS at Digital Cooperation. ISA is used for scheduling customer orders, and IMACS is an expert system for capacity planning and inventory management. Fox and Smith (1984) developed the ISIS system which formalizes the various scheduling constraints in the systems knowledge base and direct the search according to the hierarchy of tasks. Chiodini (1986) presented a real time replanning system interfaced with a material requirements planning (MRP) system. The system minimizes the impact of the perturbation and maintains production stability. Fukuda, Takeda and Hayashi (1986) developed an expert system concept for production control. Parunak (1987) developed YAMS, a planning and control system, to illustrate the distributed AI methodology. CASCADE is another system developed by the same author for material handling applications.

Borne and Fox (1984) developed TRANSCELL for cell management. Süer and Dagli (1992) developed a prototype knowledge-based system for solving single machine scheduling problems.

5.3 PROBLEM DEFINITION

The system to be analyzed is composed of assembly operations where all of the operations are manual or require light weight and low cost tools and equipment or they are performed by robots.

More than one assembly line may be in operation at any given time. The number of lines and the number of workstations in a line may vary with time. However, once a model is assigned to a line, the number of workstations (operators, robots) remains unchanged until the entire batch is completed. Simultaneous assembly of a model on more than one line is not allowed.

There are n models to be assembled (available at time zero) with a total of R resources. The maximum number of lines is limited by the number of resources since at least one resource has to be assigned to each model. Therefore, the problem is to assign n different models to a varying number of lines with a varying number of workstations in each line over a period of time T, such that the total number of available resources, R, is not exceeded. The limitations can mathematically be shown as:

$$1 \leqslant NL_t \leqslant R \quad t = 1, 2, 3, \ldots, T$$

$$0 \leqslant WS_{i,t} \leqslant R \quad t = 1, 2, 3, \ldots, T; i = 1, 2, 3, \ldots, n$$

$$\sum_{i=1}^{n} WS_{i,t} \leqslant R \quad t = 1, 2, 3, \ldots, T$$

where T is the planning horizon considered, n is the number of models, R is the total number of resources, NL_t is the number of lines at time t and $WS_{i,t}$ is the number of resources allocated to model i at time t.

There are two possible cases:

1. $n > R$: the number of models is greater than the total number of resources. Not all of the models can start at time zero. Some models are assembled only after another is completed and resources become available. In that sense, the net effect may be considered similar to that of multi-model lines.
2. $n \leqslant R$: the number of models is less than or equal to the total number of resources. Mathematically, it is possible to start all models at time zero. Therefore, there may be several lines working on different models at the

same time. The net effect may be considered similar to that of mixed-model lines since all products may be obtained at the same time or within a very short period of time. However, this may not be the best alternative when line efficiencies are considered. Real world working environments are in most cases similar to those represented in case 1.

Six line efficiency sensitive rules (RESMAX, RESMIN, FMAX, FMIN, TOTMAX and TOTMIN) are suggested and their performance is compared against two measures, i.e. average flow time and makespan, under six different schedule generation policies (delay, delay/modified, delay stop, delay stop/modified, non-delay and non-delay/modified). The schedule generation policies aim to obtain the highest possible efficiency for each product considering the fact that lower average flow time and makespan values can be attained.

5.4 PERFORMANCE MEASURES

The performance of these rules are measured against two criteria often used in scheduling research:

1. Average flow time: time that a model remains in the system is known as flow time. In other words, it is defined as the time elapsed between when a model becomes available and when it is completed. The average for all models is calculated and used as a measure.
2. Makespan: the time that the last model in the system is completed is known as makespan.

5.5 BACKGROUND

In this section, first cycle time versus line efficiency and cycle time versus number of workstations are analyzed, and the priority matrix is derived. The priority matrix forms the basis on which the scheduling rules are developed. Dagli and Süer (1986) used the priority matrix to develop a mathematical programming model to find the optimal allocation of resources with the objective of maximizing the overall efficiency in a previous work.

5.5.1 Cycle time versus line efficiency

Cycle time is the time allocated to each workstation to complete the operations assigned to it. Station time (time required to perform the assigned operations) cannot exceed the cycle time. If all of the workstations have a station time equal to cycle time, then the line efficiency is 100%. However, in practice it is almost always less than 100%.

Station and line efficiencies are defined as follows:

$$SE_i = \frac{ST_i}{CT} \quad LE = \frac{\Sigma^{ns} SE_1}{NS}$$

where SE_i is the efficiency of workstation i, ST_i is the station time for workstation i, CT is the cycle time, LE is the line efficiency and ns is the number of workstations.

As the cycle time increases, the line efficiency decreases for a given number of workstations as shown in Fig. 5.2.

5.5.2 Cycle time versus number of stations

It is obvious from the discussion above that the highest line efficiency is achieved with the lowest cycle time possible for a given number of workstations. When the number of workstations varies, then there is another best cycle time value corresponding to it. This relation is best shown in Fig. 5.3.

When the cycle time is CT_1, the number of workstations required is NS_5. As the cycle time increases, the number of workstations remains the same, thus lowering the line efficiency until the cycle time becomes $CT_2 - e$. As soon as the cycle time takes a value of CT_2, the required number of workstations becomes NS_4. The same behavior is also observed for other values. Therefore, the relation between cycle time and the number of workstations takes the form of a step function. The best cycle times (the highest line efficiency) for alternative line configurations $NS_1, NS_2, \ldots, NS_k$

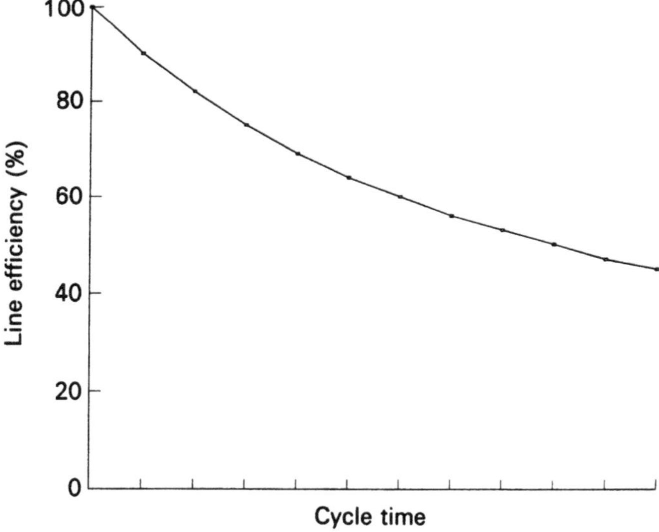

Fig. 5.2. Relation between cycle time and line efficiency.

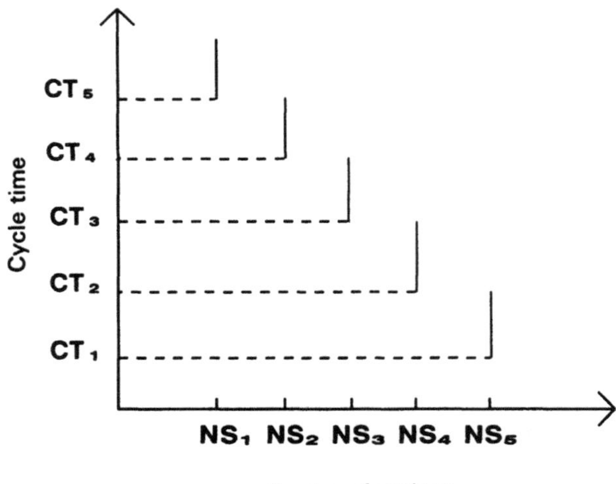

Fig. 5.3. Relation between cycle time and number of stations.

are CT_k, $CT_{k-1}, \ldots, CT_1$, respectively, and are used to prepare the priority matrix described in the next section.

5.5.3 Priority matrix

The priority matrix is formed using the following steps:

1. Find CT_{ij} for all i and j.
2. Calculate LE_{ij} for all i and j.
3. Order LE_{ij} in the decreasing order of efficiency for all i: $LE_{i[1]} \geqslant LE_{i[2]} \geqslant \cdots \geqslant LE_{i[k]}$.
4. Determine $NS_{i[j]}$ values corresponding to $LE_{i[j]}$ for all i and j.
5. Set $a_{ij} = NS_{i[j]}$ for all i and j.

where k is the number of alternative line configurations, CT_{ij} is the best cycle time (highest line efficiency) for model i when alternative j is used, LE_{ij} is the line efficiency for model i when alternative j is used, $LE_{i[j]}$ is the jth highest line efficiency for model i, $NS_{i[j]}$ is the number of workstations corresponding to the jth highest line efficiency for model i, a_{ij} is the element of the priority matrix (same as $NS_{i[j]}$) and n is the number of models.

The best number of workstation values (those that yield the highest line efficiency among alternatives) for each product are placed in the first column of the priority matrix. Similarly, the second best number of workstation values are placed in the second column, etc. The general format of a priority matrix is presented in Fig. 5.4.

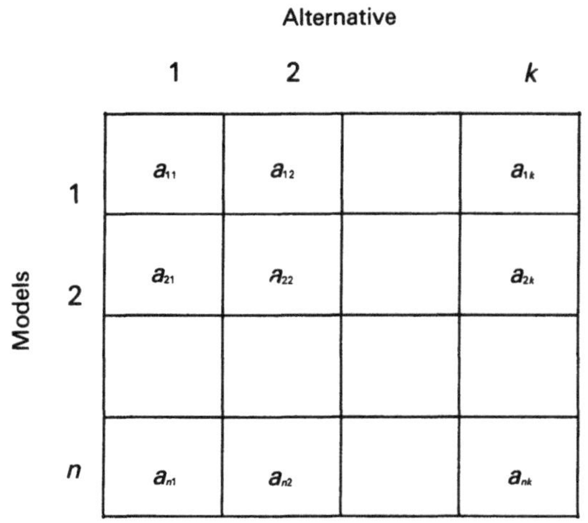

Alternative

	1	2		k
1	a_{11}	a_{12}		a_{1k}
2	a_{21}	a_{22}		a_{2k}
n	a_{n1}	a_{n2}		a_{nk}

Models

Fig. 5.4. Priority matrix.

5.6 RULES

The rules used in this study take the priority matrix into account in assigning models to assembly lines. Süer (1990) described the resource allocation rules and the algorithms in detail along with single machine and parallel machine scheduling approaches to the same problem.

5.6.1 Rule RESMAX (maximum resource)

The selection criterion (SC_{ij}) is the number of resources (workstations) needed to assemble a model ($SC_{ij}=a_{ij}$). The models with the highest number of resources (highest SC_{ij} values) are given higher priority. This rule works well if the objective is to minimize the makespan.

5.6.2 Rule RESMIN (minimum resource)

The selection criterion is the number of resources (workstations) needed to assemble a model ($SC_{ij}=a_{ij}$). The models with the lowest number of resources are given highest priority. This rule is usually good for minimizing the flowtime.

5.6.3 FMAX (maximum in-line time)

The selection criterion is the in-line time of a model and it is calculated by multiplying the cycle time by the demand ($SC_{ij}=CT_{ij} * D_i$). The models with the highest in-line time are given higher priority. This rule performs well if the objective is to minimize the makespan value. It considers the net time required to produce a model.

5.6.4 FMIN (minimum in-line time)

The selection criterion is the in-line time of a model and it is calculated by multiplying the cycle time by the demand ($SC_{ij} = CT_{ij} * D_i$). The models with the lowest in-line time are given higher priority. This rule is the best for minimizing flowtime, since it operates very similar to the shortest processing time (SPT) rule.

5.6.5 TOTMAX (maximum total time)

The selection criterion is the total assembly time for each model and is calculated by multiplying the cycle time times the demand and the number of resources ($SC_{ij} = CT_{ij} * D_i * a_{ij}$). The models with the highest total assembly time are given higher priority. Experimentation has shown that this rule performs well if the objective is to lower the makespan value.

5.6.6 TOTMIN (minimum total time)

The selection criterion is the total assembly time for each model and is calculated by multiplying cycle time times the demand and the number of resources ($SC_{ij} = CT_{ij} * D_i * a_{ij}$). The models with the lowest total assembly times are given higher priority. This rule performs well if the objective is to minimize the flowtime.

5.7 SCHEDULE GENERATION TECHNIQUES

There are three basic schedule generation techniques used in this study, i.e. non-delay, delay and delay stop. If the non-delay technique is used, a model is assembled using a line formed out of available resources, preferably with the highest line efficiency configuration. In the case of the delay technique, a model is held until the resource requirement in the first column of priority matrix (a_{i1}) becomes available. Therefore, only the first column of the priority matrix is used for the delay schedules. However, if there are resources available and some other models are to be assembled, resources are assigned to other models as long as their resource requirement is satisfied. The delay stop technique differs from the delay technique in the sense that a model is held until the resource requirement in the first column becomes available without assigning any other model to the available resources.

Delay, delay stop and non-delay schedules may further be explored by a slight modification applied to the last unassigned model. The maximum number of resources are assigned to the last model, regardless of the corresponding line efficiency hoping to reduce both makespan and the average flow time. The model is held until the maximum number of resources becomes available.

Flow charts of the algorithms for the schedule generation techniques are given in the next section.

5.8 ALGORITHMS

The algorithms for six schedule generation techniques are given in this section with brief explanation of each.

Notation:

t	time counter.
j	column (alternative) counter.
k	number of alternatives.
S	set of unassigned models.
U	set of unassigned models that are temporarily not considered.
R_t	resource (number of resources) available at time t.
SC_{ij}	the value of selection criterion for model i and alternative j.
i	model counter.

5.8.1 Non-delay algorithm

The search process for a model begins in the first column of the priority matrix since the line configurations in the first column have the highest line efficiencies. For the RESMIN, FMIN and TOTMIN rules the minimum values of SC_{ij} are selected, whereas for the RESMAX, FMAX and TOTMAX rules the maximum values of SC_{ij} are selected.

When the model r with the min(max) S_{r1} value is obtained, its resource requirement is compared with the available resources. If there are enough resources available, the model is assigned to a_{r1} resources. Then, the model is excluded from the search and the available resources are reduced. If the available resources are reduced to zero, time is advanced to the next completion time and the available resources are increased. Otherwise, the algorithm checks whether all of the models have been assigned.

If there are not enough available resources, the model is temporarily placed into another set U and the search for another model begins. If there is not any feasible model that can be assigned, the algorithm switches to the next column (reluctantly since line efficiencies are decreasing) and all of the models temporarily placed in U are placed back in S. The search for a new model begins in the new column.

Once the algorithm considers all of the columns, an indication will be made that the available resources are not sufficient for assigning any of the remaining models. Henceforth, time is advanced to the next completion time, available resources are adjusted, sets are rearranged and the search begins from the first column one more time. The flow chart for the algorithm is given in Fig. 5.5.

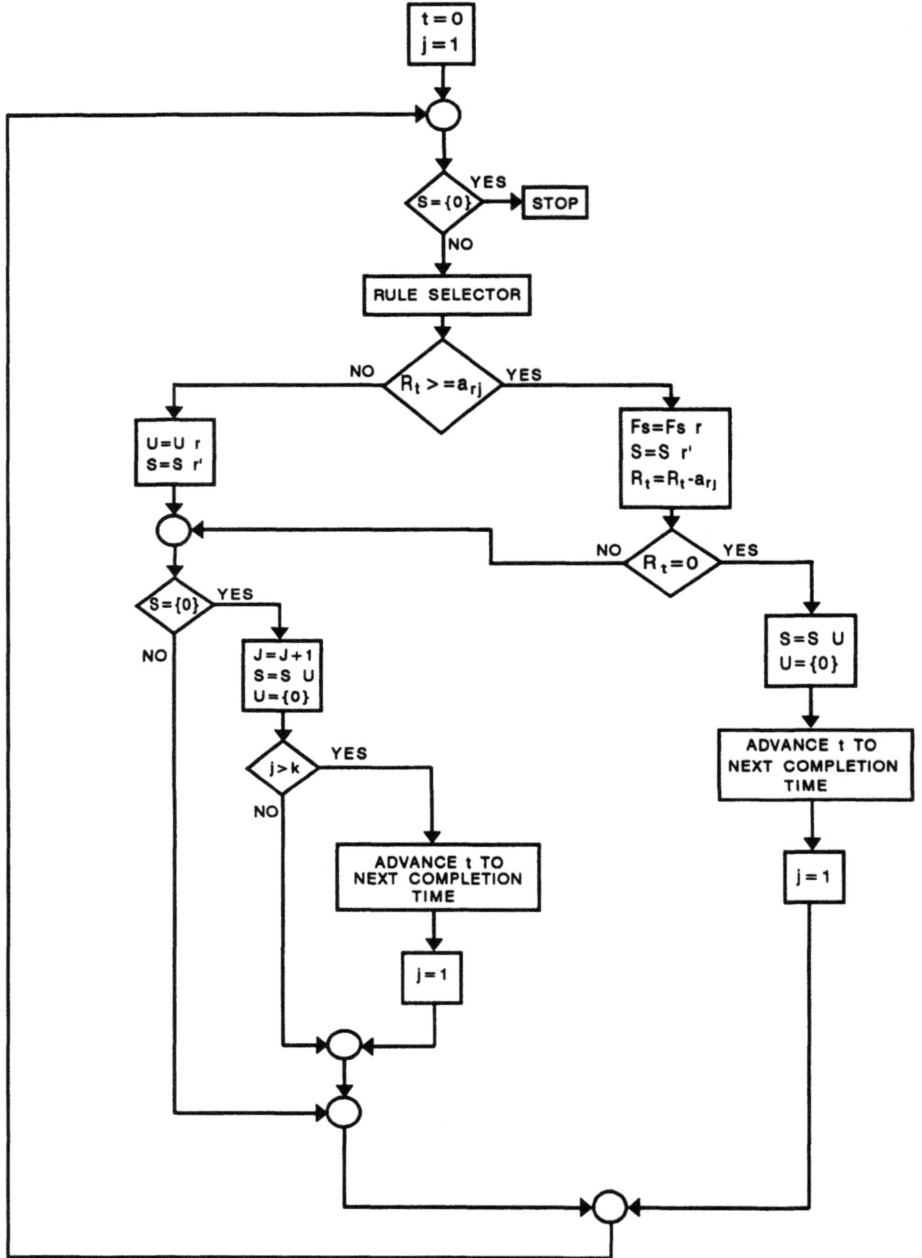

Fig. 5.5. Flow chart for the non-delay algorithm.

5.8.2 Non-delay algorithm/modified

A modified version of the non-delay algorithm is very similar to the original form except that the maximum resource configuration is used for the last unassigned model in set S. The starting time for the model is delayed until the maximum resources required become available. The flow chart for the modified non-delay algorithm is presented in Fig. 5.6.

5.8.3 Delay algorithm

The search for a model is always carried out considering the first column of the priority matrix.

When the model r with the min(max) S_{r1} value is obtained, its resource requirement is compared with available resources. If there are enough resources available, the model is assigned to a_{r1} resources. Then, the model is removed from set S and the available resources are reduced. If the available resources are reduced to zero, time is advanced to the next completion time and the available resources are increased. Otherwise, the algorithm checks whether all of the models have been assigned.

If there are not enough available resources, then the model is temporarily placed in another set U and the search for another model begins. If there is not a feasible model that can be assigned, then the algorithm advances time to the next completion time and increases the available resources by the number of resources released. The models temporarily removed from set S are placed back into S and the search for a new model restarts. The entire loading process is summarized in Fig. 5.7.

5.8.4 Delay algorithm/modified

A modified version of the delay algorithm is similar to the original form. However, the restriction on the use of the first column resource requirement is relaxed for the last unassigned model in set S and maximum resource configuration is used regardless of the line efficiency and the column. The start time of the model is delayed until maximum resource required becomes available. The details of the algorithm are represented in Fig. 5.8.

5.8.5 Delay stop algorithm

The search for a model is always carried out considering the first column of the priority in this technique as well. The difference between the delay algorithm and the delay stop algorithm is that if a selected model requires more resources than available, the search for another model starts in the former, whereas no other model is assigned until the selected model is assigned in the latter. Time is advanced to the next completion time and the resources available are increased. Resources available versus resources

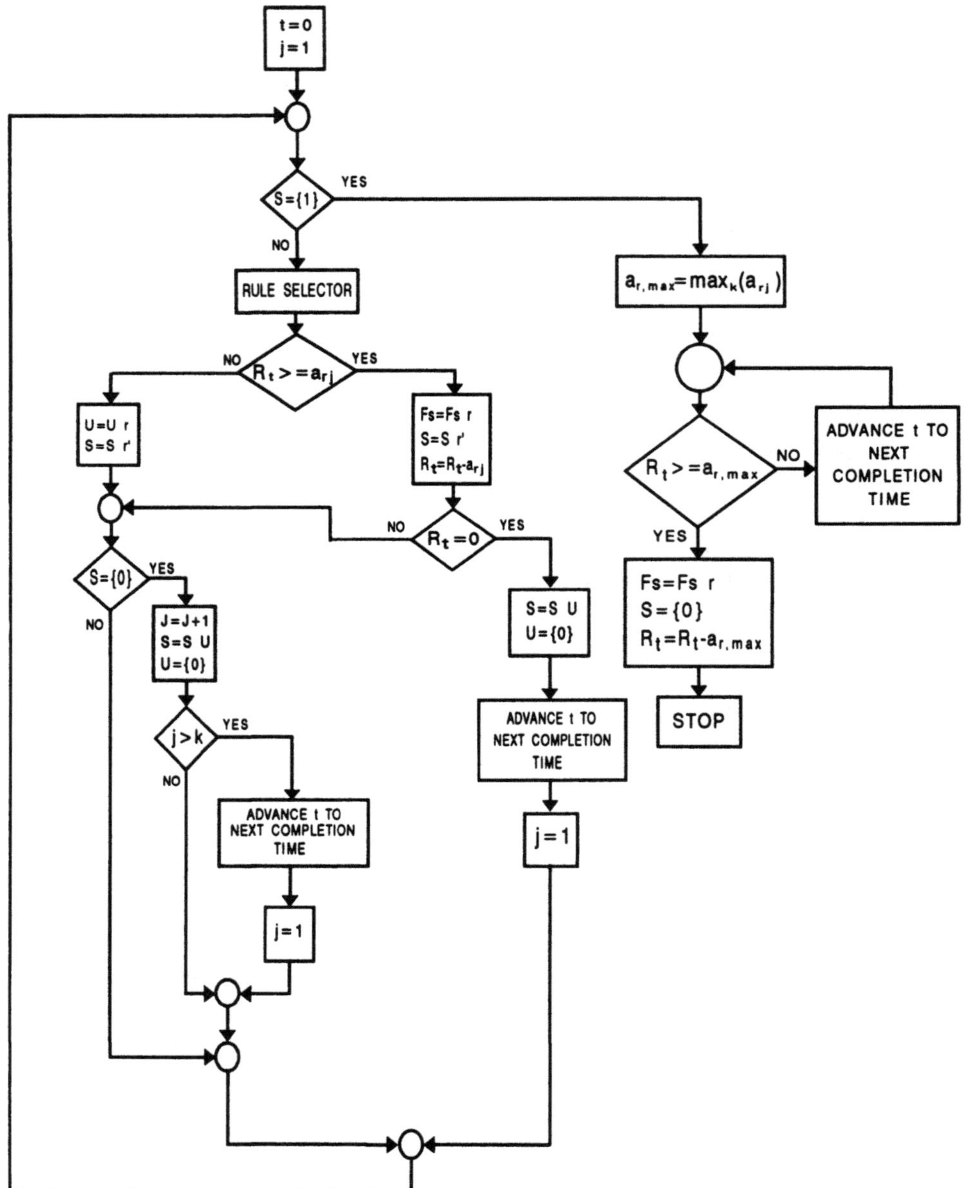

Fig. 5.6. Flow chart for the non-delay algorithm/modified.

required are compared one more time, etc. The steps of the algorithm are represented in Fig. 5.9.

5.8.6 Delay stop algorithm/modified

A modified version of the delay stop algorithm is similar to the modified

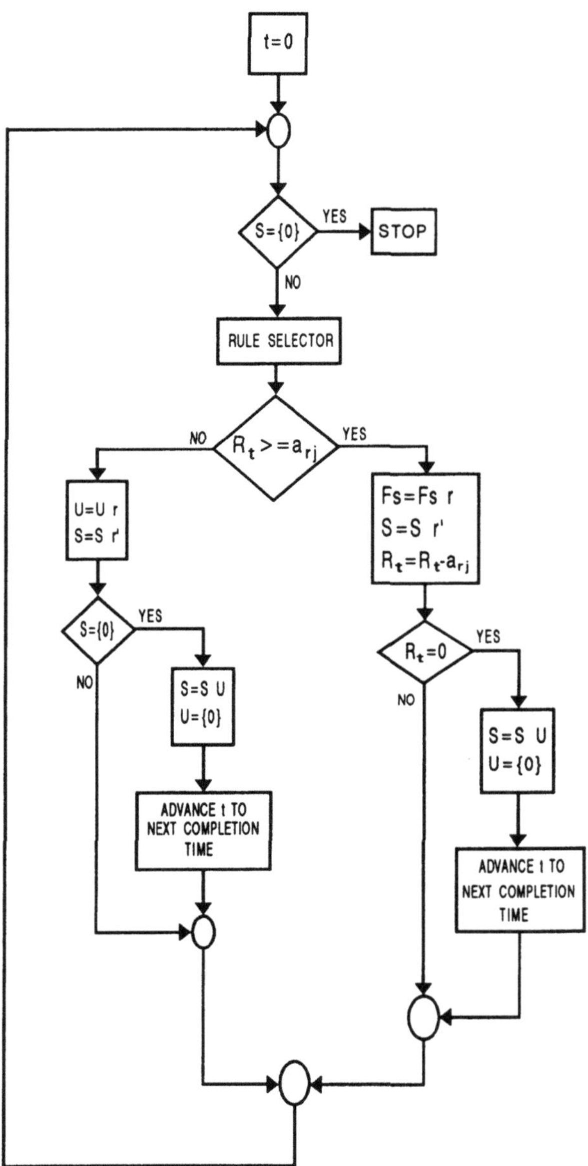

Fig. 5.7. Flow chart for the delay algorithm.

delay algorithm. The maximum resource configuration is used for the last unassigned model regardless of the line efficiency and the column. The start time of the model is delayed until the maximum resources required become available. The flow chart for the entire process is shown in Fig. 5.10.

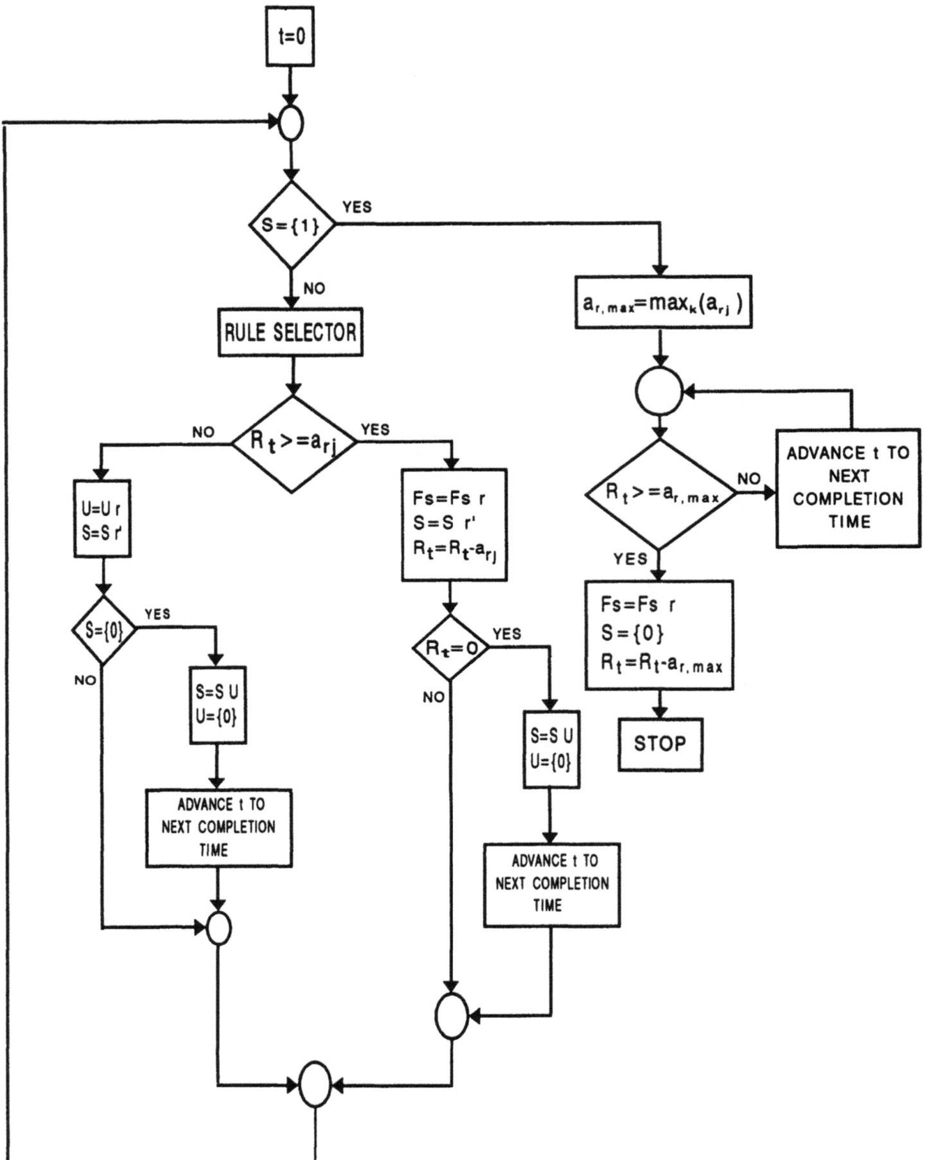

Fig. 5.8. The delay algorithm modified.

5.9 AN EXAMPLE

Six models and three line-configuration alternatives (one-resource, two-resource and three-resource) are considered with the following priority matrix. Demand for each product is assumed to be 100 units. The priority

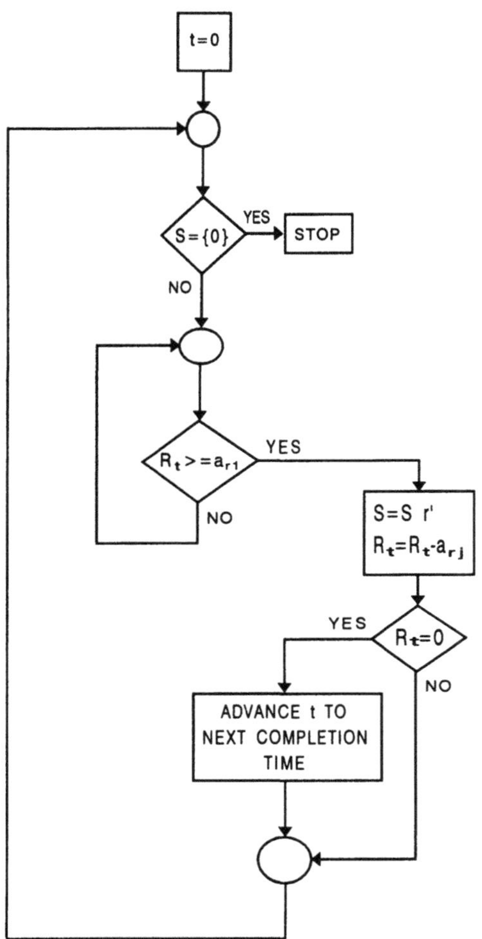

Fig. 5.9. Flow chart for the delay/stop algorithm.

Table 5.1. Priority matrix for the sample problem

Model	*No. of resource for alternative*		
	1	*2*	*3*
1	1	3	2
2	3	2	1
3	1	2	3
4	2	3	1
5	1	3	2
6	2	1	3

matrix for the sample problem is given in Table 5.1. The cycle times for the sample problem are given in Table 5.2. The average flow time and makespan values for the sample problem are summarized in Tables 5.3 and 5.4, respectively.

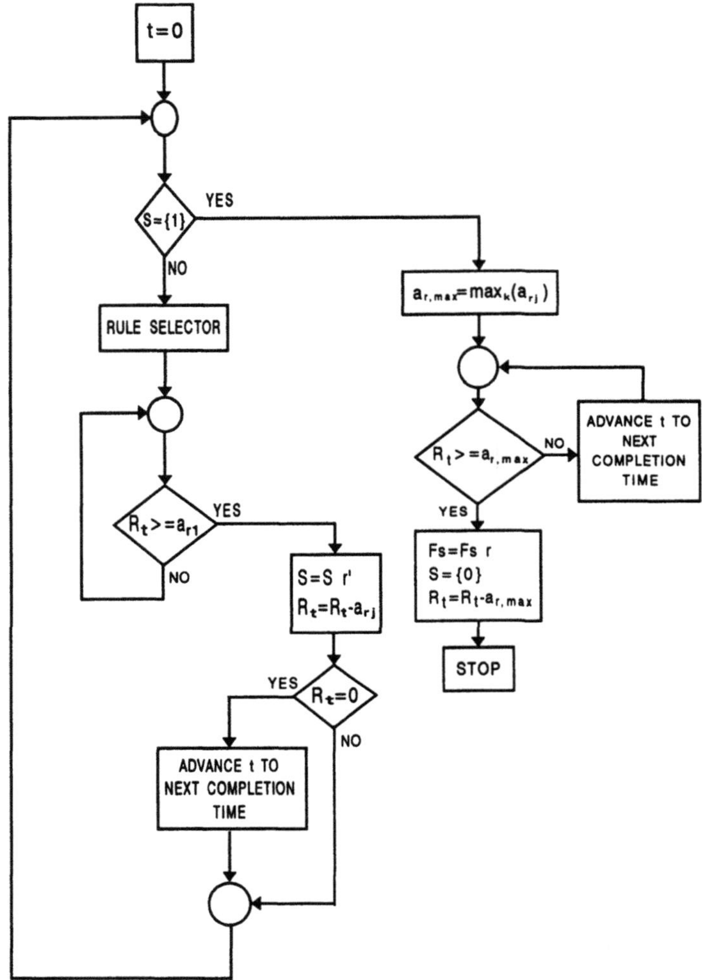

Fig. 5.10. Flow chart for the delay/stop algorithm modified.

Table 5.2. Cycle times for the sample problem

	Cycle time for alternative		
Model	*1*	*2*	*3*
1	0.224	0.075	0.117
2	0.070	0.110	0.230
3	0.095	0.055	0.040
4	0.120	0.090	0.280
5	0.150	0.055	0.090
6	0.190	0.400	0.140

5.10 EXPERIMENTATION PERFORMED

The rules of the knowledge base were determined based on the empirical results obtained by extensive experimentation. The programs used for experimentation with the resource allocation rules and the algorithms were developed using Fortran. The factors considered for the experimentation are explained below:

1. Number of models (two levels)
   (a) five models
   (b) 10 models
2. Demand (three levels)
   (a) low, 75 units for all models
   (b) medium, 100 units for all models
   (c) high, 125 units for all models
3. Number of alternatives (two levels)
   (a) two alternatives
   (b) three alternatives
4. Efficiency (two levels)
   (a) 1.0, 0.90 (two alternatives); 1.0, 0.90, 0.80 (three alternatives)
   (b) 1.0, 0.80 (two alternatives); 1.0, 0.85, 0.70 (three alternatives)
5. Priority matrix (two levels)
   (a) one to three resources per model (uniform distribution)
   (b) one to five resources per model (uniform distribution)

   In generating the priority matrix, the repetition of the same resource value for the same model is not allowed. Therefore, the uniform distribution is revised after each resource value has been determined for a model.
6. Cycle times (five levels)
   (a) 1–5 min per unit (uniform distribution)
   (b) 1–10 min per unit (uniform distribution)
   (c) 3–5 min per unit (uniform distribution)
   (d) 4–10 min per unit (uniform distribution)
   (e) 6–10 min per unit (uniform distribution)

   The uniform distributions mentioned above are used to generate the cycle times for the highest line efficiency alternative only. The remaining cycle time values for other alternatives are calculated by considering the line efficiencies and the resource figures in the priority matrix.
7. Resource levels (seven levels)
   (a) max resource level in priority matrix $= X$
   (b) $X+1$
   (c) $X+2$
   (d) $X+3$
   (e) $X+4$
   (f) $X+5$
   (g) $X+6$

Table 5.3. Average flow time values for the sample problem

Rules	*Algorithms*					
	Delay	*Delay/ modified*	*Non-delay*	*Non-delay/ modified*	*Delay stop*	*Delay stop modified*
RESMIN	26.07	26.07	27.07	26.82	26.07	26.07
RESMAX	22.40	22.13	22.40	22.13	27.15	25.98
FMIN	21.23	20.90	21.23	20.90	22.82	23.50
FMAX	24.98	24.98	26.22	24.98	25.65	25.65
TOTMIN	24.07	23.23	27.07	26.82	25.73	24.90
TOTMAX	26.07	26.07	27.15	26.07	27.98	28.47

Table 5.4. Makespan values for the sample problem

Rules	*Algorithms*					
	Delay	*Delay/ modified*	*Non-delay*	*Non-delay/ modified*	*Delay stop*	*Delay stop modified*
RESMIN	47.50	47.50	43.00	41.50	47.50	47.50
RESMAX	38.00	36.40	38.00	36.40	41.40	41.40
FMIN	38.00	36.00	38.00	36.00	41.40	45.50
FMAX	38.00	38.00	45.40	38.00	38.00	38.00
TOTMIN	47.50	42.50	43.00	41.50	47.50	42.50
TOTMAX	35.50	35.50	42.00	35.50	41.00	41.00

The total number of runs made is 1680 (=2*3*2*2*2*5*7). In each run, 36 possible combinations of the rules and the algorithms were tested against two performance measures, thus generating a total of 120 960 data points to analyze and form the knowledge base.

5.11 THE STRUCTURE OF THE RULES

The most important conclusion reached after analyzing the results is that the demand level does not affect the results to any degree. Therefore, it is not included in the knowledge base as long as the same demand value is assumed for each product. In addition to this, the two objectives are treated independently since they are not compatible with one another.

After these preliminary observations, the structuring of the rules continued by grouping the results in 80 categories. A category is defined for each possible combination of the levels of all of the factors except for demand and available resources (two product levels * two alternative levels * two efficiency levels * two priority matrix levels * five cycle time levels).

In each category, 36 combinations of the rules and the algorithms were analyzed for the resource level X, and the best combination(s) of the rules and the algorithms (with minimum result for the performance measure under consideration) was determined. The same procedure was repeated for the remaining $X+1, \ldots, X+6$ resource levels. Having completed the analysis for seven resource levels, the best combination(s) with the highest frequency of minimum results was chosen to be the overall best result in this category and added to the knowledge base. The procedure is repeated for the next performance measure as well and more knowledge is generated.

The remaining categories were also analyzed in the same manner. Some of the categories produced the same or similar results. Having considered the entire knowledge generated and the similarities among different categories, the rules were structured to represent the knowledge base for the selection of the resource allocation rules and the algorithms. This study can be considered as a medium-size project according to the classification made by Yazdani (1989).

If the user does not know the answer to a question and enters 'unknown' the system will consider all of the possible answers for the question and then proceed accordingly. Pederson (1989) has discussed handling uncertainty in knowledge-based systems.

5.12 COMPUTATIONAL REQUIREMENTS

The knowledge-based expert system was developed using an IBM compatible PC 386 and a rule-based expert system shell, M.1 v3.0 [see Harmon and King (1985) for other available expert system shells]. The knowledge base developed is included in the Appendix.

ACKNOWLEDGMENTS

The authors would like to thank graduate students Marcos Ortega and Luis Mañan for their help in the preparation of this chapter.

REFERENCES

Ben-Arieh, D. (1986) Knowledge-based control system for automated production and assembly, in *Modelling and Design of Flexible Manufacturing Systems*, (ed. A. Kusiak), Elsevier, New York, pp. 347–68.

Bensama, E., Bel, G. and Dubois, D. (1988) OPAL: a multi-knowledge-based system for industrial job-shop scheduling. *International Journal of Production Research*, **26**(5), 795–819.

Borne, D.A. and Fox, M.S. (1984) Autonomous manufacturing: automating the job-shop. *IEEE Computer*, **17**(9), 76–86.

Bruno, B., Elia, A. and Laface, P. (1986) A rule-based system to schedule production. *IEEE Computer*, **19**(7), 32–40.

Chang, F.C. (1985) A knowledge-based real-time decision support system for job shop scheduling at the shop floor level. *PhD Dissertation*, Department of Industrial and Systems Engineering, Ohio State University.

Chiodini, V. (1986) *An Expert System for Dynamic Manufacturing Rescheduling*. Symposium on Real Time Optimization in Automated Manufacturing Facilities, National Bureau of Standards, Gaithersburg, MD.

Dagli, C.H. and Cihangirli, M. (1988) *Prototype Expert System for Job Scheduling in Flexible Manufacturing*. Proceedings of the 1988 International Industrial Engineering Conference, pp. 418–23.

Dagli, C.H. and Süer, G.A. (1986) *Scheduling for Flexible Layout*. Proceedings of the 17th Decision Sciences Conference, Nebraska.

Erchler, J. and Esquirol, P. (1986) Decision-aid in job-shop scheduling: a knowledge based approach. *Proceedings of the 1986 IEEE International Conference on Robotics and Automation*, San Francisco, CA, pp. 1651–6.

Fox, M.S. and Smith, S.F. (1984) ISIS: a knowledge based system for factory scheduling. *Expert Systems Journal*, **1**(1), 25–49.

Fukuda, T., Takeda, S. and Hayashi, M. (1986) *Distributed Expert Systems for Production Control*. 1986 International Industrial Engineering Conference, Institute of Industrial Engineers, Dallas, Texas, pp. 222–8.

Harmon, P. and King, D. (1985) *Expert Systems*, John Wiley & Sons, New York.

Miller, R.K. (1984) *Artificial Intelligence Applications for Manufacturing*, SEAI Technical Publications, Madison, GA.

Morton, T.E. and Smunt, T.L. (1986) A planning and scheduling system for flexible manufacturing, in *Flexible Manufacturing System: Method and Studies*, (ed. A. Kusiak), North-Holland, Amsterdam, pp. 151–64.

O'Connor, D.E. (1984) Using expert systems to manage change and complexity in manufacturing, in *Artificial Intelligence Applications for Business*, (ed. W. Reitman), Ablex Norwood, New Jersey, pp. 149–58.

Parunak, H.V.D. (1987) Distributed AI systems, in *Artificial Intelligence: Computer Integrated Manufacture*, (ed. A. Kusiak), IFS, Kempston, Bedford, UK. Springer, New York.

Pederson, K. (1989) *Expert Systems Programming*, John Wiley & Sons, New York.

Shaw, M. (1986) A pattern-directed approach to FMS scheduling, in *Flexible Manufacturing System: Operations Research Methods and Applications*, (eds. K.E. Stecke and R. Suri), Elsevier, New York, pp. 545–54.

Shen, S. and Chang, Y. (1986) An AI approach to schedule generation in a flexible manufacturing system, in *Flexible Manufacturing Systems: Operations Research Models and Applications*, (eds. K.E. Stecke and R. Suri), Elsevier, New York, pp. 581–92.

Süer, G.A. (1990) *Manpower Allocation to Assembly Lines.* ORSA/TIMS Joint National Meeting, Philadelphia, Pennsylvania.

Süer, G.A. and Dagli, C.H. (1992) *Knowledge-Based System for Single Machine Scheduling.* Proceedings of the 14th Conference for Computers and Industrial Engineering, Cocoa Beach, FL.

Wild, R. (1972) *Mass-production Management*, John Wiley & Sons, New York.

Yazdani, M. (1989) Building an expert system, in *Expert Systems: Principles and Case Studies.* Chapman & Hall, New York.

CHAPTER 6

An intelligent shop management system for production supervision

Gary P. Moynihan

6.1 INTRODUCTION

One of the most challenging problems in industry today is to improve the productivity of the manufacturing shop floor. If a coherent control of the shop floor is to be achieved, and its associated productivity improved significantly, then the key to the solution lies in understanding the function of the production supervisor.

Production supervisors are the individuals in a manufacturing organization responsible for planning, organizing, directing and controlling the activities of the shop floor operators. In a very real sense they are the first line of manufacturing management. By directing the work group as a team, the production supervisor has the actual responsibility for ensuring that production schedules are met, quality maintained, costs controlled and work flows smoothly through the supervisor's area.

It has long been recognized that the supervisor's heavy task load might impede manufacturing productivity (Taylor, 1947). The advent of the computer provided one source of assistance. Previous research in shop floor support systems has frequently concentrated on purely algorithmic methods, which have the tendency to generate large volumes of data in the form of printouts. Analysis of the 1981 National Survey of Supervisory Management Practices revealed, however, that these systems have been largely useless to the supervisor, who has to make spot decisions based upon heuristic and intuitive techniques (LaForge and Bittel, 1983). A mixture of heuristic and algorithmic solutions are required to better support the production supervisor.

Artificial intelligence applications have been developed to utilize these types of techniques in a manufacturing environment (Bel *et al.*, 1989;

Ben-Arieh and Moodie, 1987; Fox, Allen and Strohm, 1982). As noted by McKay, Safayeni and Buzacott (1986), these applications have had limited effectiveness on the shop floor. This may be due to a misunderstanding as to the planning and control functions of a production supervisor, and the erroneous belief that they are limited solely to shop floor dispatching.

Despite its criticality, little work appears to have been done regarding analyzing the planning and control aspects of this specific domain. Perhaps the primary reason for this is the difficulty inherent in its investigation. The problem domain spans many traditional specialties (e.g. decision theory, shop floor control methodology, resource planning, cost control techniques). Instead, research has focused on such supervisory functions as leadership and worker motivation (Broadwell, 1985; Christenson, 1982; Fechter and Horowitz, 1988). The production supervisor appears to be the forgotten variable in the manufacturing productivity equation.

The approach of this research was to focus on this planning and control problem area, and from it formulate a generalized model regarding the supervisor's informational needs and decision-making processes. A software system, reflecting this model, was then developed to support the production supervisor. This approach is consistent with recent trends in information technology and business process redesign (Davenport and Short, 1990).

6.2 METHOD AND SCOPE

Three alternative knowledge acquisition approaches were considered to support this process. The first alternative involved interviewing selected production supervisors. The interviewing of domain experts is an accepted knowledge acquisition technique. Analysis of this alternative determined that interviews with supervisors from this researcher's (then) company would result in industry-specific and company-specific information. Blind mail questionnaires to other companies were considered to expand the survey base. However, it was thought that this would yield an insufficient quantity of responses to be statistically meaningful. The credentials of these domain experts would also be uncertain.

The second alternative recommended the use of the Delphi Technique. In this case, the questions would be targeted at known experts from academe and industry. The primary advantage of this alternative was the certainty of the experts' credentials. This alternative was also discarded due to doubts that a sufficient quantity of responses could be obtained. An initial mass mailing resulted in zero responses.

The third alternative was an intensive literature search. Although little work has been done regarding the supervisory planning and control function *per se*, a great deal of literature is available on associated topics, e.g. planning and scheduling models. The challenge of this alternative was to analyze these related data for applicability to the shop floor environment, then to assimilate them into an overall framework.

A total of 188 heuristics were incorporated into the expert system. The exact facets of production supervision vary from industry to industry and from company to company. These varying practices preclude guaranteeing the true generality of the model. Domain dependencies may still exist. The model will provide a general framework and baseline for further development. The value of this strategy will be discussed later in this chapter.

6.3 DEVELOPMENT OF THE SUPERVISORY MODEL

The production supervisor exists as the focal point of all activities that bear on the work of the supervisor's department. Based on preliminary work done by Moore (1969), and as expressed in Fig. 6.1, the supervisor is at the center of a vortex of information, constantly interfacing with a variety of functions on a multitude of issues. Managing an organization can be interpreted as managing the opportunities in the environment. According to Ginter, Rucks and Duncan (1985), a series of normative models have been developed to describe the management process. Translation of their work to a manufacturing environment provides the basis for the general process model portrayed in Fig. 6.2. It is within the framework of this overall manufacturing management process that the more specific supervisory model must be constructed.

The specific nature of the industrial environment drives the detailed processes within the manufacturing management model. The supervisor represents the management level closest to the particular manufacturing technology, product and shop floor conditions. Development of the supervisory model requires an understanding of these environmental ramifications on the more general manufacturing management process. Manufacturing may be partitioned into three major categories: process industry, repetitive industry and job shops. The primary characteristics of these three partitioned environments are portrayed in Table 6.1.

A summary view of the supervisory model is depicted in Fig. 6.3. The supervisory planning and control function is comprised of three primary components: resource allocation and scheduling, situation assessment, and projection and replanning. Each component, in turn, represents subordinate models. This planning and control triad revolves around a set of shop parameters specific to the company and industry. Clearly stated in the construction of the model is access to the factory information system. Each primary model component accesses and utilizes data from the other components as well as from the factory data bases and data collection systems. The Factory Information System thus provides a conduit to data from the specific manufacturing environment.

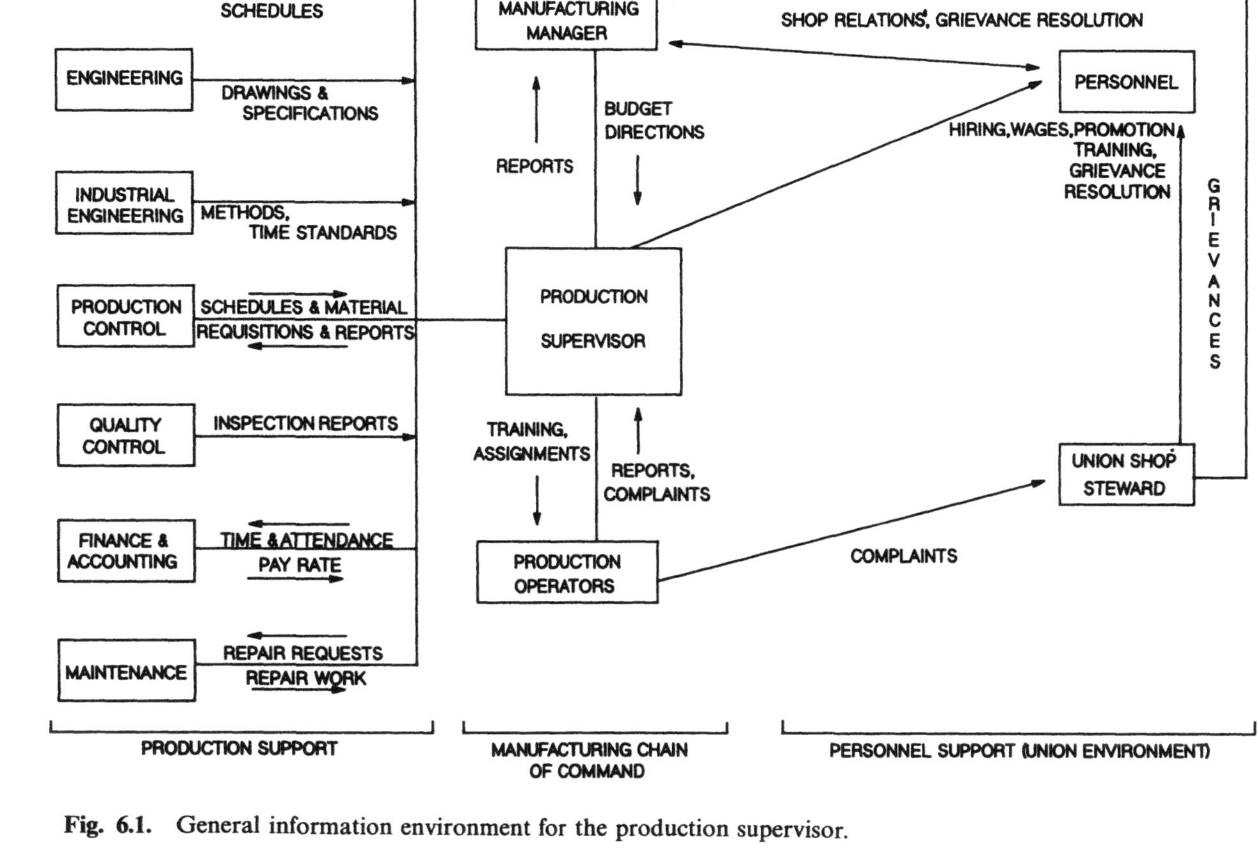

Fig. 6.1. General information environment for the production supervisor.

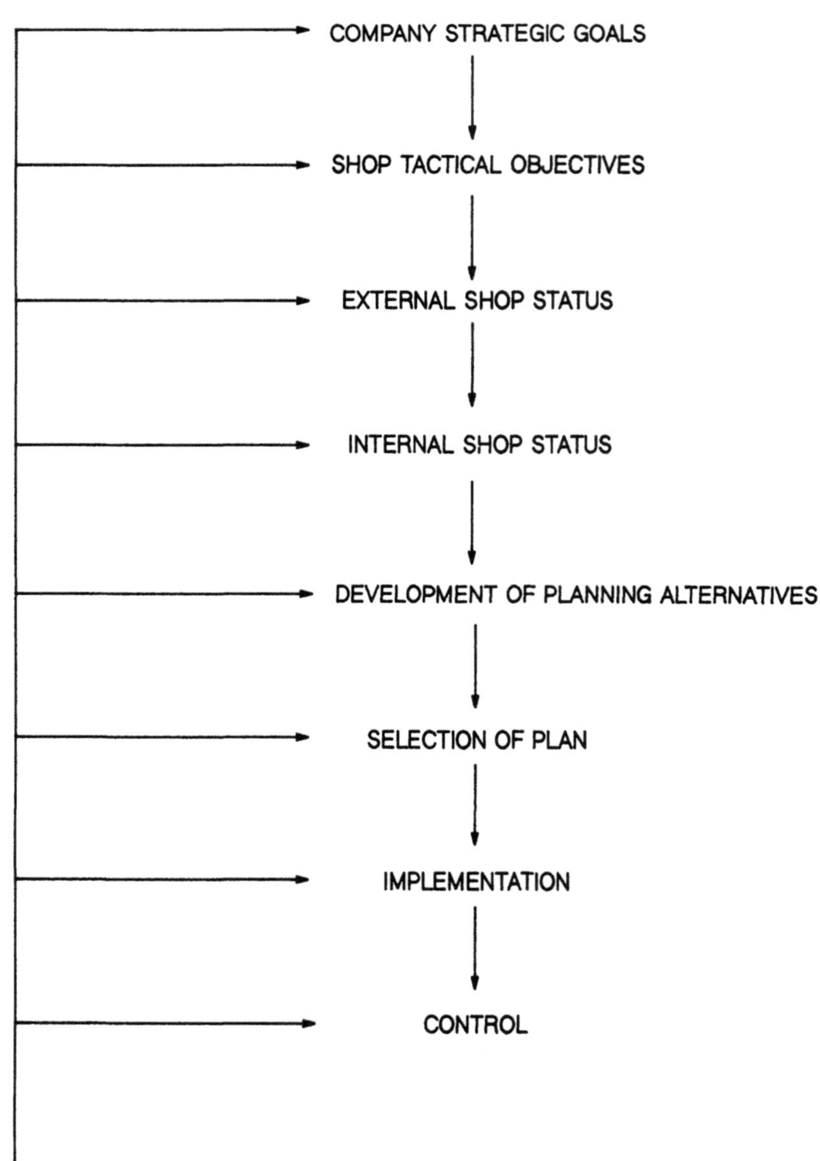

Fig. 6.2. General model of the manufacturing management process.

6.4 DESCRIPTION OF THE PROTOTYPE SYSTEM

A prototype expert system was developed as proof of the underlying conceptual model of the production supervisor's planning and control function. It provides a linkage of facts (regarding conditions within the manufacturing area) and supervisory heuristics. A rule-based system was selected as the

Table 6.1. Major industrial environment characteristics

Characteristic	*Manufacturing Environment*		
	Process	*Repetitive*	*Job shop*
Components/parent	Nondiscrete	Discrete	Discrete
Volume of production	High	High	Low-medium
Routings	Fixed	Fixed	Variable
Relative queue lengths	Short	Short	Long
Shop floor scheduling	Minimal required	Detail dispatching required	Detail dispatching required
Set-up considerations	Minimal	Minimal to variable	Significant set-up
Control	Flow-rate oriented	Flow-rate oriented	Part number/work order
Process time per step	Variable	Short	Variable
Equipment capacity	Dedicated	Dedicated	Nondedicated multipurpose
Operators	Semiskilled/skilled	Semiskilled	Skilled

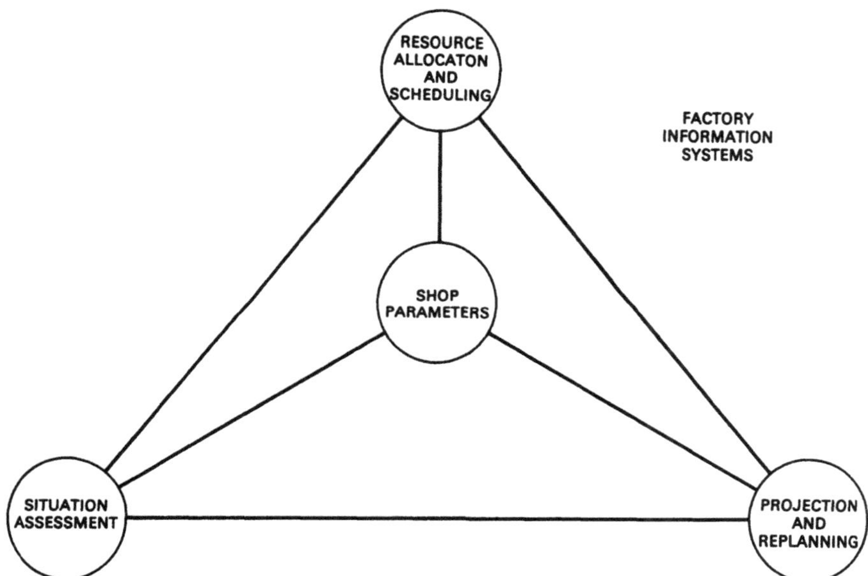

Fig. 6.3. The planning and control triad.

knowledge representation paradigm. The knowledge structures obtained during this research were input to the personal consultant plus (PCPLUS) expert system shell. PCPLUS utilizes the abbreviated rule language (ARL). The ARL-based heuristics are provided in the Appendix. A series of specialized functions were programmed directly in Scheme and interfaced with the shell.

In practice, the system would be interlinked to the manufacturing plant's data bases and data collection system. For the purposes of this prototype, it was necessary to emulate these factory information systems by utilizing a series of input files. These were developed using DBase III Plus. The PCPLUS expert system shell accesses the DBase files to support activation of heuristics within its rule base. Primarily those input files requiring data manipulation by the system utilize DBase. Other input files, only requiring access and read capabilities, exist as standard ASCII text. The software was loaded on an IBM PC AT microcomputer.

The expert system was structured into a series of modules consistent with the general model. Common functions were grouped into a system root frame. (In PCPLUS, a frame is defined as an informational structure within the knowledge base, and should not be confused with the object-oriented paradigm.) Within each of these major modules, the heuristics and algorithms are organized into a series of logic cells (see Fig. 6.4) to represent the functionality discussed in the following sections.

6.5 SHOP PARAMETERS

The shop parameters define the supervisor's particular application and adapt the general model to the specific manufacturing environment. The shop parameters consist of both fixed and transient forms of local data. The

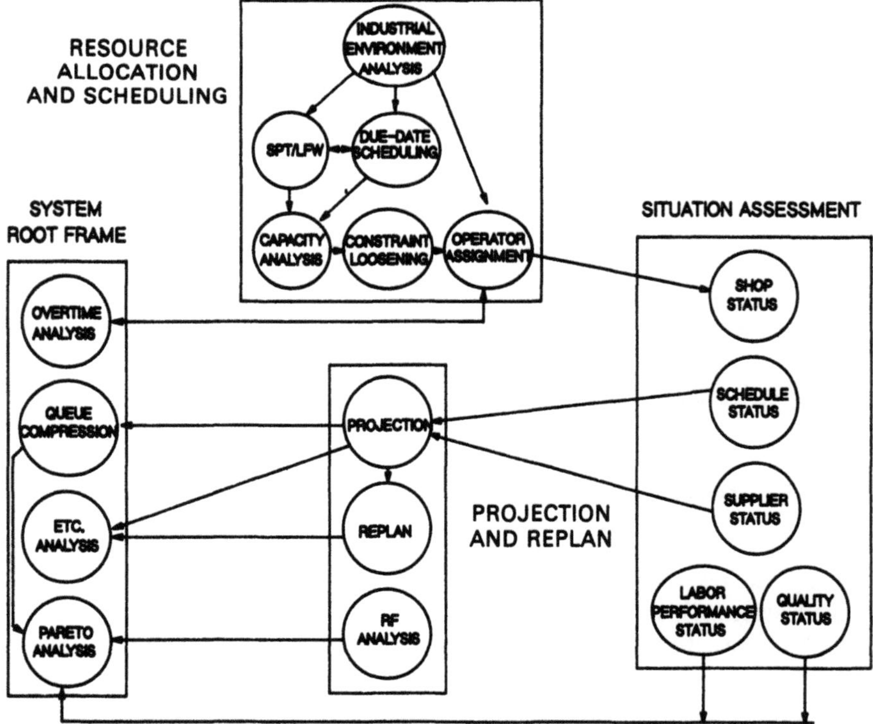

Fig. 6.4. Summary level system flow diagram.

interaction of the shop parameters with data from the factory information system serves to organize the activation of the rules in the planning and control triad's knowledge base.

Fixed data within the shop parameters identify the specific industrial environment (i.e. dedicated production line, simple job shop or assembly job shop). The fixed data also denote whether the shop is union or non-union and if the shop employees are paid on a piecework basis.

Transient data serve to fine tune the general rule activation sequence determined by the fixed data. In a sense, the transient data perform many of the functions of a supervisor's notebook. This repository contains alternate manufacturing plans that would identify appropriate 'out-of-step' operation sequences. These operation sequences would be utilized when attempting to schedule and allocate work to constrained resources. Operators' vacation plans would also be included in order to project their impact on long-range shop production plans. Another type of transient data would denote warning limits regarding quality problems. Supervisory preference for machine/workstation allocation, and operator preference regarding job assignment would also be included to focus properly the knowledge base.

6.6 RESOURCE ALLOCATION AND SCHEDULING

The resource allocation and scheduling module is intended to support routine planning for shift start-up and subsequent operator reassignments and overtime assignments. A summary diagram of the module is shown in Fig. 6.5. The shift start-up option provides assignments for all shop personnel, while the other options require identification of the specific operators to be assigned. Operator reassignment is utilized on an exception basis, whenever an operator requires a new task or completes a previous one. Normal use of the operator reassignment option is contingent on the time remaining in the shift, the span time of the operation step and the size of the work order. The overtime assignment option is used to plan the scope of work beyond normal working hours. Processing is similar for all three options, however the overtime assignment logic initially activates the overtime analysis cell resident in the projection and replanning module.

The initial phase of the module checks the status of jobs, workstations and operators as identified in the factory data collection system. Based upon the type of manufacturing, as identified in the shop parameters file, either direct assignment of jobs, modified shortest processing time (SPT) or due date scheduling is used. Although these rules are sub-optimal approaches, they are well understood by production supervisors, in general (Baker, 1984; Broadwell, 1984; Becker, 1974). As such, the model and resulting system were judged to be more acceptable by the potential user community.

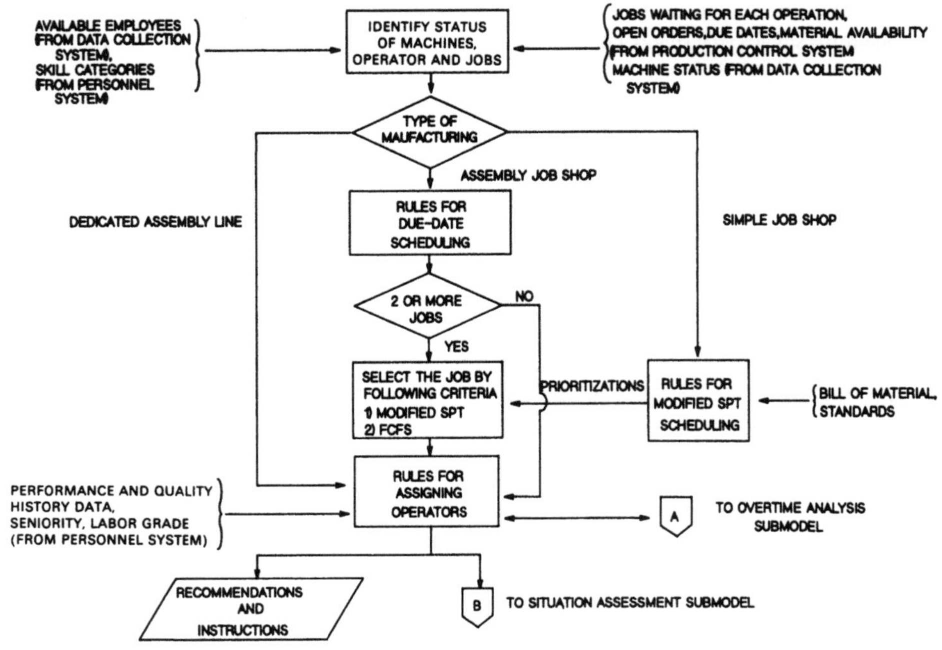

Fig. 6.5. Resource allocation and scheduling module.

The dedicated production line presents a single product flow. This is a straight-forward situation that allows the direct job assignment of operators. Conway is frequently credited with doing the foundation analysis of dispatching heuristics during the 1960s (Conway, Maxwell and Miller, 1967). Building upon Conway's work, Russell and Taylor have identified SPT as being the most effective dispatching heuristic in a simple job shop environment and due date scheduling to be most effective in an assembly job shop (Russell and Taylor, 1984). Effectiveness, as defined in their study, was based on mean flow time, mean job tardiness and percent of jobs tardy.

For the due date heuristic path, slack values are calculated. Calculation of a slack time infers a calculation of an overall operation time at the step level:

$$\text{operation time} = \{[(\text{set-up} + (\text{run} * \text{lot size}) * \text{RF})/\text{crew size}]/60\} + \text{queue time} + \text{move time} \quad (6.1)$$

Set-up and run are conventional types of labor standards. RF designates the application of realization factor (which is discussed later in this chapter). Move and queue times are conventionally in units of hours and do not require division by sixty in the algorithm. The step times are accumulated to the part level, yielding a total part time remaining. Slack times are then calculated and sequenced in ascending order:

$$\begin{aligned}\text{slack time} = &(\text{work order completion date} - \text{today's date})\\ &-(\text{work order completion date}\\ &-\text{total part time remaining})\end{aligned} \tag{6.2}$$

Based upon the due dates indicated by the Production Control system, if there is a job with negative slack it is scheduled for immediate production. If two or more jobs have negative slack, the higher priority is assigned to the job with more negative slack value.

SPT-based scheduling has been noted for reducing mean flow time, mean wait time and mean lateness (Salvendy, 1982). When combined with consideration for least forward work (LFW), SPT also maintains the smoothest work flow. This is another prime supervisory consideration. LFW prioritizations using the bill of material and labor standards information will also aid in reducing work in progress (WIP) in the supervisor's domain of responsibility. The higher the indenture level for a part (or the higher the step completed with a specific part), the higher the priority given to the job to be worked. This higher priority is analogous to the increased value-added as a product nears completion. Sequencing parts in the order of this prioritization will reduce the overall WIP costs for the shop. This methodology incorporates some of the work of Scudder and Hoffman regarding cost-based rules for job shop scheduling. In their simulation study of manufacturing shops, the VLADRAT (VaLue ADded RATio) rule proved to be the most effective at minimizing WIP levels over all utilization levels (Scudder and Hoffman, 1985). VLADRAT is functionally equivalent to LFW. LFW was used in the module, however, to maintain the heuristic on a level meaningful to the production supervisor. As with the due date heuristic path, logic is incorporated to limit job lateness.

Capacity analysis is normally accomplished by the production control system. In order to best utilize the output of this analysis at the production supervisor's level, a series of heuristics have been incorporated into this portion of the module. Similar heuristics, specifically related to flexible manufacturing systems, were first developed by Nakamura and Salvendy (1988). Identification of any potential constraints provides the supervisor with a guide to further smooth the overall workflow in the manufacturing department. In a production area at or near capacity, these potential constraints become actual bottlenecks. Even in an industrial environment below capacity, a delay at one of these potential constraints (e.g. due to tool breakage) might create a production chokepoint for the entire shop. Special consideration needs to be given to these key resources by the supervisor.

Based upon the identification of the constraints and/or the imperativeness of the original work order, a series of rules are activated to expedite the overall flow of work in the shop. These heuristics may combine or split work orders (depending upon the specific situation) or access alternate operational plans from the Shop Parameters file.

Once the work orders have been sequenced, heuristics are used to assign the operators to the jobs. These same heuristics provide the core logic for the overtime assignment and operator reassignment functions, within resource allocation and scheduling domain. Consideration of the individual operator's training certification, performance and quality history, and any union constraints are included. Recommendations are also included for operator assignment when no production work is available.

Whenever possible, the best operators are focused on the constrained resources and the high value operations. The term 'best' is based on a combination of the operator's performance and quality history. Higher value operations involve higher indenture level parts with lower remaining operational time. Due to the accrued value-added of these parts, the supervisor wishes to minimize scrap and rework by assigning this work to operators with the best quality rating. Since these higher indenture parts are also closer to completion, the supervisor wishes to finish them as soon as possible in order to attain the production quota. As a result, historical labor performance data are combined with quality indices to create an average quality performance (AQP) rating. The AQP is defined as (total units completed – units rejected)/productive time, for all units completed by the employee for this operation. Due to frequent labor contract constraints, the AQP is only used to support operator assignment in non-union environments.

6.7 SITUATION ASSESSMENT

The situation assessment module (Fig. 6.6) provides a control mechanism for the supervisor to evaluate his planning. The production supervisor is alerted if any of the parameters exceed their warning limits. Recommendations for action are then issued. For example, one function resident within this module provides a series of alerts regarding delays on the shop floor. The urgency and magnitude of the problem are also defined. The underlying function of this module is to provide the necessary information so that the production supervisor can make an intelligent decision on where to focus his attention within a very limited timeframe.

The monitoring of labor performance (classically defined as labor standards versus labor actuals) is fundamental to the production supervisor function. In the module, the labor performance logic measures labor efficiency at the operator/shop and the part number/product levels. Historically, labor reporting systems have buried shop inefficiencies within the operator's performance rating. The identification and segregation of non-productive time (due to delays or rework) will solve this problem.

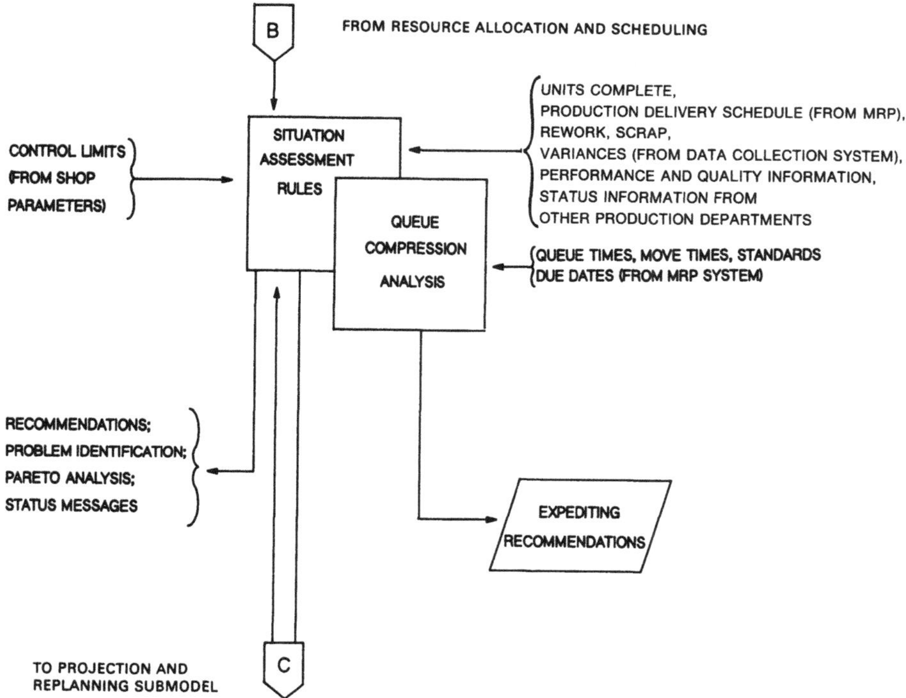

Fig. 6.6. Situation assessment module.

Non-productive time is identified through the use of variance transactions, input by the operators and accessed from the factory data collection system.

In this module, operator performance is based only on productive time (the time spent actually performing work on the product). The shop performance, also known as cost performance, is based upon the total of productive time plus non-productive time. The supervisor then has the ability to determine the sources of inefficiency within his shop and the means to quantify it. Recommendations are displayed for improving poor performance levels.

The capability to measure labor performance on a product level also exists within the situation assessment module. As with operator/shop performance, data are obtained from performance history. Consistently poor performance for a product, regardless of operator, may indicate an equipment problem (if dedicated only to that product) or the need for an engineering design review. This is functionally equivalent to a part that cannot be held to quality inspection tolerance.

Quality may be defined as the degree to which a company's design specifications have been met (Crosby, 1979). The price that the supervisor pays for non-conformances (i.e. failures to meet specifications) entails an increase in units of product to meet the same delivery quota. A ripple effect occurs due to these added units, increasing the risk of achieving the scheduled output while inflating the work in process inventory. The quality status function, resident in the situation assessment module, is designed to provide visibility to the sources of quality defects within the supervisor's area.

Traditional supervisory practice often is to wait until the quality inspector issues an alert, before making any readjustments to the process. This reactive approach results in many non-conformances being produced prior to identification. A more proactive approach, informally used by some supervisors, can provide a means to limit the quantity of these non-conforming units. This approach involves setting informal quality warning limits on key operations. These limits, based primarily on experience, may be used to monitor and project overall activities within the shop, not just for the individual operation. For example, by reviewing the inspection results for a particular part in a job shop, the supervisor may be able to estimate when a grinding wheel will have to be replaced and plan the shift accordingly. To support this task, the warning limits are accessed from the shop parameters file. In the absence of informal warning limits based on supervisory experience, formal quality warning limits may be substituted if thought significant for this application. Further research in this aspect of the system would include access and utilization of statistical process control (SPC) data.

Historical quality data are also required to support the supervisory monitoring function. The concept of selection criteria has been incorporated into the situation assessment module, in order to support the labor performance and quality functions. Selection criteria provide a means of management by exception, displaying all operators performing outside an indicated range. Although the selection criteria are envisioned to target low performance and quality ratings, exceptionally high peformance can also indicate a problem. This may be due to inaccurate labor standards, incorrect method of operation or that the operator is 'gaming the system'.

Pareto analysis is a technique of arranging data according to significance or importance, then selecting the relatively few elements that account for a disproportionate share of the measurements. Focusing on these key elements allows the supervisor to achieve the greatest gain within his limited time. Incorporation of a Pareto analysis feature into the overall supervisory model provides a method to indicate those areas with the highest potential for productivity improvement. Linkage to the labor performance and quality analysis functions identifies the causes for low ratings and directs the supervisor towards solving those problems from which the most benefit can be derived.

One important function of the production supervisor is to expedite the flow of work, as necessary within the shop. Material handling time is a critical, and often overlooked, factor of a product's overall span time on the manufacturing shop floor. Queue compression analysis is embedded in the situation assessment module to control this. A predicted completion time for a work order's remaining operations is calculated based on labor standards and historical move and queue times from the factory data bases. Realization factors are then applied. If the predicted completion time and the due date are the same or later, then no expediting is necessary. If the due date is earlier than the predicted completion time, normal material handling times are insufficient and expediting is recommended (Robbins and Rabbi, 1988).

Construction of the supervisory planning and control triad explicitly depicts interaction between the modules. The situation assessment module works in concert with projection and replanning to provide the supervisor with visibility to the shop's and to upstream department schedules. Within the supervisor's own department, this facility allows capability to determine if production quotas can be met based upon current conditions within the shop. Visibility to supplier departments provides the supervisor with information on upstream workloads and delays. It also provides the capability to determine the potential impact of upstream schedule slippages on the supervisor's own production schedule. Heuristics and algorithms embedded in the projection and replanning module can then be utilized by the supervisor to formulate a recovery plan.

6.8 PROJECTION AND REPLANNING

The projection and replanning module is used for variable horizon planning and 'what-if' analysis by the supervisor. The supervisor's locus of planning tends to be near-term. For a group supervisor (defined as responsible for a single work area), 38% of the planning tasks are for that day, 40% are for the week ahead and 15% are for 1 month ahead (Kirkpatrick, 1987). For the section supervisor (defined as responsible for multiple work areas within a department), the planning horizons are partitioned to be 15, 20 and 25%, respectively (Kirkpatrick, 1987). Heuristics and algorithms, resident in the projection and replanning module, provide the capability to plan over these horizons of primary interest and to view the potential ramifications of that plan on the production schedule (see Fig. 6.7).

An important aid in the function of projection/replanning is the use of realization factors. These are multipliers that when applied to labor standards result in a forecasted labor time (United States Department of Defense, 1983). The realization factor is determined by analyzing the trend of the historical labor performance data. There is an intuitive appeal to give more weight to recent labor performance observations. To support this, an exponential smoothing algorithm was incorporated into the projection and replanning module to provide a means of trend analysis. The resulting realization factor is then incorporated in the projection calculation.

Heuristics within the situation assessment module interact with those in projection and replanning to allow review of the significant components of the realization factor. This review is enhanced by a data descent capability.

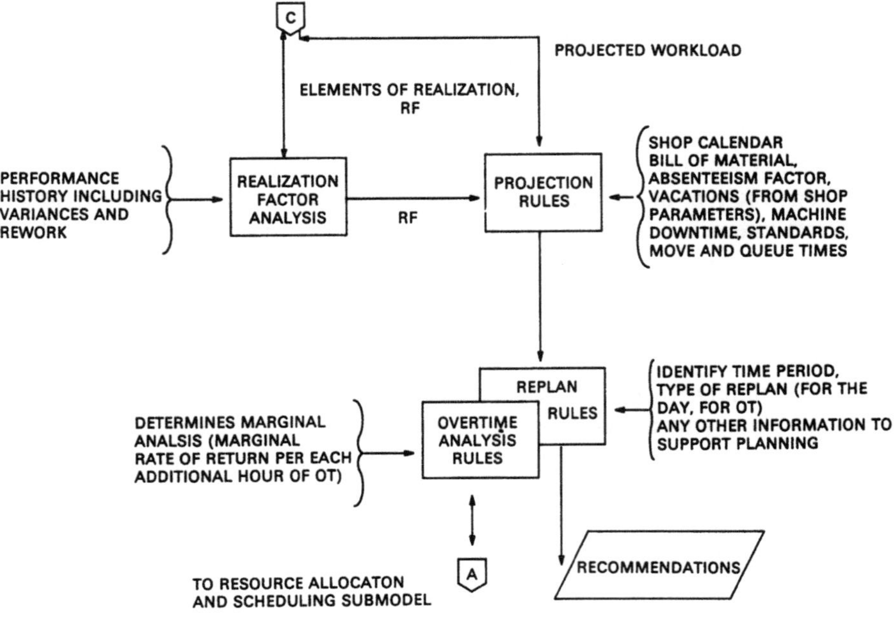

Fig. 6.7. Projection and replanning module.

The concept of data descent, sometimes referred to as report-tiering or drill-down, provides visibility to the next level in the hierarchy of detail. In the case of realization factors, the data descent capability identifies the elements of realization and those contributing workstations for a specified element of realization.

Logic embedded in the projection and replanning rule base calculates the anticipated workload on the shop in order to predict, in advance, the match between this load and the capacity required to meet it. In a dedicated production line environment, the generation of future workload is calculated by multiplying the quantity of product to be delivered by the labor standards. An estimate to complete (ETC) is then calculated by subtracting standards earned from this original standards quantity and multiplying the result by a realization factor. Comparison of the ETC with the available capacity, in units of hours, suggests whether the shop is on schedule and if the production quota can be achieved. The available capacity of the shop is calculated considering the number of operators, the days remaining in the planning horizon, the number of shifts currently working in the shop, planned vacations and the shop's historical absenteeism factor.

For job shop projection, many of the heuristics embedded in the queue compression rule base, within the situation assessment module, are activated to calculate a total remaining part time. A similar comparison of total remaining part time to available capacity is transacted to determine if production is on schedule.

When production is identified as being behind schedule, the supervisor requires the capability to simulate a recovery plan and project its probable outcome (Siegel, 1987). Replan heuristics, embedded in this module, allow the capability to test out these recovery plans. Modification to the number of shifts and/or the number of operators provides the driver to the recovery plan. Changing the shift configuration takes into account scheduled overtime. Modification to the number of operators can account for any additional loan-in or temporary personnel.

The final component of the production and replanning module is overtime analysis. The function of this component is to determine if the assignment of operators to overtime work is worthwhile, based upon a marginal analysis of the manufacturing department's overtime history. A marginal rate of return, per each overtime hour, is calculated and conveyed to the production supervisor. A marginal rate of return, in this situation, may be defined as the increase in production that occurs for each additional hour of overtime. On a detail analysis level, an average production during overtime is calculated for each employee, based upon cumulative history. Based upon these values, the supervisor can decide to proceed with the overtime assignment.

This overtime analysis logic is also automatically activated by the overtime assignment rules in the resource allocation and scheduling modules. If overtime is still requested after viewing the marginal analysis, the overtime analysis and the overtime assignment components work in concert to provide a series of heuristic-generated recommendations. This process builds upon the work of Scudder (1985) regarding evaluation of overtime policies.

6.9 CONCLUSIONS

Repeated studies have reported that supervisors indicate the least confidence in, and most problems with using, existing information systems (Kulonda and Moates, 1986; LaForge and Bittel, 1983). This research has contributed a general model of the production supervisor's planning and control function and accompanying decision-making heuristics and algorithms. An expert system has been constructed which encompasses both the theoretical and the practical knowledge reflected in the model.

Experts frequently have difficulty analyzing the steps taken to arrive at a specific conclusion. Once knowledge is incorporated into a model or expert system, it provides a foundation from which others can expand and clarify the nature of the expertise. As it is fine tuned, the problem-solving techniques are transformed from an art to a science. This initial identification and analysis of the supervisor's function will act as a basis for further refinements, in a more industry-specific direction. As Eli Whitney stated in 1821, 'One of my primary objects is to form the tools so the tools themselves shall fashion the work and give to each part its just proportion, which when accomplished, will give expedition, uniformity and exactness to the whole' (Jackman, 1984).

REFERENCES

Baker, K. (1984) Sequencing rules and due-date assignments in a job shop. *Management Science*, **30**, 1093–8.

Becker, C.H. (1974) *Plant Manager's Handbook*, Prentice-Hall, Englewood Cliffs, NJ.

Bel, G., Bensana, E., Dubois, D. *et al.* (1989) A knowledge-based approach to industrial job-shop scheduling, in *Knowledge-Based Systems in Manufacturing*. Taylor and Francis, London.

Ben-Arieh, D. and Moodie, C.L. (1987) Knowledge based routing and sequencing for discrete part production. *Journal of Manufacturing Systems*, **6**, 287–97.

Bittel, L. (1968) *What Every Supervisor Should Know*, McGraw-Hill, New York.

Blackstone, J.H., Phillips, D.T. and Hogg, G.L. (1982) A state-of-the-art survey of dispatching rules for manufacturing job shop operations. *International Journal of Production Research*, **20**, 27–45.

Broadwell, M. (1984) *The Practice of Supervising*, 2nd edn, Addison-Wesley, Reading, MA.

Broadwell, M. (1985) *Supervisory Handbook: A Management Guide to Principles and Applications*, John Wiley & Sons, New York.

Christenson, C. (1982) *Supervising*, Addison-Wesley, Reading, MA.

Conway, R.W., Maxwell, W.L. and Miller, L.W. (1967) *Theory of Scheduling*, Addison-Wesley, Reading, MA.

Crosby, P. (1979) *Quality is Free*, Mentor Executive Library, New York.

Davenport, T.H. and Short, J.E. (1990) The new industrial engineering: information technology and business process redesign. *Sloan Management Review*, 11–25.

Fechter, W. and Horowitz, R.B. (1988) The role of the industrial supervisor in the 1990s. *Industrial Management*, **30**, 18–19.

Fox, M.S., Allen, B. and Strohm, G. (1982) *Job Shop Scheduling: An Investigation in Constraint-Directed Reasoning*. Proceedings of the National Conference on Artificial Intelligence, Carnegie-Mellon University, Pittsburgh.

Ginter, P., Rucks, A.C. and Duncan, W.J. (1985) Planners' perceptions of the strategic management process. *Journal of Management Studies*, **22**, 581–6.

Jackman, M. (1984) *The Macmillan Book of Business and Economic Quotations*, Macmillan, New York.

Kirkpatrick, T. (1987) *Supervision*, Kent Publishing, London.

Kulonda, D. and Moates, W.H. (1986) Operations supervisors in manufacturing and service sectors in the United States: are they different? *International Journal of Operations and Production Management*, **6**, 21–34.

LaForge, R.L. and Bittel, L. (1983) A survey of production management supervisors. *Production and Inventory Management*, 99–113.

Maxwell, W., Muckstadt, J., Thomas, L.J. and Vander Eecken, J. (1983) A modeling framework for planning and control of production in discrete parts manufacturing and assembly systems. *Interfaces*, **13**, 92–9.

McKay, K., Safayeni, F. and Buzacott, J. (1986) Job shop scheduling theory: what is relevant? *Interfaces*, **18**, 84–9.

Melnyk, S.A. and Carter, P.L. (1986) Scheduling, sequencing and dispatching: alternative perspectives. *Production & Inventory Management*, **27**, 58–67.

Moore, F. (1969) *Manufacturing Management*, Richard D. Irwin Inc. Homewood, IL.

Nakamura, N. and Salvendy, G. (1988) An experimental study of human decision-making in computer-based scheduling of flexible manufacturing system. *International Journal of Production Research*, **26**, 567–82.

Robbins, J.H. and Rabbi, M.F. (1988) Supporting the shop: successes and failures in production systems. *Industrial Management*, **30**, 6–12.

Russell, R. and Taylor, B.W. (1985) An evaluation of sequencing rules for an assembly shop. *Decision Sciences*, **16**, 196–212.

Salvendy, G. (1982) *Handbook of Industrial Engineering*, John Wiley & Sons, New York.

Scudder, G.D. (1985) An evaluation of overtime policies for a repair shop. *Journal of Operations Management*, **6**, 87–98.

Scudder, G.D. and Hoffman, T.R. (1985) Composite cost-based rules for priority scheduling in a randomly routed job shop. *International Journal of Production Research*, **23**, 1185–95.

Sharit, J., Eberts, R. and Salvendy, G. (1988) A proposed theoretical framework for design of decision support systems in computer-integrated manufacturing systems: a cognitive engineering approach. *International Journal of Production Research*, **26**, 1037–62.

Siegel, S. (1987) Simulation of scheduling rules helps decision-making on various objectives in manufacturing plant. *Industrial Engineering*, 40–4.

Taylor, F.W. (1947) *Scientific Management*, Harper & Row, New York.

Texas Instruments, Inc. (1986) *Personal Consultant Plus User's Guide*, Texas Instruments, Inc., Dallas, TX.

United States Department of Defense (1983) *Military Standard 1567A*.

CHAPTER 7

Intelligent systems for conceptual design of mechanical products

Qun Wang, Ming Rao and Ji Zhou

7.1 CONCEPTUAL DESIGN AUTOMATION

7.1.1 Conceptual design

Development of industrial techniques is closely related to the application of computers in engineering design, integrated computer-aided design (ICAD) technology has evolved as a new generation of design techniques. It also paved the way for implementing computer-integrated manufacturing (CIM). However, the key issue to accomplish the objective is conceptual design automation (Wang *et al.*, 1990).

Conceptual design is a very important but difficult target in CAD. As we know, product quality, reliability, production cost and productivity depend on not only the quality of components/parts, but also on the conceptual design and the coordination of the layout design of components, i.e. the quality of design synthesis.

In the past few decades, computers have been extensively used in optimization, finite element analysis, reliability design, computer graphics and simulation for detail in engineering; moreover, there has been relatively little achievement in the conceptual design stage. In other words, their use has been limited almost exclusively to purely algorithmic solutions and they cannot handle non-numerical or non-algorithmic information. As a result, there are many difficulties in the real applications of computers in engineering design. We find that one of reasons is not implementing the automation of the conceptual design stage (design synthesis), which therefore hinders the development of ICAD and applications of CAD. Obviously, it is very important to develop an integrated distributed intelligent environment to improve the quality and efficiency of product conceptual design.

7.1.2 Conceptual design automation

Conceptual design of mechanical products consists of two main aspects. (1) Conceptual design: it needs to conceptually determine specifications, performances, functions and structure parameters of a product, and select product structural forms, materials and configuration. All design results will provide numerical and symbolic information for the following detail design. (2) Layout (or structure) design: it performs the task of placing all parts and components. Clearly, these two aspects are usually ill-structured problems, which deal with non-numerical or non-algorithmic information, and are not amenable to purely algorithmic computation (Rao, Jiang and Tsai, 1989). The methodology to solve these ill-structured problems is thinking, reasoning and decision making on the basis of special domain knowledge and expert's experience. Therefore, conceptual design could not be handled by conventional CAD techniques, and is suitable for the use of expert system techniques.

Expert systems provide a programming methodology for solving ill-structured problems that are difficult to handle by purely algorithmic methods. An expert system is also an intellectualized computer program that acquires the knowledge of human experts and applies it to solve real world problems by reasoning and decision making. Such a problem-solving strategy is similar to the thinking activity of experts during the conceptual design stage. In order to implement the automation of conceptual design and to develop a powerful integrated distributed intelligent environment for conceptual design, it is imperative to use artificial intelligence (AI) techniques and combine them with conventional CAD techniques such as data bases, computer graphics, mathematical algorithms, solid modeling, system analysis, optimization, etc. (Manochetti and Seireg, 1987; Wang, Zhou and Yu, 1989a,b; Yu, Zhou and Wang, 1987).

7.1.3 Characteristics of conceptual design

The experience gained from building expert systems for solved problems has shown that their power is very apparent when the problem at hand is sufficiently complex (Rao, Tsai and Jiang, 1988; Rao, Jiang and Tsai, 1988). However, in engineering design, many practical and successful numerical computation packages, e.g. the optimization package, are already available. We agree that AI should emphasize symbolic processing and non-algorithmic inference, but it should be noted that the utilization of numerical computation will make expert systems more powerful in dealing with engineering design problems (Rao, Tsai and Jiang, 1988; Rao, Jiang and Tsai, 1988).

Having reviewed the existing expert systems, we feel that most of them were developed for specific purposes. Many were implemented with LISP-based tools and production rules were used to represent domain expertise.

In terms of applications, such expert systems can only process symbolic information and heuristic inference. A lack of numerical computation and uncoordinated single applications limit their capability to solve real design problems.

It can be shown that conceptual design is a hybrid engineering problem of performing numerical computation and symbolic inference alternately. In other words, problem solving depends on not only inference but also the algorithm. As reported recently, thousands of practical expert systems have been established in manufacturing engineering in the US and Germany. Unfortunately, most of them are used in fault diagnosis and production planning, and only a few in engineering design. In addition, hundreds of the tools (or shells) for building expert systems are available in the software market; however, many of them are only available for the special purposes of diagnosis and planning problem solving, and are not suitable for design. Obviously, design problem solving is different from others. It is very demanding to build AI development tools and develop expert systems for design. Compared with other types of expert systems, the development of design expert systems is confronted mainly with the following problems.

(*a*) *Multiplicity of design results and uncertainty of objectives*

Generally speaking, problem solving in diagnosis is a 'multiple input/single output' problem solving pattern, i.e. one conclusion (output) can be inferred from some evidence with an inference engine (of course, sometimes more than one). In contrast, design problem solving is 'single input/multiple output', i.e. a few results meeting the same requirements may be obtained at the same time. As a result, two obstacles may be involved in design: large decision space and comprehensive evaluation of the quality of design. The problem is how to find out all acceptable schemes that satisfy design requirements and how to choose the effective decision-making method to pick up the best one from all acceptable schemes.

(*b*) *Multiple levels and objectives of design tasks*

Obviously, a design needs to perform various subtasks that lie on different levels. For instance, the design for a machine tool involves many aspects such as transmission, hydrostatic circuits, electric circuits, power utility, operation, etc. These design tasks may be implemented on different levels and controlled by meta-knowledge (Rao, Jiang and Tsai, 1989). Thus we will face new problems, i.e. how to automatically resolve and plan a design task, how to represent the relationship among subtasks, how to solve event conflicts, and how to choose the appropriate problem-solving strategy to match a subtask.

(c) *Intelligent design environment for alternately performing computation and inference*

Conventional expert systems and tools emphasize symbolic processing and non-algorithmic inferences. Since design problem solving needs not only symbolic reasoning but also numerical calculation, the intelligent distributed intelligent environment should be able to call a variety of existing analysis and simulation packages, and to exchange information with a data base management system at any time.

(d) *Multiplicity of knowledge representation and problem-solving strategy*

Product design deals with various problem-solving methods and knowledge representation forms. For example, it often employs reasoning, calculation, table look up and graphics. During the development of such expert systems, we have to keep the segregation of the knowledge base, data base and control strategy to allow users to efficiently organize different models and domain expertise, because each of these components can be designed and modified separately.

(e) *Structure problem solving and geometry knowledge representation*

The final results of product design, including those from conceptual design and detail design stages, should be ultimately represented on the drawings that involve 80% of design information. In fact, it is inevitable that design deals with various geometric information, and implements structural and layout designs that touch upon the representation and inference of space knowledge. Compared with historical symbolic inference, space inference is more difficult. The problem is how to describe and cope with the geometric, functional and topological information of bodies.

(f) *Complexity of redesign and combinatorial explosion of the problem*

Redesign is the inevitable obstacle to design problem solving. When the results are unsatisfactory to the requirements from customers, the integrated intelligent design environment (IIDE) has to carry out redesign. Obviously, with an increase of the system size and problem complexity, redesign will be much more difficult. The problem is how to store and apply the failing information to direct redesign, how to select an optimum problem-solving strategy when multiple tasks conflict and how to implement expertise knowledge.

7.2 PROBLEM-SOLVING STRATEGY

7.2.1 Design–analysis–evaluation–redesign (DAER) model

In this paper, we describe a very effective and practical model for conceptual design, called the DAER model. The model is only a summary of the current situation of engineering design and not a new design methodology (Dixon, 1983). In fact, it authentically reflects the design expert's thought in solving real engineering problems and can handle the problems that need to deal with the empirical knowledge designs. This IIDE is developed on the basis of the DAER model.

Design actually is an art and a creative process (Douglas, 1988). Therefore, designers might try to approach design problems in much the same way as a painter develops a painting. In other words, designers' original design procedures should correspond to the development of a pencil sketch, where designers want to suppress all but the most significant details of the design, i.e. they want to discover the most important parts of an object to be designed which determine the final performances and price of a product. An artist next evaluates the preliminary painting and makes modifications, using only gross outlines of the subjects. Similarly, designers want to evaluate their first guess at a design and generate a number of design alternatives. In this way, designers hope to generate a 'reasonable-looking' rough product design before they start adding much detail.

To sum up, a design procedure can be divided into two stages: (1) generating a preliminary design scheme, and (2) repeating the procedure of 'analysis–evaluation–redesign' until all requirements and restraints are met and the design results are feasible. The DAER model follows the procedure. As Fig. 7.1 indicates, the DAER model will put forward a preliminary design scheme (synthesis) according to customers' requirements and marketing investigation, then analyze and evaluate the scheme, and finally make a decision regarding acceptability in terms of different decision criterion. If the scheme is deemed not acceptable, redesign will be performed. All the parts of the DAER model are indepenent modules with different functions.

(a) *Preliminary design*

Preliminary design is usually used to design a new product. Problem-solving strategies used in the preliminary design stage include the following:

- Generating a scheme on the basis of comparison with the past types.
- Generating a scheme that meets all requirements and constraints.
- Generating a scheme that only meets part of the constraints.
- Randomly generating a scheme.

These strategies can be used singly or in combination. The process of preliminary design depends on the specific domain knowledge and experts' experience. The fact should be neglected that some design issues are in the redesign area to improve old products. In this case, the preliminary design results are equal to the parameters of the existing products. These results can be analyzed directly without loading the preliminary design module, because the results already include enough data and information for the analysis module.

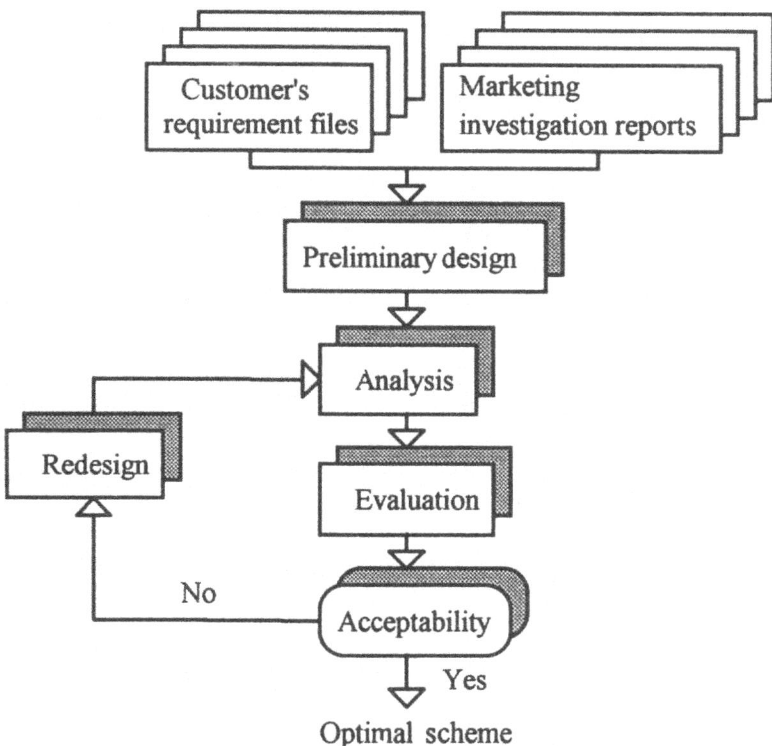

Fig. 7.1. The DAER model.

(b) Analysis

The purpose of analysis is to provide more detailed data and information so that evaluation can be implemented. Generally speaking, two types of information are produced in a design process: one is symbolic information directly from reasoning and the other is numerical information from computation. The analysis module will use various modern design methods such as finite element analysis, optimization and reliability, as well as different mathematical formula (computation models).

(c) Evaluation

Evaluation is a comprehensive consideration of all the design targets to choose an optimal design scheme. It requires not only statistics models but also experts' experience.

(d) Acceptability

This is a Yes/No test to determine whether all requirements and design constraints have been met.

(e) Redesign

Redesign will use the feedback messages from earlier analysis and evaluation, and then generate a modified design scheme.

7.2.2 Problem definition

To better understand and implement the DAER model, we will present a common problem-solving strategy for conceptual design based on the DAER model. First of all, it is necessary to introduce a few definitions and terminology.

Definition 1. Concept is defined as the abstract description of the natural property of an object. Each concept has its own identifier to be refereed as C_i.

Definition 2. Concept space is a set that includes all concepts in a specific domain. These concepts are organized in a specific order and hierarchy to behave the certain relations between them. Concept space is written as CS. Needless to say, $C_i \in CS$.

Definition 3. Function concept is the abstract description of the features of system functions. It is expressed as FC_i.

Definition 4. Function concept space is a set that includes all functions in a specific domain. It is written as FCS. Similarly, $FC_i \in FCS$ and FCS $\in CS$.

Definition 5. Structure concept is the abstract description of the essence of component structures. It is represented as SC_i. There exists such a relationship that $SC_i \in CS$.

Definition 6. Structure concept space consists of all structure concepts. It is also a subset of CS. It is denoted as SCS. $SC_i \in SCS$ and $SCS \in CS$.

Definition 7. Effective concept is the concept that satisfies the application environment and objectives as well as key constraints. Its notation is EC_i.

Definition 8. Effective concept space (ECS) consists of all effective concepts. $EC_i \in ECS$ and $ECS \in CS$.

Definition 9. Effective function concept is a function concept that satisfies that application specifications and constraints. It is denoted as EFC_i.

Definition 10. Effective function concept space ($EFCS$) incudes all effective concepts. $EFC_i \in EFCS$ while $EFCS \in FCS$.

Definition 11. Effective structure concept (ESC) is a structure concept that satisfies the application environment, objectives, constraints and effective function concepts. $ESC_i \in SCS$.

Definition 12. Effective structure concept space ($ESCS$) consists of all effective structure concepts. $ESC_i \in ESCS$ and $ESCS \in SCS$.

Definition 13. Design pattern is a tree structure that consists of nodes and arcs. Each node represents effective structure concept. Each arc indicates an 'AND' relation of nodes or an 'OR' relation of a single mode. It is denoted as DP_i.

Definition 14. Design pattern set (DPS) is equivalent to $ESCS$. Each design pattern represents a design scheme. Therefore, a pattern set is also a scheme set that meets application environment and design specifications. $DP_i \in DPS$.

7.2.3 Problem solving strategy

The problem-solving strategy of conceptual design based on the DAER model can be described in the following five stages (see Fig. 7.2).

Stage 1 is a problem definition stage for design tasks (from application environment and purposes to functions). Functions to be used are chosen from the expertise function memory (it can be viewed as a part of the knowledge base) according to the application environment and purposes provided by customers. For example, site location has an impact on the conceptual design of mechanical products because utilities available on site, such as cooling water temperatures, will depend on the geographical location. The knowledge to define functions is shallow knowledge (Kapp *et al.*, 1989).

The first step in problem solving can be expressed as follows:

$$\text{STEP } 1 = \left(FCS,\ \sum_{i=1}^{n} EFC_i | S_i \text{ and } T_i,\ EFCS \right) \tag{7.1}$$

where S_i and T_i $(i = 1, 2, \ldots, n)$ represent specifications and constraints provided by users. The objective in the above expression is to find an EFC_i to satisfy S_i and T_i in FCS, then to combine these EFC_i into $EFCS$.

Stage 2 is an effective conceptual design stage (from function to structure). The structures to execute functions are selected from structure memory (which can be viewed as another part of the knowledge base). The communication between functions and structures is not a 'one-to-one' mapping. Such a 'multiple-to-multiple' mapping configuration (see Fig. 7.3) indicates that a function can be realized with many different structures and a structure may possess many functions, e.g. the function for cooling can be implemented by several structures: cooling water, air, oil or others. This 'multiple-to-multiple' mapping makes the pattern design alternatives more completed and diversified. Each design alternative has a design scheme. If there is more than one design alternative (pattern), we need to select an optimal (or near optimal) one. In the case that no design alternatives are available, new design techniques will be used (since no existing structure can be used for the needed functions). If the existing design alternatives fail to satisfy application requirements, the design has to be improved. The knowledge to formulate effective structure concepts is heuristic knowledge which can be represented by heuristic rules.

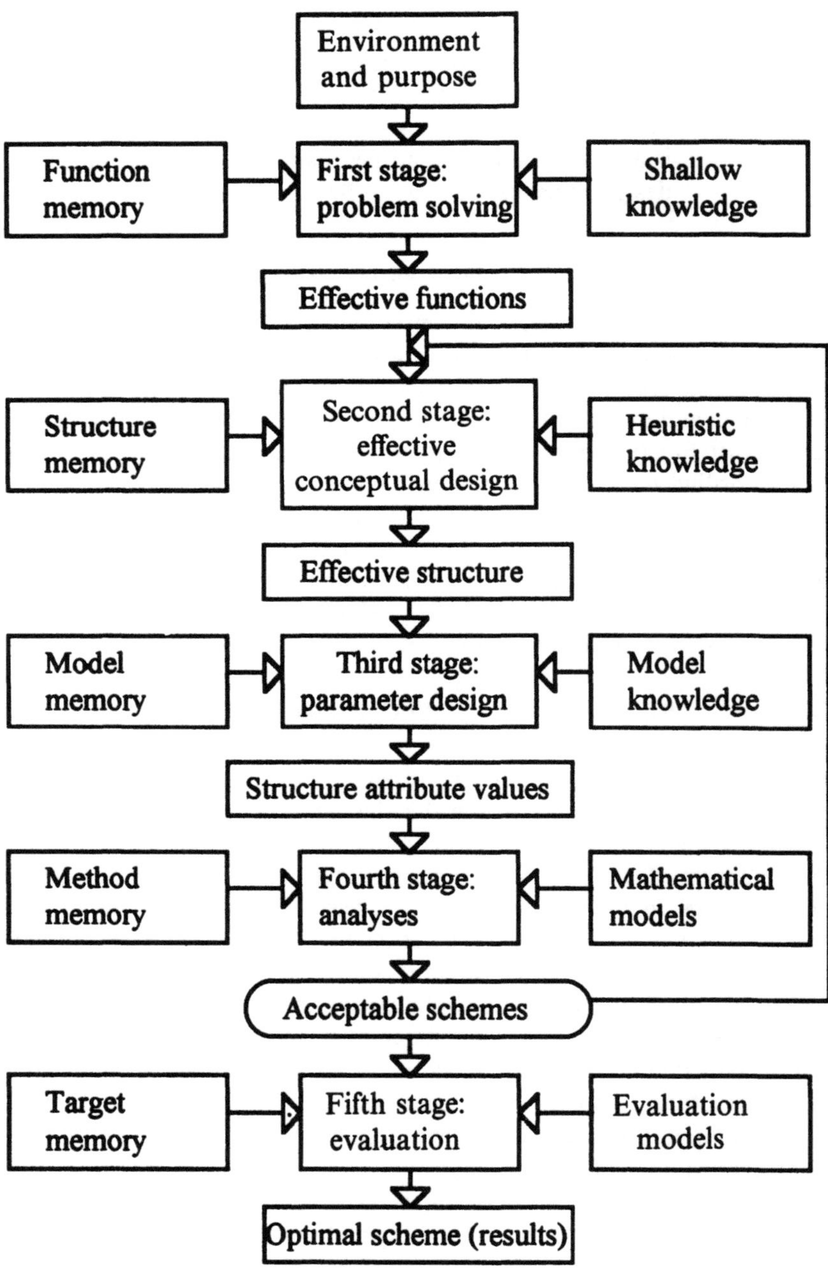

Fig. 7.2. Problem-solving strategy.

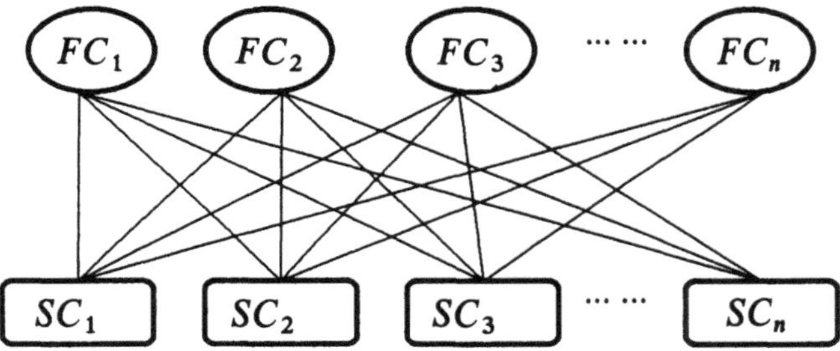

Fig. 7.3. Configuration of multiple-to-multiple mapping.

The following expression is the second step of problem solving:

$$\text{STEP } 2=\left(SCS,\ \sum_{i=1}^{n} ESC_i|S_i \text{ and } T_i,\ ESCS\right) \tag{7.2}$$

This expression presents how to select all ESC_i to satisfy S_i, T_i and EFC_i in SCS, then to combine these ESC_i into $ESCS$.

In general, $ESCS$ is a set of design patterns (alternatives). It is necessary to resolve them into individual patterns such that each individual pattern is one design scheme. The process is called as scheme resolving. It can be expressed as the following algorithm:

$$\text{RESOLVING}=\left(ESCS, OP,\ \sum_{i=1}^{n} DP_i\right) \tag{7.3}$$

where OP stands for a set of resolving operations. The purpose in the above expression is to divide ESCS (or DPS) into several DP (or design schemes) with OP.

Stage 3 is a parameter design stage (from structure to parameter). At this stage, the detailed description of structures can be completed by using the design models stored in model memory according to the characteristics of effective structure concepts. The knowledge to determine structural attributes is deep knowledge, i.e. model knowledge, which is represented by object-oriented frames.

The third step can be shown below:

$$\text{STEP } 3=\left(\sum_{i=1}^{n} DP_i, OPF,\ \sum_{i=1}^{n} ADP_i\right) \tag{7.4}$$

where, OPF represents an operating set of frames, and ADP_i is a design pattern with attributes and values (parameters). The expression indicates that it converts DP_i into ADP_i by OPF.

Stage 4 is an analysis stage (from parameter to analysis). Because functions and structures share a 'multiple-to-multiple' mapping configuration, numerous design schemes are usually produced. After parameters are given, all design schemes will be analyzed by selecting numerical computation methods (e.g. statistic analysis, optimization, etc.) from the method memory. Conventional CAD techniques can be utilized here.

The fourth stage of problem solving can be expressed:

$$\text{STEP } 4=\left(\sum_{i=1}^{n} ADP_i, OM, \sum_{j=1}^{m} ADP_j\right) \tag{7.5}$$

where OM represents an operating set of analysis methods and $n \geqslant m$. The algorithm is used to analyze every design scheme so that the feasible schemes are selected from ADP_i to satisfy the requirements of analyses. Usually a few schemes (or patterns) are omitted and only the practical schemes are kept.

Stage 5 is a final stage for comprehensive evaluation (from analysis to evaluation). According to analysis data, a proper evaluation target system from target knowledge memory and a comprehensive mathematical model from evaluation models will be chosen to evaluate the selected practical scheme. Techniques of fuzzy mathematics and system engineering are used in evaluation.

The fifth step of the proposed strategy can be represented as:

$$\text{STEP } 5=\left(\sum_{j=1}^{m} ADP_j, OE, ADP^*\right) \tag{7.6}$$

where OE is an operating set of evaluations. This algorithm intends to find the best scheme among practical candidate patterns by using evaluation OE. ATP^* is an optimal design to be sought.

Each stage in the problem solving strategy is very important. It combines numerical calculation (such as mathematical modeling, optimization and scheme analysis) with symbolic reasoning (knowledge representation and model handling as well as scheme evaluation) to accomplish the objectives in every stage.

7.3 SYSTEM CONFIGURATION

A good problem-solving strategy must match a good program structure to ensure the quality and efficiency of the software. In this chapter, an IIDE has been developed on the basis of the meta-system architecture (Rao, Jaing and Tsai, 1989). The meta-system is a large-scale knowledge development environment. It consists of several symbolic reasoning systems, numerical computation packages and graphics programs. The module techniques are used to implement the IIDE. Each module performs different functions. For example, the task definition module provides a window to input information. Users can define design tasks, application environments, purposes and specifications through the window. Also IIDE contains a function design module, a structure design module, a parameter design module, an analysis

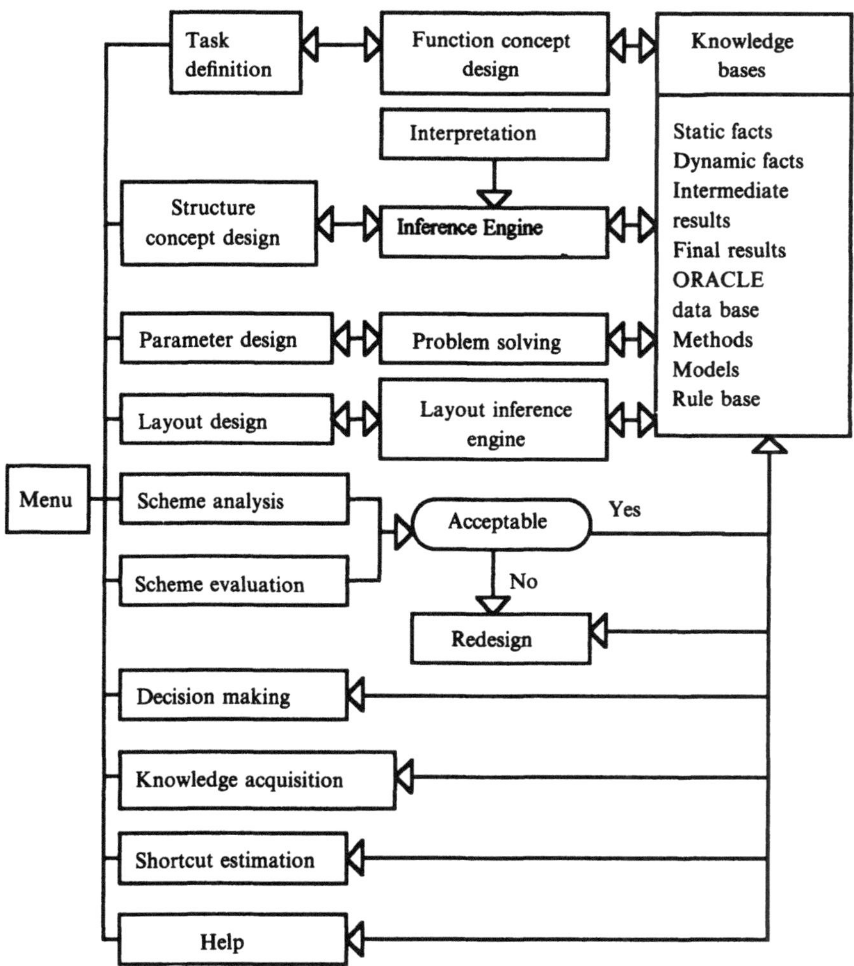

Fig. 7.4. Control structure of the IIDE.

and evaluation module, etc. (see Fig. 7.4). As a subsystem, each module may be written in different languages and used independently. They are under the control and management of the meta-system. This structure simulates human reasoning behavior in engineering design such that it can be used as a general framework for developing an applied integrated intelligent system to accomplish conceptual design. The following descriptions briefly describe their functions and characteristics.

7.3.1 Menu management system

It can guide users to select modules and observe the performance of the modules. Since the menu system employs a tree structure (each subsystem has its own sub-menu) and an object-oriented programming technique, users can select and run modules according to the contents on screen.

7.3.2 Task definition module

This is a window to input information. Users can define design tasks, application working environments, purposes and specifications through the window. The information that is normally available at the initial stages of design is given.

7.3.3 Function concept design module

This functions as the first step of problem solving. Its purpose is to further expand facts and information in the knowledge base due to the existing specifications to reduce interactive contents and determine some of the global variables (parameters) that might be shared by many unit operations. In general, global variables are not equally well understood. It is obvious that the entire system must be considered when determining these global variables. However, it is also very difficult for a beginning designer. A user never ensures that he/she receives correct and complete information. If some important data are missed, the module should supply these data using domain knowledge.

7.3.4 Structure conceptual design module

This can select the type of scheme and potential structural configuration components to satisfy the functions and constraints. This module works in the second step and provides parameter design with a variety of features. Its knowledge base includes expert exprience and heuristic knowledge which are represented as heuristic rules. Its inference employs constraint reasoning (Wang *et al.*, 1990).

7.3.5 Parameter design module

This functions similar to the third step of design. With the structures obtained from concept design and facts provided by users, attributes and attribute values (parameters) of the system/product to be designed are determined. Parameter design usually deals with local variables by short-cut calculations. Local variables are well understood by engineers, because these variables only associate with subsystem/component design, rather than the overall system or other subsystem. The knowledge of parameter design is expressed by the data structure of an object-oriented frame that is operated by problem solver.

7.3.6 Layout design module

Based on the geometrical parameters and information, the layout design module determines the position and orientation of each component and produces a drawing design. Another purpose is to perform simulation for testing the geometric interference between bodies.

7.3.7 Scheme analysis module

This functions in the fourth step. Based on the results from structural, parameter and layout design, this module analyzes each preliminary design scheme and provides data for evaluation. There are two paths in the module. If there exist satisfactory specifications, then the system selects the next module. Otherwise, local redesign is required. In addition, the module can call the existing analysis and simulation software packages.

7.3.8 Scheme evaluation module

This module is equal to the fifth step of problem solving (comprehensive evaluation). Its purpose is to evaluate comprehensive functions. In other words, all schemes entering the evaluation module are practical ones that are different from each other only in quality. The evaluation uses comprehensive evaluation models, fuzzy mathematics and system engineering techniques. Indices systems (or targets, such as operability, maintainability, manufacturing cost, etc.) and weights are selected by domain experts. There are two paths in the evaluation module. If the evaluating results satisfy the specifications provided by users, then the information is sent to the decision-making module. Otherwise, a global redesign will be performed. Generally speaking, it is difficult to evaluate the quality of a design scheme because most targets are vague and uncertain.

7.3.9 Decision-making module

In general, multiple schemes are generated as design results. During a mechanical system design, the index system solicits various opinions from different domain experts. Thus, conflicting solutions are generated from different expert opinions. The decision-making module will pick a best scheme among these schemes.

7.3.10 Interpretation module

This connects with the inference engine and provides the interpretation for concepts and reasoning paths in order to help users understand items, concepts and then to manage the system.

7.3.11 Inference module

The inference engine usually performs specific tasks and formulates knowledge representation. Obviously, in order to solve large hybrid engineering design problems, a simple inference engine that provides only one reasoning technique is unsatisfactory. The integration of different inference mechanisms is required in solving real world problems.

7.3.12 Problem solver

This operates the method knowledge base using a variety of problem-solving strategies, including reasoning, calculation, table look-up, curve observation and analogy. In conceptual design, these engineering design methods are often used. For example, reasoning can determine empirical coefficients, table look-up or curve observation provides numerical information (from a handbook) and analogy can choose better structure types.

7.3.13 Layout inference engine

Layout inference is much more complicated than symbolic inference because designing the engine will consider the shapes, positions, orientations of a body to be assembled as well as space constraints and interferences of bodies.

7.3.14 Static facts base

This stores task definitions and specifications provided by users. The information in the static facts base contains the essential conditions (constraints) for function and structure concept design. The facts in the static facts base are expressed with vector lists.

7.3.15 Dynamic facts base

In a reasoning process, users must continuously provide more detailed facts and data which are stored in the dynamic facts base. In our integrated intelligent design system, the dynamic facts base only associates with an inference engine and supports the interpretation module.

7.3.16 Intermediate result base

This stores intermediate results in the processes of symbolic reasoning and numerical computation. There are two purposes for setting up the intermediate result base. First of all, these results are used in a continuous reasoning process. Secondly, when a design fails, a backtrack (redesign) will be performed using the information stored in the intermediate result base.

7.3.17 Final results base

This stores all acceptable schemes. The decision making module refines the best one among those stored in the final results base.

7.3.18 Method base

This records the structure descriptions for all problem-solving methods. Each parameter or structure attribute has its own specific methods to be generated through reasoning, table look-up, analogy, calculation, etc. The method base is operated in many different ways. Separated from the problem solver, it partially describes the problem solving procedure. When required to handle a new parameter, users may add description to the method base without changing the problem solver. All methods are described with an object-oriented frame.

7.3.19 Rule base

This consists of many files organized as production rules. Each file is generated by the knowledge acquisition module to perform a specific subtask.

7.3.20 Model base

This model base stores various parametric models needed in part assembly. The parameters consist of space position (x, y, x) and orientation (v_x, v_y, v_z). A mechanical model (part or component) can be placed on a suitable position if the key parameters and a transformation matrix are provided.

7.3.21 Oracle

This is a commercial data base under the UNIX operation system and functions as a center for exchange information. The data base is a bridge between product design and manufacturing in the CIM environment.

7.3.22 Optimization base

This provides seven optimization techniques, such as linear optimization, non-linear constraint optimization, discrete optimization, etc. The optimization programs provide common optimal algorithms. An optimal model of mechanical design described in the model base can automatically link the programs and generate an executable file. The results obtained by running the file will be stored in the intermediate result base.

7.3.23 Shortcut estimate module

Scheme estimating is the important part of product design and is also the foundation of scheme evaluation and decision making. Analysis and appreciation of product cost and profit rate before manufacturing have a practical significance to direct production.

7.3.24 Knowledge acquisition module

This can be used to operate the rule base which consists of many files. Each file is a collection of rules for special problems. The knowledge acquisition module can maintain all files by skimming, adding, deleting and modifying rules in a file.

7.3.25 Redesign module

Redesign needs to use the messages from earlier analysis and evaluation as its input to improve the product design quality and find the best solution.

7.3.26 Help

The help module brings up a context sensitive help window explaining each of the system commands. The help window can be selected at any time from any module.

7.4 FUNCTION AND STRUCTURE CONCEPTUAL DESIGN

7.4.1 Function conceptual design

Function concept (definition 3) design, as first-stage reasoning, aims to expand the user's specifications into design conditions and restraints. In general, a user may fully describe the specifications and the working environment, but he/she can hardly translate these specifications into design conditions and restraints. That is to say, a user never ensures that his/her information is correct and complete. Some important data may be missing or may be in a too narrow range. For example, during the design of a chemical plant, where the plant is located has an impact on conceptual design because the utilities available on a site (e.g. cooling water temperatures) will depend on the geographical location. Similarly, the costs of raw materials will reflect the transportation costs, depending on where these materials are produced. The function concept design makes functional descriptions more specific and supplies some missing information.

Function concept design needs to use the shallow knowledge. In our system, shallow knowledge is defined as simple cause/effect knowledge (Kapp *et al.*, 1989), which is represented as heuristic rules without including mathematical formula or computation in context. All the conclusions that satisfy the restraints should be triggered and all effective concepts should be added to the fact base.

7.4.2 Structure concept design

Structure concept (definition 5) design, as second-stage reasoning, is the main content of the IIDE. The module can extract effective structure concepts from the structure concept space. Each concept is a specific symbol corresponding to a structural type, a structural concept or a structural alternative. Second-stage problem solving can be divided into five steps: establishment of structure concept space, establishment of the reasoning network, building of a knowledge base, restraint reasoning and scheme resolving (or scheme decomposition).

(a) Structure concept space

The structure concept space consists of all possible alternatives of parts/components and structure types. The space can be described by an AND/OR tree. The tree root represents a product/system to be designed and the leaves (or nodes) stand for its components and possible structure alternatives. The arcs (branches) of the tree represent the specifc hierarchical relationships among concepts (or nodes).

In the concept representation, if there exists an AND relationship between a node and all its subnodes, the node can be implemented only if all the subnodes are triggered (or performed). On the other hand, if there exists an OR relationship between a node and all its subnodes, the node can be carried out only if any one subnode is triggered. For example, Fig. 7.5 shows part of a structure concept space in a wheel loader. A typical wheel loader consists of several sybsystems such as a hydraulic subsystem, a transmission subsystem, a brake subsystem, a working device, etc. Here, if we only consider designing transmission, the subsystem is broken down into four independent parts in terms of transfer transforming ways: hydraulic, fluid, mechanical and electric transmission. When we design fluid transmission, four tasks will be performed, i.e. selecting engines, and designing the clutch coupling, torque converter and gear box. Thus a structure concept space (or a concept tree) may consist of hundreds or thousands of concepts or nodes. Finally, it should be noted that the tree structure depends on expert design thinking, i.e. each domain expert has his/her own structure concept tree. Therefore, when building a structure concept space, an intelligence engineer should fully consider the differences between experts and be able to coordinate the conflicts between them.

(b) Establishment of a reasoning network

When all function concepts, specifications and restraints (they are always viewed as the premises of rules) that trigger the nodes are linked to each concept in the structure concept space, this space (or tree) will be translated into a concept reasoning network. Since a structure concept may be a

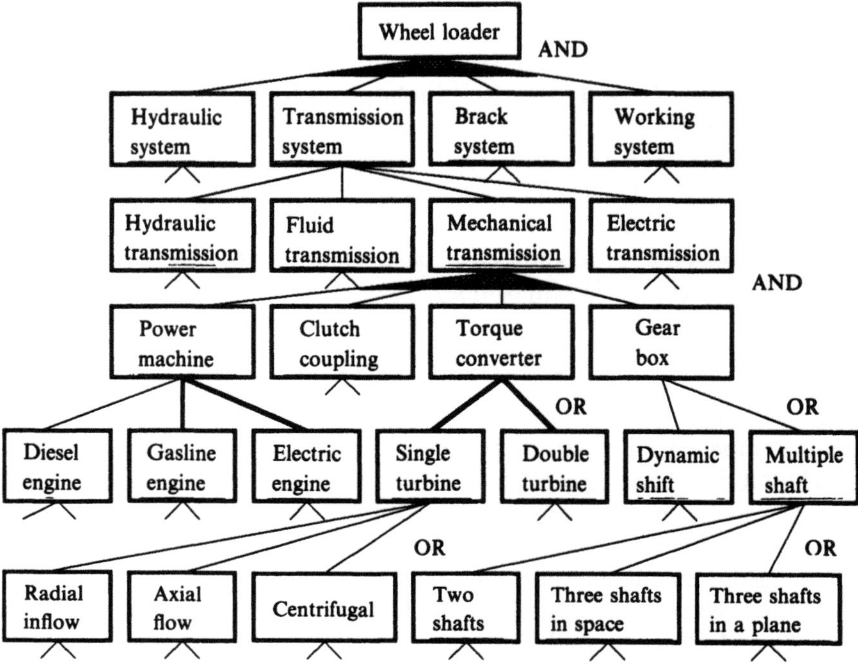

Fig. 7.5. Part of a wheel loader structure concept space.

premise of a concept node at another branch, the reasoning network becomes in reality an AND/OR graph. Without losing generality, we suppose that C_i represents a structure concept and E_i expresses a restraint condition (or a premise), Fig. 7.6(a) shows a simplified structure concept space from Fig. 7.5 and Fig. 7.6(b) demonstrates a concept reasoning network that includes a few restraints and functions.

(c) *Building the rule base*

The rule base of structure concept design consists of three aspects: production rules, the relationship list between rules and nodes, and the record list of goal nodes.

Rule: In this integrated intelligent design system, rules are defined as follows:

$\langle$Rule$\rangle$::=(IF{$\langle$Premise$\rangle$}$^+$THEN {$\langle$Conclusion$\rangle$}$^+$)
$\langle$Premise$\rangle$::=({$\langle$Function$\rangle$}*{$\langle$Element}$^+$)
$\langle$Function$\rangle$::=AND, OR, NOT, +, −,..., any predicate or operator
$\langle$Element$\rangle$::=$\langle$Vector Element$\rangle$|$\langle$Attribute Value$\rangle$
$\langle$Value$\rangle$::=$\langle$Symbol$\rangle$|$\langle$Number$\rangle$

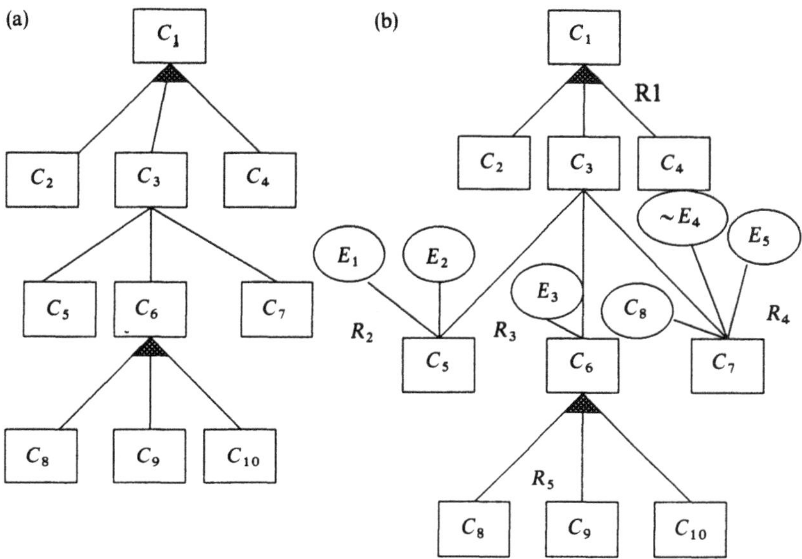

Fig. 7.6. (a) A simplified structure concept space; (b) reasoning network for conceptual design.

⟨Conclusion⟩::=({⟨Statement⟩}$^+$|{⟨Action⟩}$^+$)
⟨Statement⟩::=(⟨Vector Element⟩)|⟨Attribute Value⟩)
⟨Action⟩::={⟨Function⟩}$^+$|{⟨operator⟩}$^+$

where { }* means optional, { }$^+$ means one occurrence at least and | means 'OR'.

The structure concept reasoning network as shown in Fig. 7.6(b) can be converted into the following rules:

(Rules
  (R_1 (IF C_1 THEN $C_2C_3C_4$))
  (R_2 (IF (AND $E_1E_2C_3$) THEN C_5))
  (R_3 (IF (AND E_3C_3) THEN C_6))
  (R_4 (IF (OR (AND C_3C_8) (AND (NOT E_4)E_5C_3) THEN C_7))
  (R_5 (IF C_6 THEN $C_8C_9C_{10}$)))

Relationship list between rules and nodes: In general, the efficiency of the search and the velocity of an inference engine are relatively low, since production rules are not a kind of structural knowledge representation. In a production system (or rule-based system), the logical relationship between rules and nodes is implied in the context. In order to improve the reasoning efficiency, the IIDE applies the representation of tree node levels to explicitly stand for the logical relationship. In terms of the location of rules, the knowledge acquisition module can automatically classify the rules linking a node into its forward rules and backward rules, i.e. each node has its forward rules and/or backward rules. Here, we define the rules that verify the node (it can be viewed as a conclusion) are forward rules of the node and the rules that can be triggered by the node (it can be viewed as a premise) are backward rules of the node. This classification can bring out the following three advantages:

- Increasing the reasoning efficiency, since the logical relationship is explicitly represented and automatically accomplished by the knowledge acquisition module before the system is run. Thus, when the inference engine verifies a node, it can quickly find out the rules concerning the node. Clearly, the inference engine need not search all rules in the rule base. A special testing shows that the efficiency can be raised 8–10 times after the logical relationship is recorded. Of course, the processing method needs more memory space.
- Improving the explanation ability to reasoning paths, i.e. the inference engine can backtrack a reasoning process using the logic relationship.
- Enhancing the function for checking the contradictoriness and redundancy among rules.

Here, we define

$$G=(C, R, Q) \tag{7.7}$$

as a concept reasoning network graph, where $C=\{c_i\}$ $(i=1,2,\ldots,I)$ is a concept node set, $R=\{r_j\}$ $(j=1,2,\ldots,J)$ is a rule set and $Q=\{\text{After, Before}\}$ represents the relationship between rules and nodes (forward and backward rules).

The rules (from R_1 through R_5) discussed above only represent the context between concept nodes, but do not give us the logical relationship between nodes and rules (i.e. nodes and arcs). To clearly illustrate this relationship in correspondence to Fig. 7.6(b), we set up a 'Before' list and a 'After' list as:

(After ($C_1(R_1R_2R_3R_4R_5)$
$C_3(R_2R_3R_4R_5)$
$C_6(R_5)$))

(Before ($C_8(R_1R_3R_5)$
$C_9(R_1R_3R_5)$
$C_{10}(R_1R_3R_5)$
$C_5(R_1R_2)$
$C_6(R_1R_3)$
$C_7(R_1R_4)$
$C_2(R_1)$
$C_3(R_1)$
$C_4(R_1)$))

Record of goal nodes:
A concept reasoning network is equivalent to a concept set needed for domain experts to solve real problems. Therefore, operation on the reasoning network should not change the existing logical relationship between concepts. For this reason, a record list about goal nodes is added to the rule base. This list can also control the depth of searching nodes. The record list for Fig. 7.6(b) is expressed as

$$\text{(GOAL } (C_2C_4C_5C_7C_8C_9C_{10}))$$

The following example can best illustrate the function of the list. For the reasoning network in Fig. 7.6(b), if no goal list is set up and when E_3 and C_1 are true, R_1, R_3, R_4 and R_5 are triggered. In this case, the result of reasoning will be restored as an effective concept space shown in Figure 7.7(a). Compared with Fig. 7.6(a), C_7 has obviously lost its original logical relationship with C_3. However, if a goal list is set up, the original logical relationship will be maintained (see Fig. 7.7b).

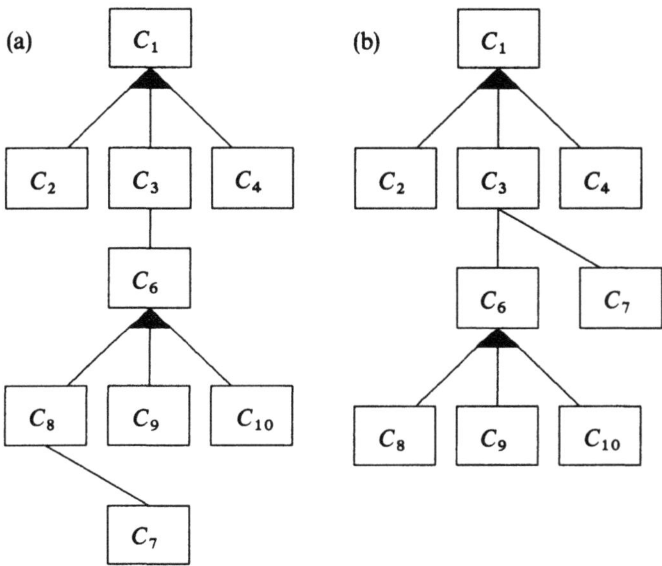

Fig. 7.7. Function of the record list of goal nodes. (a) Losing its logical relationship; (b) keeping its logical relationship.

(d) Restraint reasoning

The purpose of conventional reasoning (i.e. first-stage reasoning) is to prove an assumption, to obtain a conclusion or a few conclusions. Such an inference engine usually employs a forward chaining control strategy (data driven) or backward chaining (goal driven). The objective of restraint reasoning is to convert a structure concept space into an effective concept space that satisfy design restraint conditions. In other words, this inference can prune the original structure concept space according to facts provided by users. So far, this inference engine has been developed in IIDE.

In symbolic processing, the verification of a single concept node (a conclusion or an assumption) is not enough; instead, all the possible concept nodes in the space should be considered. The important criterion to an unsuccessful reasoning is this situation when an AND node or none of the OR nodes are not triggered. In this case, no design schemes or patterns will be produced. The restraint reasoning employs the bidirectional control strategy (combination of forward chaining and backward chaining). That is, the sub-nodes are expanded by using forward chaining reasoning to find out backward rules of these nodes, and then push them into the stack. The backward chaining reasoning can verify a specified concept node.

Before we introduce the restraint reasoning algorithm, we define

Stack	to record expanding nodes
Static Facts	to store facts associated with design
Intermediate Results	to record intermediate results and asked facts
Final Results	to record the final effective concept space

Restraint reasoning algorithm

1. Give C_1 (generally, it is the design goal, e.g. a wheel loader).
2. Push C_i in the Stack, Static Facts and Final Results.
3. Check whether Stack is empty. If it is empty, print or show the effective concept space, and then, exit. Otherwise
4. Pop out C_i from Stack and then find out the backward reasoning rules of C_i and all the forward nodes concerning the rules (i.e. all subnodes of C_i).
5. If C_i has a subnode (i.e. C_i is not in the goal node record), then
   (a) if the subnode is not a good node, push C_i into Stack and go to (3), else
   (b) the subnode is a goal node, continue the following.
6. If C_i has no subnodes or C_i is a goal node, verify the premises of forward rules concerning C_i:
   (a) try to match C_i with the restraints in Static Facts, Intermediate Results and Final Results and record the rules triggered, and then go to (3). Otherwise
   (b) consult a user, and push the facts obtained from him/her into Intermediate Results. If C_i is verified, push C_i into Final Results and record the rules triggered, and then, go to (3). Otherwise
   (c) if C_i is not verified, push this negative conclusion ($\sim C_i$) into Intermediate Results, give up C_i, and go to (3).

Figure 7.8 is an example of explaining the symbolic operation based on Figure 7.6(b) and the result of the reasoning is shown in Fig. 7.7(b).

Step 1: Push C1 the tacks.

ST	SF	MR	FR	UR
C_1	E_3		C_1	
	C_1			
	$\sim E_4$			

Step 2: C1 is expanded.

ST	SF	MR	FR	UR
C_1	E_3	C_1		
C_4	C_1			
C_3	$\sim E_4$			
C_2				

Step 3: C_2 is verified.

ST	SF	MR	FR	UR
C_1	E_3		C_1	R_1
C_4	C_1		C_2	
C_3	$\sim E_4$			

Step 4: C_3 is expanded

ST	SF	MR	FR	UR
C_1	E_3		C_1	R_1
C_4	C_1		C2	
C_3	$\sim E_4$			
C_7				
C_6				
C_5				

Step 5: C_5 is not verified.

ST	SF	MR	FR	UR
C_1	E_3	$\sim E_1$	C_1	R_1
C_4	C_1	$\sim E_2$	C_2	
C_3	$\sim E_4$	$\sim C_5$		
C_7				
C_6				

Step 6: C_6 is expanded and verified.

ST	SF	MR	FR	UR
C_1	E_3	$\sim E_1$	C_1	R_1
C_4	C_1	$\sim E_2$	C_2	R_3
C_3	$\sim E_4$	$\sim C_5$	C_6	R_5
C_7		$\sim E_5$	C_8	
			C_9	
			C_{10}	

Step 7: C_7 and C_4 are verified.

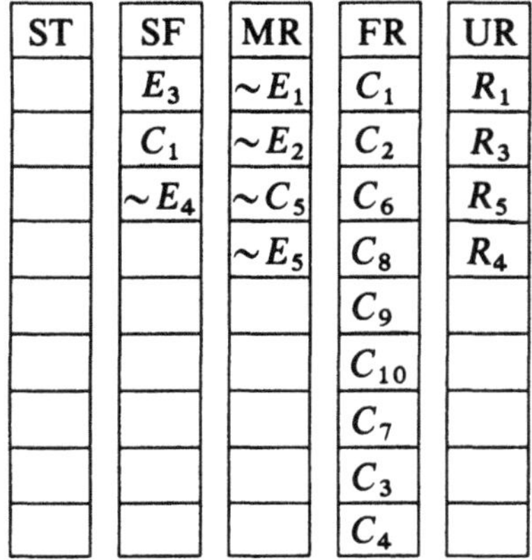

ST	SF	MR	FR	UR
	E_3	$\sim E_1$	C_1	R_1
	C_1	$\sim E_2$	C_2	R_3
	$\sim E_4$	$\sim C_5$	C_6	R_5
		$\sim E_5$	C_8	R_4
			C_9	
			C_{10}	
			C_7	
			C_3	
			C_4	

ST = stack, SF = static facts, MR = intermediate results,
FR = final results, UR = used rules.
Known: E_3 is true, E_1, E_4 and E_5 are false.

Fig. 7.8. A simplified example for explaining the restraint reasoning.

(e) Scheme resolving (or separation)

Generally speaking, an effective structure concept space is an assemblage of a number of design schemes. For example, Fig. 7.7(b) can be divided into two acceptable schemes as shown in Fig. 7.9. The number of acceptable schemes in the effective concept space is given by

$$NM = \prod_{i=1}^{n} t_i \tag{7.8}$$

where n is the number of OR nodes on the effective concept tree and t is the number of branches i.

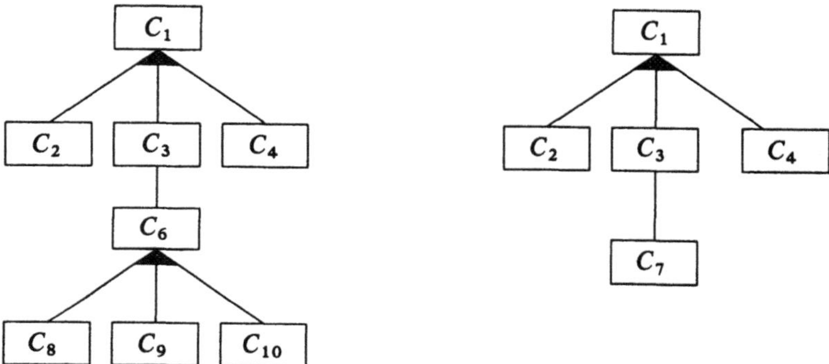

Fig. 7.9. Scheme resolving.

We may imagine that there exist a vast amount of schemes when n and t increase. However, because the restraint reasoning used domain experts' knowledge, the effective concept space can be confined to an ideal extent.

Scheme resolving is thus to divide an effective concept space (a tree) into a few subtrees containing only AND nodes and a single OR node. Each subtree can be served as an acceptable scheme. For example, the effective concept space as shown in Fig. 7.7(b) can be expressed as the following list:

$$\text{(schemes } (C_1(C_2C_3(C_6(C_8C_9C_{10})C_7)C_4)))$$

After the resolving operation, this list is separated into two subtrees (or sublists) as shown in Fig. 7.9:

$$\text{(scheme 1 } (C_1(C_2C_3(C_6(C_8C_9C_{10}))C_4)))$$
$$\text{(scheme 2 } (C_1(C_2C_3(C_7)C_4)))$$

Let us now discuss the algorithm of scheme resolving. An effective structure concept space is defined as:

$$\text{Schemes} = (EC, UR, Q) \tag{7.9}$$

where $EC = \{ec_i\}$ $(i = 1, 2, \ldots, I)$ represents a set of effective concept nodes, $UR = \{ur_j\}(j = 1, 2, \ldots, J)$ represents a set of the rules triggered and

$Q = \{\text{After Before}\}$ represents the logical relationship between the effective nodes and the triggered rules.

Scheme resolving algorithm
(Scheme, as a variable list, records all acceptable schemes)

1. Push Final Results data into Stack.
2. Verify whether Stack is null. If it is, separately display all acceptable schemes in Scheme and then exit. Otherwise
3. Pop a list Li out of Stack and look for OR nodes from the bottom of Li.
   (i) If an OR node is found and it has more than two branches, Li should be divided into sublists from this node. Push the sublist into Stack and then go to (2).
   (ii) If no OR node is found, push Li into Scheme and go to (2).

7.5 PARAMETER DESIGN

The purpose of parameter design is to determine the attributes and attribute values of structure types (or structure concepts). Structural parameter design will accomplish the detailed description of the system or product's structure, and provide necessary data and information for quantitative analysis and qualitative evaluation of design schemes. Parameter design is not a single symbolic reasoning problem. It also needs to use numerical computation, table look-up and curve observation. Therefore, its knowledge representation should not only express problem-solving methods but also consider the description of problem-solving processes; and not only employ a few problem-solving strategies but also solve the conflicts between different methods. This is why many methods may be chosen to solve the same engineering problem, which one is better often depends on the assumptions of the application environment and experts' private knowledge. As a result, we should fully consider this characteristic when designing the knowledge base, data structure and control strategy of the parameter design module.

7.5.1 Knowledge structure of parameter design

The knowledge associated with parameter design is called the deep knowledge or the model knowledge (Kapp *et al.*, 1989). In the IIDE, an object-oriented data structure (or frame) is employed to describe the deep knowledge. The objects (frame names) usually represent structure concepts, structure types of equipment tokens. The assemblage of all frames is called the method base for parameter design, which can be executed with a problem solver. Since the method base is independent of the problem solver, it can be easily updated without the adjustment of the problem solver.

7.5.2 Frame structure

Frames is a special data structure for representing domain knowledge in the field of AI applications. One of its advantages is to enhance the capacity for describing attributes of an entity (through facets and slots of a frame) and the logical relationship between an entity and others (through is_a slot or has_a slot). The IIDE here makes use of frames to describe problem-solving methods and problem-solving processes.

Generally speaking, a frame is comprised of a frame name, several slots and values. The slots represent attributes of the object, and values express attribute descriptions as well as a frame name stands for a specific object. In the IIDE, three formulated frames are developed to define problem solving strategies: symbolic reasoning, numerical computations and table look-up/curve observation. Their structures are defined as follows:

Numerical computation structure

Frame name: (string) x x x

1. Knowledge source: (string) x x x
2. Setting date: (string) x x x
3. Task level: (string) x x x
4. Task content: (list) x x x
5. Solution condition: (list) x x x
6. Search path: (list) x x x
7. Inquiry content: (list) x x x
8. Computation formula: (list) x x x
9. Related frame: (list) x x x
10. Explanation: (string) x x x
11. Default method: (list) x x x

Symbolic reasoning structure

It also needs to add two slots in the frame description as follows:

12. Rule set: (string) x x x
13. Inference engine type: (string) x x x

Table look-up structure

For the table look-up frame, table slot and interpolation must be given:

14. Data table: (string) x x x
15. Interpolation method: (string) x x x

The above three frames are already defined in the IIDE. If a new problem-solving method is required, a new frame structure should be defined. The functions of the slots are briefly explained:

1. Knowledge source. The slot records the source of the knowledge, and it may be a book, an item of literature reference or a domain expert.

2. Setting date. It describes when the frame or the problem solving method is set up to make is easier for users to modify and update the method base.
3. Task level. It represents the task hierarchy, i.e. the relationship of structure concepts. Ordering the number of task levels, which are dependent on structural concept space, is automatically implemented before parameter design.
4. Task content. It describes the content of the problem to be solved, such as what parameters were used to describe the current problem.
5. Solution condition. It specifies the parameters related to the current problem to be processed, i.e. the parameters in the slot are the premises. In addition, the slot also stores the conditions for triggering formula in the computation formula slot.
6. Search path. It is developed to illustrate the order of searching Static Facts, Dynamic Facts, Intermediate Results and Final Results.
7. Inquiry content. It asks users for information when some conditions cannot be found in fact bases and data bases.
8. Computation formula. This slot stores the analytical formula or numerical computation models.
9. Related frames. It describes the logical relationship between the current frame (or structure type) and other related frames.
10. Explanation. It records the descriptions about variables, concepts and problem-solving strategies. Textual information can be directly shown in English on the screen if necessary.
11. Default method. This slot can provide another problem-solving method or default solutions when a chosen method in the computation formula slot fails to solve the problem.
12. Rule set. The slot records the file name of a rule set, which may be called if necessary to use symbolic reasoning strategy for solving the current problem.
13. Inference method. It stores the file name of an inference engine. Since different problems may require different reasoning strategies, the IIDE provides four reasoning strategies.
14. Data table. It functions to record the name of data tables, and the data and information in the tables are stored in a file or data bases.
15. Interpolation method. It stores the name of interpolation methods that may be called during the process of problem solving.

The analogy is a usable design methodology, which is often employed in conceptual design or design synthesis. Thus, the IIDE provides the function. Users may obtain statistical formulae from the design and manufacturing data in the existing products, and put the formula in the default method slot. If we cannot get the problem solutions from the mathematical models, it is reasonable to use statistical methods to obtain an approximate solution.

7.5.3 Method base

The method base consists of all frames corresponding to structural types, parameters or empirical coefficients. Because the frames are inter-related to each other, there exists a certain relationship among the frames. In the IIDE, this relationship is simply treated as a kind of level relation or dendritic relation. Therefore, one frame only depends on the other vertically.

The levels of frames are automatically generated by scheme solving. For instance, two acceptable schemes are generated for Fig. 7.9. In scheme 1, there are eight frames on four levels on which the problem solving relies. The problem solver can manipulate the frames from bottom to top.

7.5.4 Problem solver

The problem solver identifies the function and structure of a frame, and then processes the slots of the frame in order of frame levels. Figure 7.10 shows the block diagram of the problem solver. The problem solving strategy has the following special functions:

1. Coupling symbolic reasoning and numerical computation. After the task is recognized, three problem solving methods can be used to solve the real problems: reasoning, calculations and table look-up.
2. Executing default method slot. If the specified method fails, the IIDE can execute the default method slot to obtain a statistical value or an approximate solution.
3. Solution provided by an expert. If all the above problem solving strategies fail, and only the user is an expert, the system will ask him/her for the problem solution.
4. Asking for expert's suggestions. If the user is an expert and his/her suggestions disagree with the solutions generated by the IIDE system, the expert has the right to modify the solutions. If the user is not a domain expert, the user has no right to do so.

7.6 SCHEME ANALYSIS

Parameter design provides data and information for analyzing product/system performance. Since the analysis contents and methods are different for special products, an intelligent environment cannot provide all analysis programs for general applications. However, it does provide a friendly internal interface that can be connected with the existing analysis models, commercial tools and computation packages. This subsystem interface can implement the data transformation and communication between the IIDE and analysis programs. In addition, the interface can call the

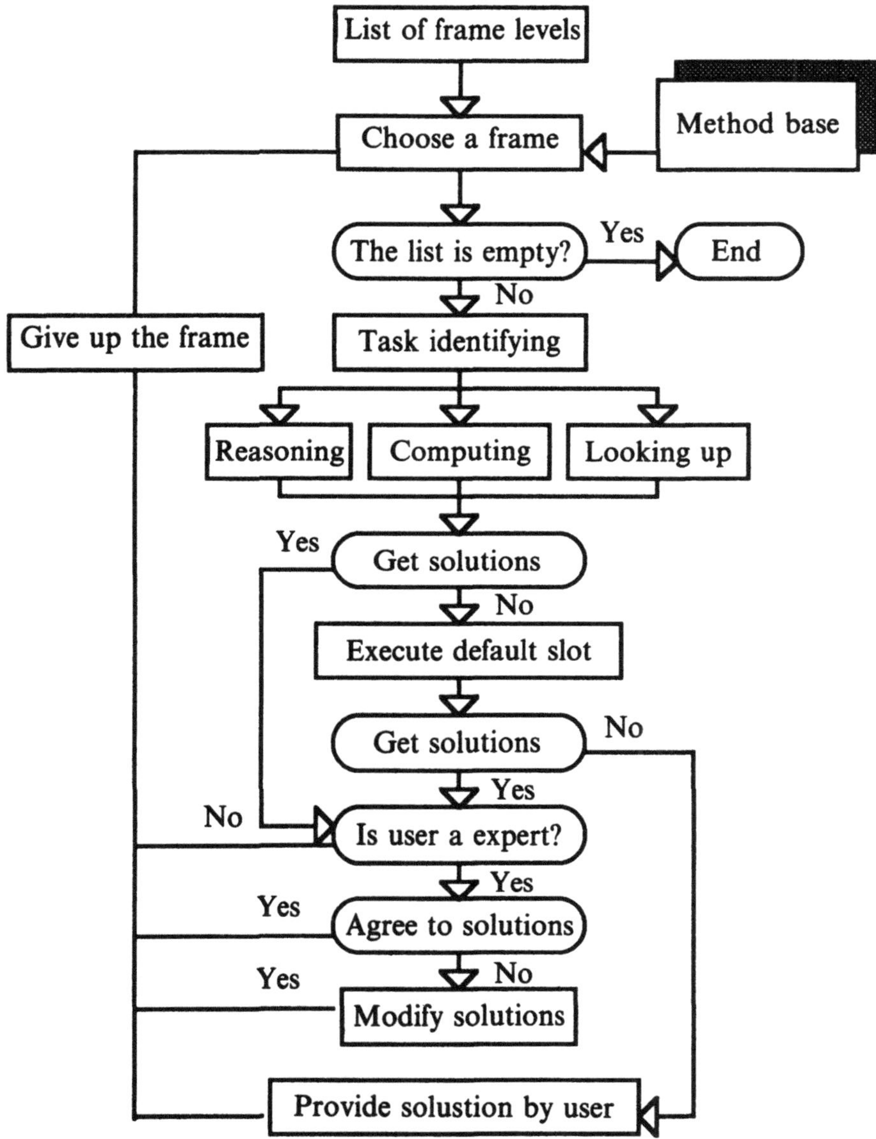

Fig. 7.10. Block diagram of problem solver.

different packages written in different languages such as C, Pascal and FORTRAN.

In the course of analysis, all generated schemes should be analyzed and the results of analysis must be stored. It should be expected that some of schemes might be eliminated through selection and competition because they cannot satisfy the criteria of design and specifications.

7.7 COMPREHENSIVE EVALUATION

The purpose of comprehensive evaluation is to determine whether the results (or schemes) of conceptual design are reasonable or practicable, and to form a true estimation of the design quality. It also provides the detail feedback information for improvement of design schemes (redesign) and an optimal backtrack point (or position) in the hope of avoiding a blind search in the new design and reducing the number of iterations. The IIDE implements a general evaluation algorithm suitable for estimating the comprehensive performances of a mechanical product/system in the early conceptual design stage. For a given design scheme, domain experts only need to give an index (target) system and the weight of each index, the evaluation module in the IIDE can perform a comprehensive estimation based on the index system and index weight.

7.7.1 Control structure of evaluation module

Figure 7.11 shows a control structure for the evaluation module and feedback redesign module. It possesses the following performances:

- It can give the final conclusion that presents the comprehensive performance of scheme using a multi-hierarchical index system.
- It can stimulate the performances of the schemes using analysis results.
- It can arrange the acceptable schemes in order of design quality with multi-hierarchical fuzzy evaluation method.
- It can provide feedback information and an optimal backtrack point for redesign models.

7.7.2 Determination of comprehensive evaluation indices

An index system is a systematized and formalized description of outer performances of a product/system to be designed, and reflects the integrated performance of the design object and the relationship between sybsystems. Here, we adopt a multi-hierarchical model (Chen, 1983). The index system is a tree structure that represents the evaluation targets horizontally and their sub-targets vertically (see Fig. 7.12). Generally speaking, the row number of targets (tree's depth) should be not more than 4, and the column number (tree width) not more than 10. This reason is that if the index system is too big, errors from various circumstances will impair the quality of the evaluation. For the specified mechanical product or system, it is very important to build its index system and to determine the index weight. The task depends on the complexity of a design problem and experience knowledge in the domain field.

Items of the index system represent the more detail descriptions of the specifications. They should be able to reflect the performance of a design

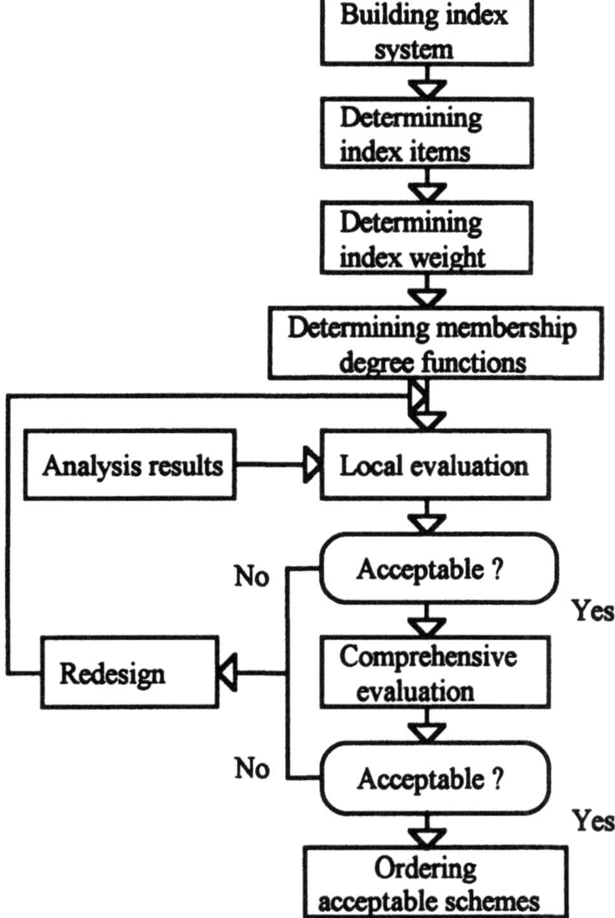

Fig. 7.11. Control structure for comprehensive evaluation.

object in the future both objectively and comprehensively. Therefore, a variety of factors with impact on the performance should be fully considered. The criterion of integrated optimum is a fundamental rule for determining the index system. A mechanical product is a complicated system where a variety of interdependent and interactive relationships exist between its subsystems. As a result, the quality of a product depends not only on the quality of its subsystems, but also mainly on the work in coordination between the product components. The acceptability should pursue the objective of whole design optimization and all items should reflect the comprehensive performance.

For a specific index system, the top items are abstract descriptions of characteristics of a whole mechanical product and the bottom items are the specific descriptions. The bottom items directly associate with design parameters generated by the analysis module. When the design fails, the

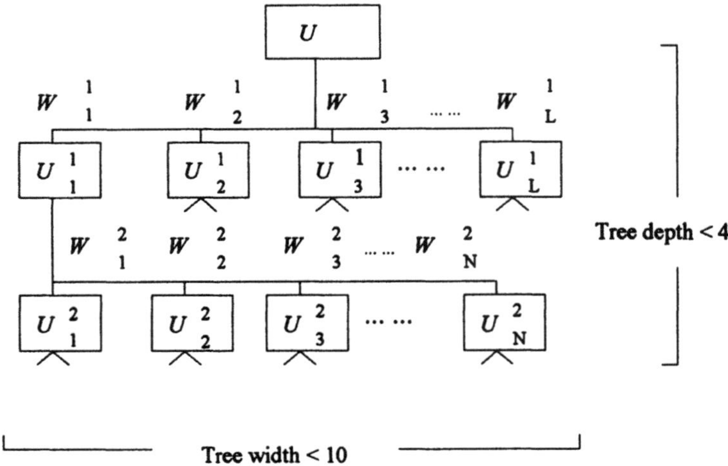

Fig. 7.12. Tree structure of the index system.

redesign module will improve the design scheme with the values of these items.

7.7.3 Determination of index weight

The index weight system is also a tree structure as shown in Figure 7.12. The IIDE uses two methods to decide index weight. One is mathematical statistics (Zhou, Duan and Wang, 1989; Zhou, Wang and Yu, 1989). The method is sometimes inapplicable for some real engineering design problems because of the difficulties in collecting statistic data, even if the model was accomplished in the software. The other is an empirical method in which index weights are directly provided by domain experts.

7.7.4 Determination of membership degree function

After the index values are given by analysis models, a designer will make a comment of 'satisfactory' or 'dissatisfactory'. This is an obscure and qualitative comment. On the other hand, the degree of 'satisfactory' for a new product design remains problematic. Therefore, the concept of membership degree in fuzzy mathematics is introduced into the evaluation module to make the quantitative evaluation of various levels of a mechanical system.

Suppose the possible value for a given index is $R \in [a, b]$ and a satisfactory value for the index is represented by A, then

$$\mu_A = f(R) \quad R \in [a, b] \tag{7.10}$$

is called the membership degree function of concept A for the index value. Many methods for determining membership degree functions have been reported (He, 1983). The IIDE selected a type of expert evaluation.

7.7.5 Comprehensive evaluation algorithm

Comprehensive evaluation refers to the qualitative estimation of a single design scheme. The evaluation procedure will be performed step by step from the lowest level of the index system up to the highest level. Finally, a summary (or conclusion) will be made on the evaluation of the first level (or the tree root) in the index system, such as the efficiency, size, maintainability and manufacturing cost.

In evaluating of each level, feasibility is first considered, i.e. the module will only answer whether the scheme satisfies all customers' requirements and design specifications on the basis of analysis results (index values) and membership functions. The evaluation is divided into two stages. The local evaluation, i.e. estimating each aspect of the comprehensive performance of a product, will be executed and then a comprehensive comment on the design scheme will be performed. The module will finally give the following information:

- Comprehensive evaluation scores.
- Feasible indices and weights.
- Dissatisfactory and satisfactory degree of indices.

In the following the comprehensive evaluation method is illustrated by an example of the third level.

Suppose the index vector for the third level (see Figure 7.12) is

$$U_3=(U_1^3, U_2^3, \ldots, U_L^3), \tag{7.11}$$

and the corresponding value for each index is

$$R_3^i \in [r_{i1}, r_{i5}] \quad i = 1, 2, \ldots, L. \tag{7.12}$$

The corresponding membership degree function for each index is

$$\mu_i(r), \quad r \in R_3^i \quad i = 1, 2, \ldots, L. \tag{7.13}$$

The weight vector of each index (see Fig. 7.12) is

$$W_3=(W_1^3, W_2^3, W_3^3, \ldots, W_L^3) \tag{7.14}$$

and the index values computed by analysis model is

$$P_3=(P_1^3, P_2^3, P_3^3, \ldots, P_L^3). \tag{7.15}$$

Algorithm for the third level evaluation

1. Calculate $U_i^3(P_i^3)$, $i=1,2,\ldots,L$.
2. If $U_i^3(P_i^3) \geqslant U_{\min}^3$, $i=1,2,\ldots,L$, then calculate $C_3^L=\Sigma_{i=1}^n W_i U_i(P_i^3)$. Otherwise, fill up the table that records error messages.
3. If $C_3^L \geqslant M_2^L$ (where M_2^L is the threshold value of the higher level), give the result of calculation. Otherwise, fill up the error table.

The general evaluation procedure of other levels is similar to the one discussed above.

7.8 DECISION MAKING

Decision making means an arrangement of acceptable schemes in a feasible order. To achieve this objective, the IIDE uses the comprehensive decision method based on multiple levels (He, 1983). If several acceptable schemes are generated in conceptual design, the decision-making module can arrange the schemes and give the comprehensive quantitative difference among them.

7.9 REDESIGN

The five stages of the problem-solving strategy can be accomplished as discussed above. However, we should consider a special situation in which all design schemes cannot meet the customers' requirements and specifications, i.e. no acceptable scheme is generated. In this case, redesign is needed on the basis of the feedback information from the analysis and evaluation module. To do so, a mathematical model of redesign on optimal feedback information has been developed in the IIDE.

The basic concept of optimal feedback is to divide the integrated performance of a mechanical product or system to be designed into various levels of indices. The corresponding membership degree of each index serves as a decision variable. These variables can construct a computing model that optimally corrects the old values of each level's indices.

Figure 7.13 shows how an optimal feedback mathematical model is established. M represents goal functions, s.t. is restraint conditions, and Opt. stands for optimal solutions. The following is an example that illustrates how to set up the second level of the optimal model under the case of index M.

Suppose that evaluation index vector and weight vector to M_{12} are, respectively

$$U_{12}=(U_{121}, U_{122}) \quad \text{and} \quad W_{12}=(W_{121}, W_{122}) \tag{7.16}$$

the corresponding membership degree vector of decision variable U_{12} is

$$D_{12}=(W_{121}d_{121}, W_{122}d_{122}) \tag{7.17}$$

and the ideal membership degree (which corresponds to the best index value) is

$$\bar{D}_{12}=(W_{121}\bar{d}_{121}, W_{122}\bar{d}_{122}) \tag{7.18}$$

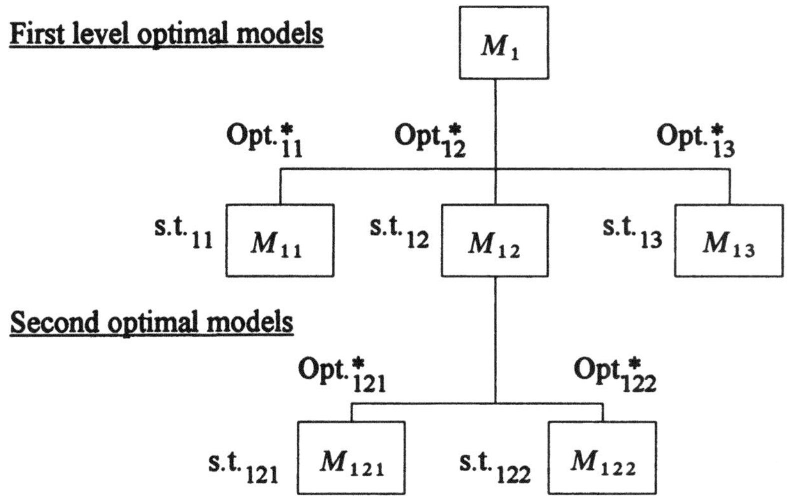

Fig. 7.13. An example for optimal mathematical models.

We have the following optimal model of feedback information:

$$\begin{gathered}\operatorname{Min} \| D_{12} - \bar{D}_{12} \| \\ s.t. \sum_{i=1}^{2} W_{121} d_{121} \geqslant d_{12}^* \\ g_{12i}^{\max} \geqslant d_{12i} \geqslant g_{12i}^{\min} \quad (i = 1, 2)\end{gathered} \tag{7.19}$$

where $g_{12i}^{\max}$ and $g_{12i}^{\min}$ are the upper and lower limits for membership degree, and d_{12}^* is the distributed value of index M_{12} that is determined by the optimal models on the first level (see Fig. 7.13). If the solution vector is

$$D_{12} = (W_{121} d_{121}^*, W_{122} d_{122}^*), \tag{7.20}$$

then D_{12}^* is the optimum distribution of M_{12} indices and the goal that should be reached by comprehensive performance of the third level indices M_{121} and M_{122}.

7.10 APPLICATION CASE STUDY

So far, we have developed three integrated distributed intelligent systems for conceptual design in the field of mechanical design using the IIDE:

- Integrated intelligent system for conceptual design of wheel loaders.
- Integrated intelligent system for conceptual design of industrial turbines.

- Integrated intelligent system for conceptual design of mechanical transmissions.

Here, CDEST (Conceptual Design Expert System for Transmission) will be discussed.

7.10.1 Functions of CDEST

CDEST is an integrated distributed intelligent system for the conceptual design of mechanical transmission, which combines expert systems with conventional CAD techniques. CDEST can be used to design the general scheme of transmissions according to customer's requirements.

CDEST can implement various designs, including power utility, shift, layout sketch and so on; and a variety of selections, such as bearing, lubricating, connecting, sealing; as well as various part parameters designs, e.g. property parameters, function parameters, geometry parameters, position and orientation parameters. In addition, the system can carry out the rigidity, strength stress analyses of important parts, the evaluation of final results (or design schemes) and shortcut estimations of manufacturing cost. The information and data (numerical and symbolic) generated by the above functions will lay the foundation for the following detailed design of parts and create good conditions for an integrated CAD system. Another function of the system is to complete the layout design, which can automatically generate rough assembly drawings and provide the interactive capability, thus allows users to modify the drawings and influence the system's decision making.

In order to maintain the knowledge base and add new knowledge, the system provides a special knowledge acquisition module which can be used to modify and expand the knowledge base. According to engineering design requirements, the system can accomplish conceptual design, layout design and shortcut estimation for various types of transmission devices.

7.10.2 Design principles of the system

The key problem of CDEST is how to combine expert systems with conventional CAD techniques and how to meet the specifications under a CIM environment (highly integrated information). Before the system is developed, the following basic principles are introduced:

- The research and development activity must satisfy the requirements of CIM software engineering.
- Two or three domain experts are designated for obtaining domain specific knowledge.
- The system must ensure the segregation of the knowledge base from the inference engines to allow users to reorganize different knowledge models and domain expertise efficiently.

- In order to efficiently maintain and upgrade the knowledge base, the knowledge acquisition subsystem should be included in the system.
- Common-LISP is used to implement the system, and process symbolic information. Numerical information will be processed by Fortran, Pascal or C.
- The system is able to exchange information with the ORACLE data base management system.
- A user-friendly environment should be provided.
- When solving the problems, the system should provide the interpretation to users conveniently and practically if necessary.
- The system can be used as a commercial intelligent tool for developing intelligent design systems.
- The system should contain various problem-solving strategies.
- The system has to be evaluated through real industrial product designs.

7.10.3 Illustration

CDEST is implemented on a SUN SPARCstation 1 plus, which is heterogeneous knowledge integration environment. Its symbolic reasoning system is developed in Common-LISP, while geometric conceptual design and simulation systems are programmed in C language. Fortran is used to develop the evaluation module, optimization package and analysis module. A graphics package (CGI) is run at a SUN graphics workstation.

The following tables and figures illustrate a designing process of the system. Here, we will describe the designing process step by step and discuss the designing results.

- First of all, several input pages about design requirements have to be filled out (Table 7.1). After the user confirms them, a 'done' button should be typed.
- The conceptual design can determine various structure forms, material, etc. Table 7.2 shows the specific matrix to present a kind of structure form. The system can automatically generate the matrix and the user can also modify it.
- In the example, after structure concept design, the system gives the user two design schemes. Figure 7.14 shows their rough layout form. Figure 7.15 is their rotational speed diagram expressing speed change. Figure 7.16 represents a more detailed structure drawing, which offers the data and information for analysis and evaluation. Figure 7.17 shows the result of side layout design. The side layout design means determining position and orientation of shafts.
- Tables 7.3–7.7 represent the results of parameter design for scheme 1, including numbers of transmission stages, shafts and gears; parameters of the selected motor; a variety of drive ratios; materials, geometric parameters, machining requirements of shafts, gears and box, etc. The

Table 7.1. Specifications of design

Tab-Surface-W (mm)	1000
Tab-Surface-W (mm)	3000
Space-Width (mm)	500
Space-Height (mm)	500
Space-Length (mm)	500
Table-Load (N)	8000
Max-Feed-V (mm/min)	1000
Min-Feed-V (mm/min)	10
Rapid-Traverse (mm/min)	0
Spindle-Power (kW)	15
Max-Spindle-V (r/min)	630
Min-Spindle-V (r/min)	50
Feed-Travel (mm)	0
Move-Part-Weight (kg)	0
Millsprindle-Max-T (Nm)	0
Cutter-Diameter (mm)	0
Mill-Head-Number	1
FB-Model	FB001
Order-No.	MUN001
Designer	Engineer
Plano-Type	Small
Feed-Box-Type	Mill-H-F
Driven-Type	T-Lead-SC
I/O-Shaft-D	Orthogonal
Mill-Head-D	Mill-Head-H
Mill-Head-P	Left

Table 7.2. The matrix for representing scheme 1

	Structure form: 1[3]–2[4]–3[3]–4[1]					
shaft 1	d_gear 3	dn_gear 2	d_gear 1	0	0	0
shaft 2	d_gear 41	d_gear 1	dn_gear 1	0	d_gear 41	0
shaft 3	d_gear 4	0	0	0	dn_gear 41	d_gear 5
shaft 4	0	0	0	0	0	dn_gear 5

results provide important information for scheme analysis and evaluation.

- Table 7.8 presents the conclusions if the results of scheme 1 analysis satisfying the conditions of the working environment. These analyses consist of stress and safety of gears, shafts, bearings and box.
- Table 7.9 shows the result of comprehensive evaluation. We specify four items as final targets, i.e. working efficiency, volume, manufacturing time and price. Similarly, we may produce the parameter design, scheme analysis and evaluation to scheme 2. Finally, better conceptual design can be determined based on the results of comprehensive evaluation.

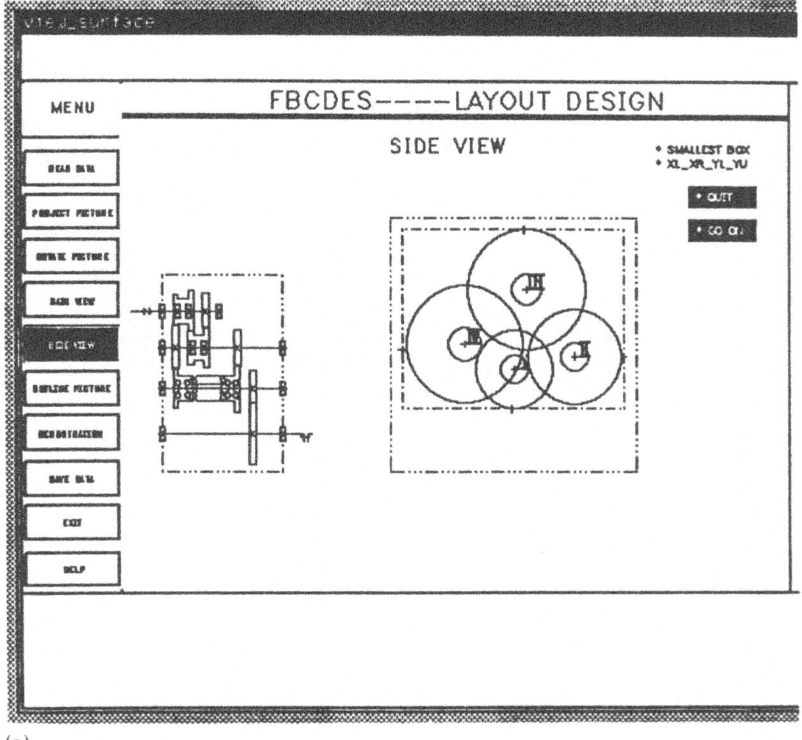

(a)

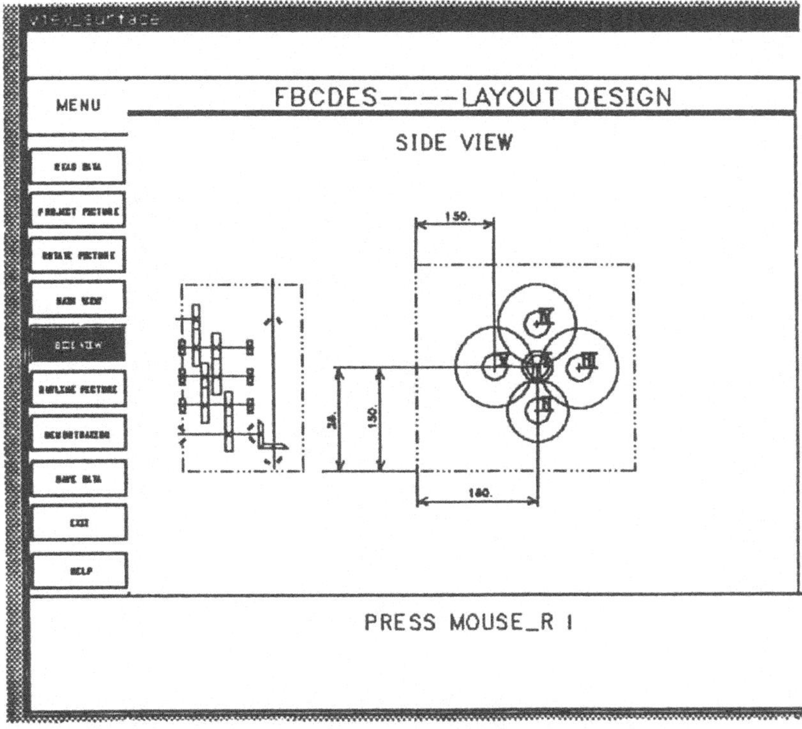

(b)

Fig. 7.14. Sketch of transmission. (a) Scheme 1; (b) scheme 2.

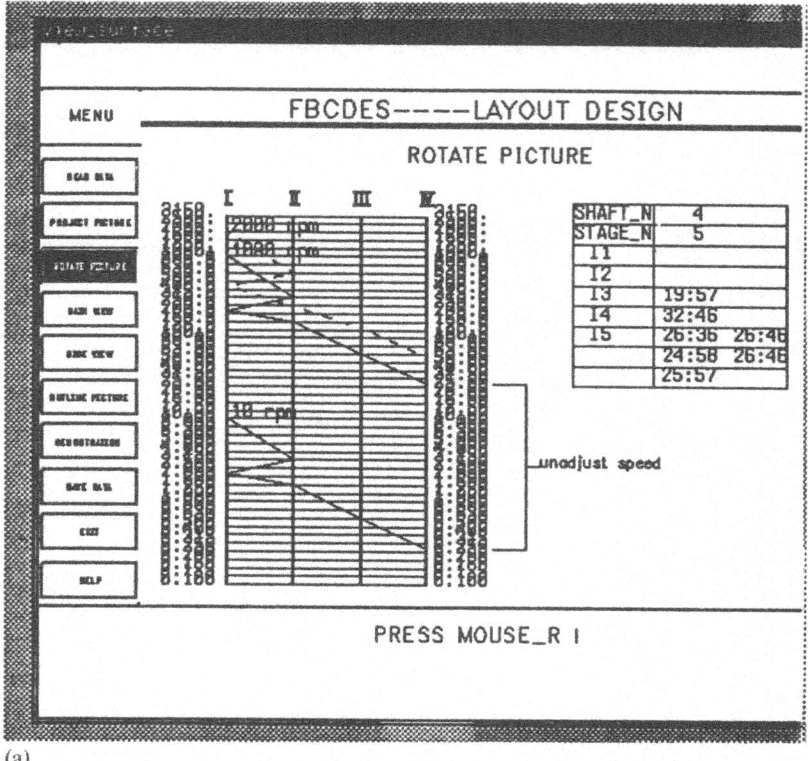

(a)

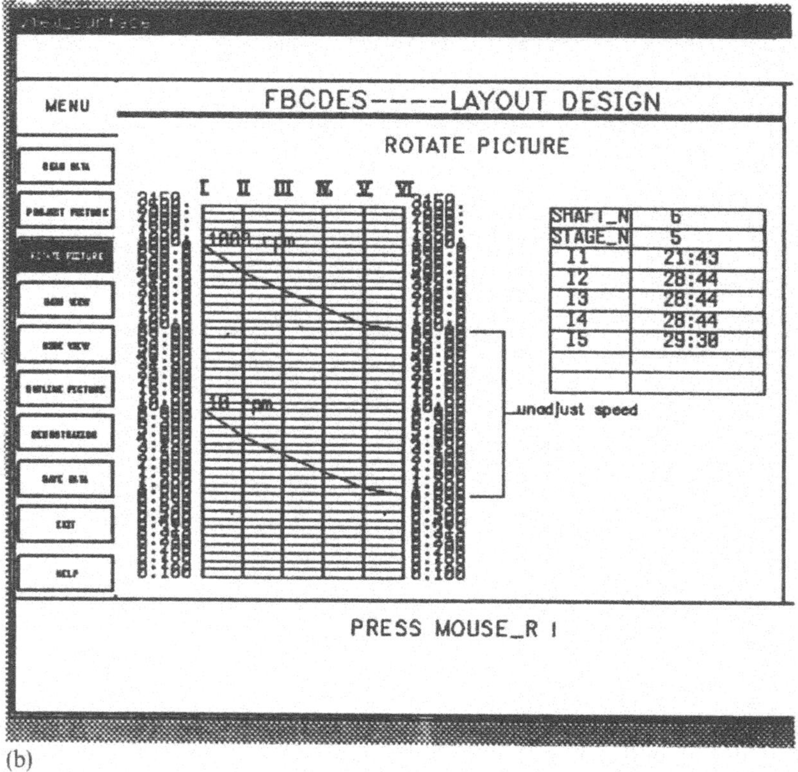

(b)

Fig. 7.15. Rotational speed diagram. (a) Scheme 1; (b) scheme 2.

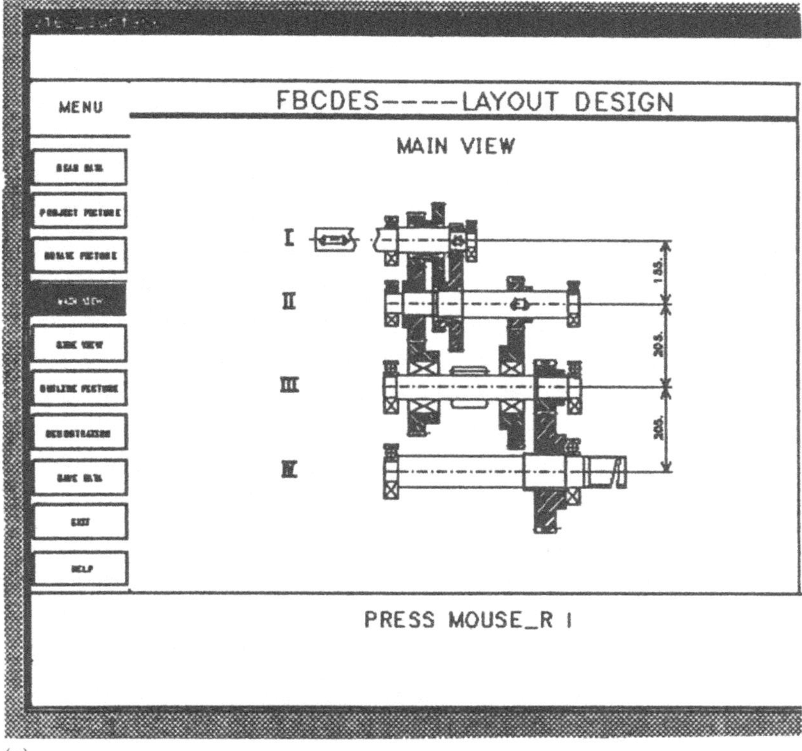

(a)

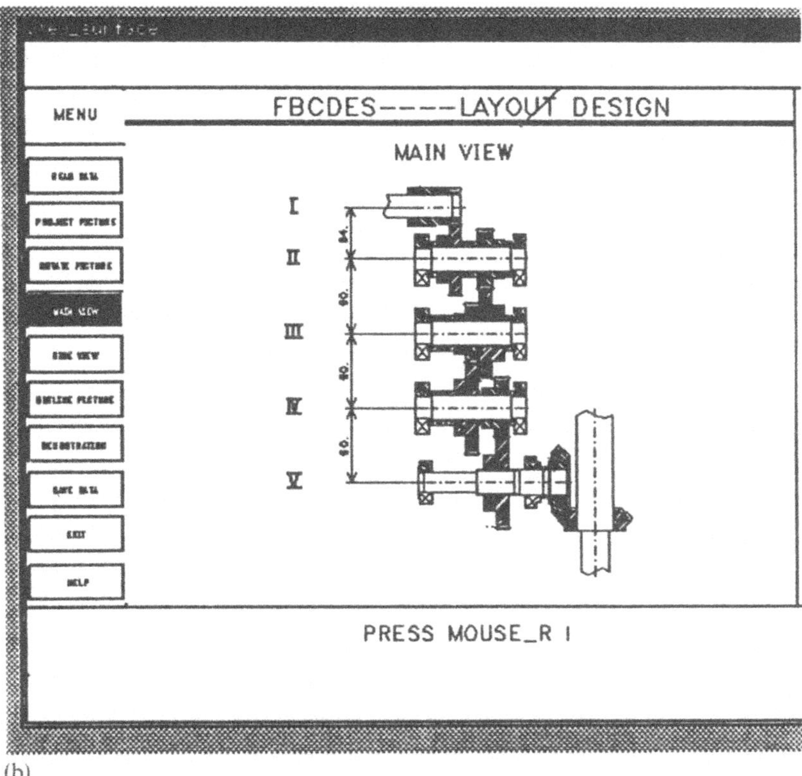

(b)

Fig. 7.16. Structure drawing. (a) Scheme 1; (b) scheme 2.

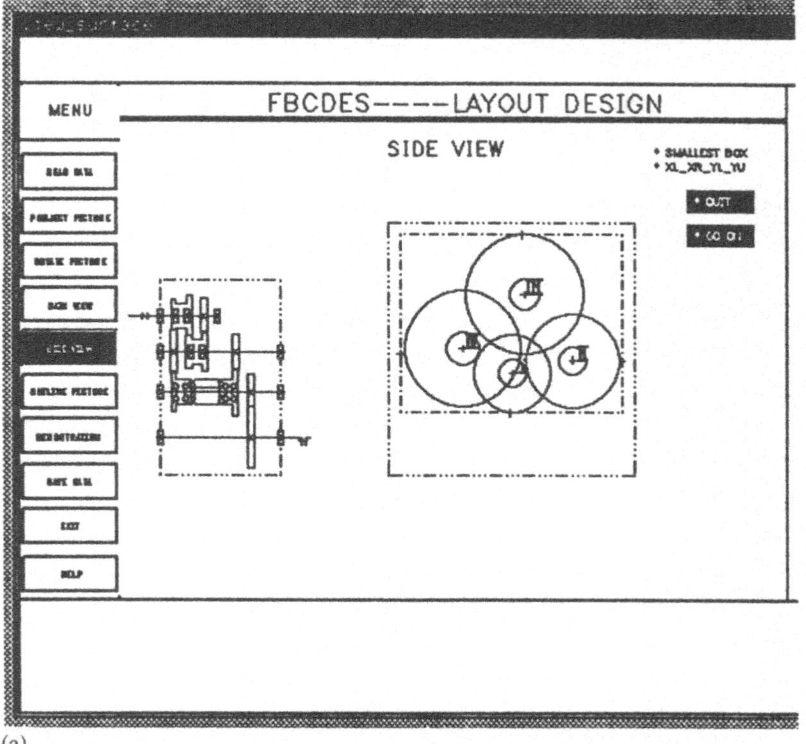

(a)

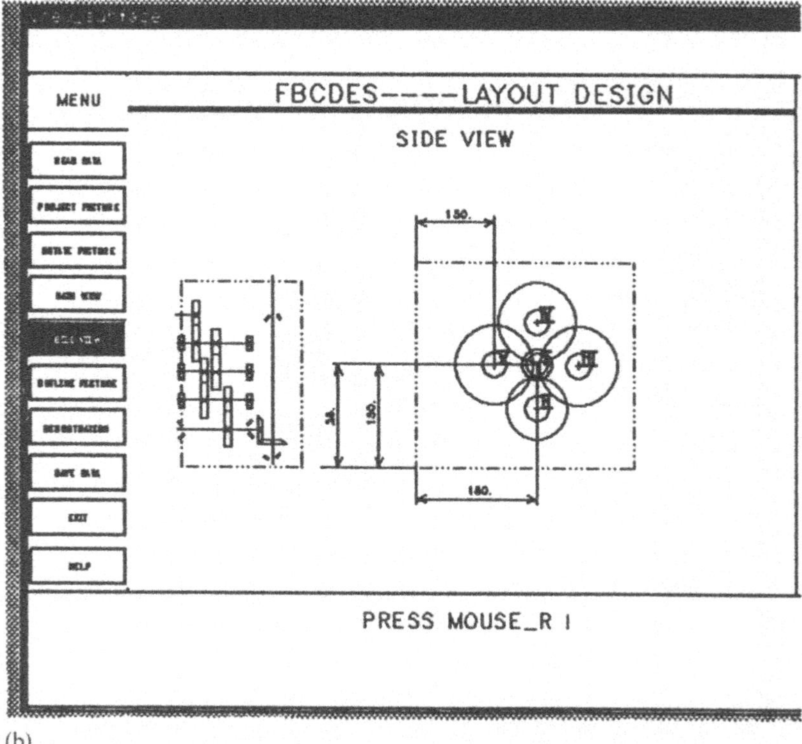

(b)

Fig. 7.17 Sectional drawing. (a) Scheme 1; (b) scheme 2.

Table 7.3. Numbers of stages, shafts and gears

Type	*Value*
Stages	5
Shafts	4
Gears	11

Table 7.4. Parameters of motor

Parameter	*Value*
Power_Source_Type	D.C. motor
Power	10
Max_Rotation (r.p.m.)	1000
Min_Rotation (r.p.m.)	10

Table 7.5. Drive ratio

Drive_Ratio	*Value*
Drive_Ratio_1	3.0
Drive_Ratio_2	3.0
Drive_Ratio_3	18/13
Drive_Ratio_4	23/18
Drive_Ratio_5	57/25

Table 7.6. Results of shafts

Shaft_No.	*Type*	*Material*	*D(mm)*	*Rotation*	*Torque*
Shaft 1	C_Shaft	45_Steel	45	2000	48.4
Shaft 2	C_Shaft	45_Steel	70	335	285.1
Shaft 3	C_Shaft	45_Steel	80	262	364.3
Shaft 4	C_Shaft	45_Steel	90	115	830.1

Table 7.7. Results of gears

G_No.	*T_shape*	*Model*	*T_No.*	*Material*	*Precision*	*Treatment*
Gear 1	SG	4	19	45CR	8_DC	teeth_G48
Gear 2	SG	4	57	45CR	8_DC	teeth_G48
Gear 3	SG	4	32	45CR	8_DC	teeth_G48
Gear 4	SG	4	46	45CR	8_DC	teeth_G48
Gear 5	SG	5	26	45CR	8_DC	teeth_G48
Gear 6	SG	5	36	45CR	8_DC	teeth_G48
Gear 7	SG	5	46	45CR	8_DC	teeth_G48
Gear 8	SG	5	25	45CR	8_DC	teeth_G48
Gear 9	SG	5	25	45CR	8_DC	teeth_G48
Gear 10	SG	5	57	45CR	8_DC	teeth_G48
Gear 11	SG	5	23	45CR	8_DC	teeth_G48

Table 7.8. Results of gear analysis

G_No.	*Stress count (Mpa)*	*Stress allow (Mpa)*	*C_safe coefficient (Mpa)*	*C_safe count (Mpa)*	*Bend_stress allowed*	*Bend_stress coefficient*	*Satisfy ?*
Gear 1	246.32	1370	354.0	22.48	354.0	15.75	OK
...	...	...	...	...	...	...	...
(omit)	...	...	...	...	...	...	...

Table 7.9. Results of evaluation

Type	*Efficiency (%)*	*Volume (m^3)*	*M_Time (h)*	*Price*
Calculated	59.56	0.0267	111.91	1892.71
Allowable	50.00	0.0250	150.00	2500.00

ACKNOWLEDGMENTS

The research is financially supported by the Natural Sciences and Engineering Research Council (NSERC) of Canada and National Natural Science Foundation (NNSF) of China.

REFERENCES

Chen, Y.Y. (1983) A mathematical model for comprehensive evaluation. *Journal of Fuzzy Mathematics*, **1**, 61–70.

Dixon, J.R. (1983) Expert system for mechanical engineering. *CIME*, **4**(10), 120–34.

Douglas, J.M. (1988) *Conceptual Design of Chemical Design*, McGraw-Hill, New York.

Gundersen, T. (1991) *Achievements and Future Challenges in Industrial Design Application.* Proceedings of the 4th International Symposium on PSE, Quebec, Canada, pp.I.I.1–I.I.31.

He, Z.X. (1983) *Fuzzy Mathematics Principle*, Science and Technology Press, Tianjin, China.

Kapp, D. *et al.* (1989) An analysis of distinction between deep and shallow expert systems. *International Journal of Expert Systems*, **2**(1), 1–34.

Kidd, P.T. (1990) Organization, people and technology: advanced manufacturing the 1990s: *Computer-Aided Engineering*, **7**(9), 149–52.

Manoocheiti, S. and Seireg, A. (1987) Computer-aided generation of an optimum machine topology for specified tasks, *CIME*, **6**(3), 10–24.

Rao, M., Jiang, J.S. and Tsai, J.P. (1988) IDSCA: an intelligent direction selector for the controller's action in multiloop control system. *Applied Artificial Intelligence*, **2**, 285–305.

Rao, M., Tsai, J.P. and Jiang, J.S. (1988) An intelligent decision maker for optimal control. *Applied Artificial Intelligence*, **2**, 19–27.

Rao, M., Jiang, T.S. and Tsai, J.P. (1989) Combining symbolic and numerical processing for real-time intelligent control. *Engineering Application of Artificial Intelligence*, **2**, 1927.

Rao, M., Cha, J. and Zhou, J. (1990) *New Software Platform for Intelligent Manufacturing*, Proceedings of the AAAI 90 Workshop Intelligent Manufactures, Boston, MA, pp. 62–5.

Wang, Q., Zhou, J. and Yu, J. (1989) *A Method of Product Conceptual Design Using AI Technology*, Proceedings of the 5th International Conference CAPE, Edinburgh, UK, pp. 200–10.

Wang, Q., Zhou, J. and Yu, J. (1989) *Decision Making System of Mechanical Product General Scheme Design*. Proceedings ICED'89, pp. 705–714.

Wang, Q., Yang, H., Zhou, J. and Yu, J. (1990) *QUINT: A Problem Solving Strategy for Mechanical System Concept Design, Part I and Part II*. Proceedings of Advances in Design Automation, Chicago, IL, pp. 331–43.

Yu, J., Zhou, J. and Wang, Q. (1987) *Mechanical Product General Scheme Design CAD Based on Expert System Technology*. Proceedings of Advances in Design Automation, Boston, MA, pp. 310–14.

Zhang, Z. and Rice, S.L. (1989) Conceptual design: perceiving the pattern. *CIME*, **11**(7), 58–60.

Zhao, H.C. (1986) *Level Analysis Methodology*, Science Press, Beijing.

Zhou, J., Duan, J. and Wang, Q. (1989) A study of evaluation subsystem for scheme design expert system. *Journal of HUST*, **17**(2), 10–24.

Zhou, J., Wang, Q. and Yu, J. (1989) Mechanical product general scheme optimization design and intelligent CAD. *Journal of Mechanical Engineering*, **3**, 56–61.

CHAPTER 8

Knowledge-based surface treatment and coating selection in product design

Chanan S. Syan

8.1 INTRODUCTION

The majority of engineering failures occur because of surface initiated effects such as wear, corrosion and fatigue. Surface treatments/coatings (T/C) offer the potential to prevent failure and often to reduce cost. T/C selection is an important part of the design for manufacture (DFM) spectrum and has been a very elusive activity to formalize. With the advent of knowledge-based systems, provision of T/C selection assistance to the designer early on in the design process has become a practical reality.

This chapter reports the development of the TESS (tribologically engineered surface selection) expert system, which is concerned with the field of surface T/Cs and their selection for optimum engineering design solutions (Syan, 1988).

8.2 DESIGN FOR ECONOMIC MANUFACTURE (DEM) AND SURFACE T/Cs

It is well known that the design of a product not only governs its performance but largely dictates how it is made. Studies by Andreason, Kahlen and Lund (1983) and later by Sheldon *et al.* (1990) showed that over 70% of the product cost is determined during the design phase, and the rest by subsequent production and other processes. The proportions of actual operating costs incurred by various departments and contributions to the product cost are shown in Fig. 8.1. The operating costs of the design department are very small compared with the influence the design activity has on total product cost. Therefore, optimization of the product design can offer the most cost-effective benefits to product manufacturing as a whole. The benefits of DEM to the industry have been recognized by researchers in the past, predominantly by Gladman (1968), Chisholm (1973) and Boothroyd (1975).

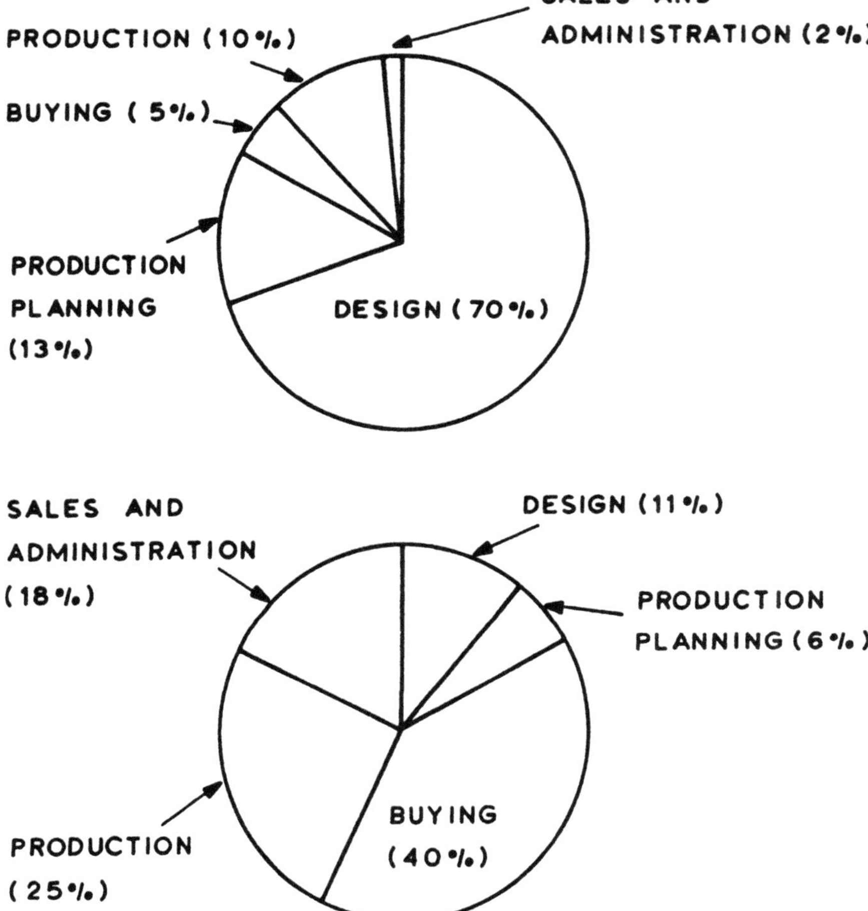

Fig. 8.1. (a) Proportion of product costs determined by various departments; (b) proportion of operating costs incurred in these departments.

In practice the design engineer relies on his own experience for most of this activity. There are occasional inputs from the production engineer, metallurgist or a value analyst, but prohibitive time penalty and cost discourage such liaison in most cases. The exceptions to this rule are few, even where the size, cost or quantity of the component is extremely large so as to warrant such expense. There are few, if any, design engineers who have such breadth of knowledge, judgmental skills and expert knowledge to carry out this activity with much confidence. Experienced design engineers working in particular domains do gain experience and hence produce mostly acceptable solutions. To be competitive, the design solutions have to be optimized both for functional and operational requirements. The concept of DEM encompasses the design process and the wider issues of manufacturing, planning, product costing, etc. The aim is to produce the design of a product to achieve optimum manufacturing efficiency and to optimize the product's competitiveness in the market.

In Table 8.1, a summary of DEM activities is given, with the sole aim of listing the major areas that need to be considered by the analysis (based on Swift, Matthews and Syan, 1985). Solutions to the DEM problem have been attempted but none have been taken up widely. The solution to the DEM problem turns out to be diffuse, hence the difficulty of formalizing it into a convenient and reliable procedure for its wider application.

Table 8.1. Some of the factors which need considering in DEM: design for economic manufacture satisfaction of function at minimum cost

Create product design concepts to satisfy functional needs
Material selection: bulk, surface treatments or coatings
Design for assembly: assembly method, ease of handling and construction
Design for part manufacture: process selection, design for processing and tooling design
Quality and reliability
Standardization and variety reduction

8.3 THE T/C SELECTION PROBLEM

Historically the situation regarding coating selection has been extremely unsatisfactory, as discussed by Syan, *et al.* (1987). Selection has been pursued on a very *ad hoc* basis. In a typical scenario a company will have accumulated experience of a few coatings or treatments, which will then have been used exclusively. Only when insurmountable problems arise would a company consider the alternatives available. Thus selection is not based on the progressive elimination of all alternatives, but rather the use of the first one that appears to fit the bill. This means that companies do not try new coatings unless they literally have to. Too often coatings are seen as a solution to a problem rather than a resource whose optimum use can prevent problems and contribute the product quality and company profits. In most situations there is a trade-off between properties to achieve a balance in the selection process. Often the selection decision would be influenced by a previous experience which may now be in the expert's subconscious.

The selection process or system must thus carry out two important tasks.

1. It must short-circuit the need for laborious experimentation to simulate the many possible wear mechanisms.
2. It must also allow the encapsulation of a vast array of case study information which can be made available to the design engineer, often just put down to 'experience' when utilized by the human expert.

The process is not, however, merely as simple as that. The human expert adviser will know what the effect of a small change in the specified operating conditions will be and will, almost subconsciously, make allowances for this. Therefore the intelligent knowledge-based system approach presents the most suitable and feasible system building option.

8.4 KNOWLEDGE ELICITATION IN TESS

There is no 'how-to-do-it' guide to knowledge acquisition. Various elicitation techniques have been used by workers in the field. The review by Wellbank (1983) and text by Hart (1989) give excellent background to the techniques used to date and the state of understanding of the knowledge acquisition domain. Work by Burton *et al.* (1987) overviews the knowledge engineering field. There are also no guidelines as to what type of knowledge elicitation technique is more suitable for a given problem domain than another.

For TESS, the knowledge acquisition task evolved throughout the duration of the project. Initially, the knowledge elicitation was via intensive interviews with the expert. This technique resembles the structured interviews as reported by Schweickert *et al.* (1987). In these interviews, the expert was asked to provide:

1. A list of variables for consideration while making a specific selection.
2. A list of outcomes, e.g. surface T/Cs and processes.
3. The knowledge and/or data used in making decisions.

The rules in the form of 'IF–THEN' form were compiled from transcripts of these interviews. These rules were also verified or modified in collaboration with the expert at a later meeting.

A table was developed containing all the variables found to be important for T/C selection for the 30 families of the T/C process. These generic T/C processes are groupings of similar processes that have identical characteristics for selection purposes. These were adopted in order to:

1. Reduce the data collection exercise to a manageable size.
2. Allow a comprehensive analysis of the T/C field without being overly restrictive on the range of T/C processes.

3. Allow future expansion of the families of processes to incorporate specific commercial processes and the knowledge/data for their ultimate choice.

8.5 TESS KNOWLEDGE-BASED SYSTEM

Surface T/C selection is an integral part of the material selection process. An ideal material and T/C selection process is illustrated in Fig. 8.2. It shows that, for a given component, the bulk requirements (e.g. strength, stiffness) need to be accommodated foremost. Only then can the surface requirements be considered with the option of surface modification via T/Cs. Since this work has only been concerned with the surface T/C selection part of the process, certain prerequisites exist for efficient use of the system. These are:

1. That the substrate material(s) have been selected predominantly using the bulk requirements.
2. That the component operates in a situation which has a likelihood of encountering wear in practice. This is essentially because the system is designed for wear-resistant surface selection. That is, the component is not merely a structural part, e.g. the casing in a gearbox.

The system is designed to offer advice to the design engineer regarding this domain. This is accomplished by:

1. Providing feasible T/C alternatives for the component.
2. Ranking of the alternatives in terms of relative costs.
3. Suggesting suppliers for the offered solutions.

Based upon the knowledge engineering study of the T/C selection problem, five major solution states were identified. These solution stages are groups of rules and data regarding aspects of the T/C selection process. The solution stages are groupings of subsets of the 16 criteria and their constituent factors (see Table 8.2).

The five solution stages are (1) operating constraints, (2) processing constraints, (3) geometrical constraints, (4) topographical constraints and (5) economic constraints.

8.6 TESS SYSTEM METHODOLOGY

The TESS selection procedure can be thought of as a five-stage transformation process. Each stage represents one of the five factor groups. The consultation starts with the hypothesis that all 30 generic coating and treatment technologies are equally suitable for the particular consultation. As the user supplies information about the particular application,

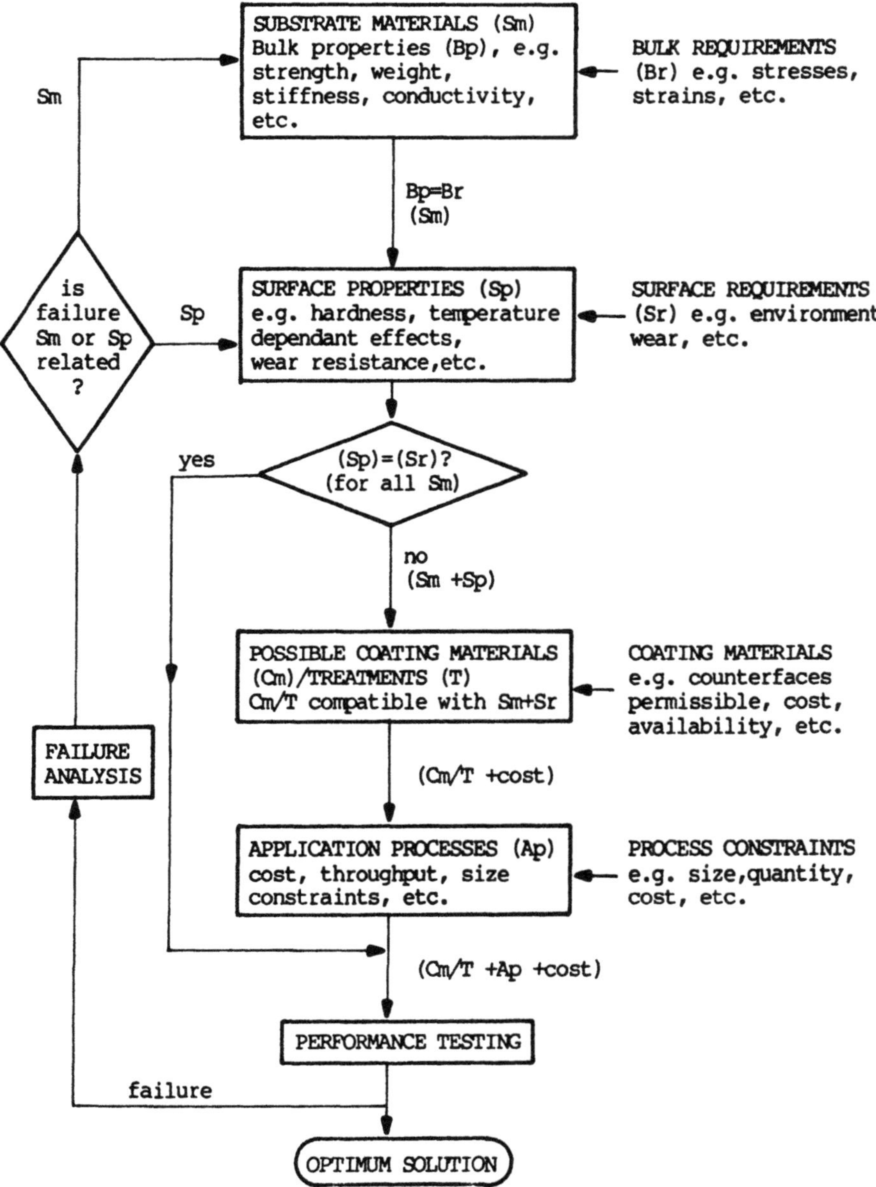

Fig. 8.2. Surface T/C selection process.

the system uses the rules in the knowledge base to reject any generic coating and treatment technologies that are unsuitable. Hence the initial hypothesis is modified as it progresses through each stage. The revised hypothesis is then carried forward, with the acquired data, to the next stage, as represented in Fig. 8.3. In Fig. 8.3 ($A_1, \ldots, A_n$) represents

Table 8.2. Factors considered under the 16 criteria

Criteria	*Factors considered*
Operating temperature	Component operating temperature Maximum allowable temperature for T/Cs
Operating environment	Component operating environment Allowable environments for T/Cs
Contact and loading types	Component/counterface contact type (e.g. line, point, area) Component/counterface relative motion (e.g. sliding, rolling) Allowable relative motions and contacts for the T/Cs
Substrate	Component material Component hardness Component surface finish Permissible materials, hardness, surface finishes for T/Cs
Counterface	Counterface material Counterface hardness Counterface surface finish Permissible materials, hardness, surface finishes for T/Cs
T/C/substrate bond strength	Loading of the component surface Allowable stress at the T/C substrate interface for T/Cs
Processing economics	Component production quantities Component cost T/C processing, quantity related, economics Relative T/C costs
Surface hardness	Component surface hardness requirement T/C surface hardness
Processing temperature	Substrate temperature constraints (e.g. tempered substrate, melting point)
Component size	Component overall size T/C processing size constraints
T/C depth	Component surface stresses (e.g. depth maximum shear) T/C depths obtainable
T/C uniformity	Component dimensional tolerances T/C re-entrant penetration capability Tolerance of T/C thickness
Component shape	Component shape (e.g. sheet, rotational, etc.) Permissible shapes (sections) Selective T/C capability
T/C surface finish	Component surface finish requirement T/C process surface finish obtainable
Pre-processing	T/C process pre-processing requirements
Post-processing	Component surface finish Post-machinable T/Cs Surface finish obtainable after post-machining

an initial hypothesis containing all the possible coating solutions ($n=30$ at present); $(B_1, \ldots, B_m)$ represents a revised hypothesis containing still possible coatings ($m<n$); $(C_1, \ldots, C(n-m))$ are the rejected coatings; and $(I_1, \ldots, I_g)$ are the data items supplied by the user during the analysis.

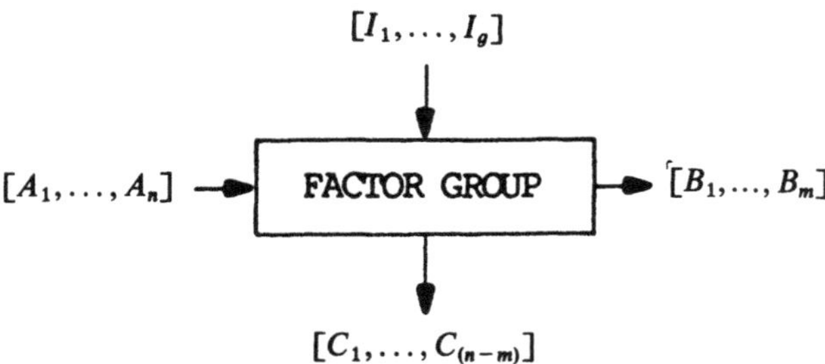

Fig. 8.3. Representation of a solution stage.

Factor group constraints represent the set operators used to transform the initial hypothesis. The whole selection process is a succession of transformations on the initial hypothesis, until all the factors have been considered using the expert knowledge within the system. Figure 8.4 illustrates the complete selection process showing the five solution stages. At each solution stage, the application information, supplied by the user, is used by the system together with the expert rules to satisfy the constraints of the 30 generic coating and treatment technologies. This process is similar to the approach taken by experts in the field. At the end of the consultation, all the possible solutions are reported, together with the data supplied by the user during the consultation.

The conclusion goals are selected using the Prolog 'Top-down' internal strategy. Rules are then scanned to find the match with head literals and then 'fired' to prove or refute the hypothesis. Backtracking control is used for preventing default rules from being fired, if previously named rules have been fired successfully. Default rules are hence only used for control after the normal system rules have been tested and are refuted.

8.7 REPRESENTATION OF RULES IN TESS

The rules derived for the five solution stages are represented as Horn clauses in the Prolog system. The logical operators 'and', 'or' and 'not' are used to form conjunctions, disjunctions and negations of antecedents. These antecedents represent the individual pieces of knowledge contained within the derived rules. The head of a clause represents the consequents

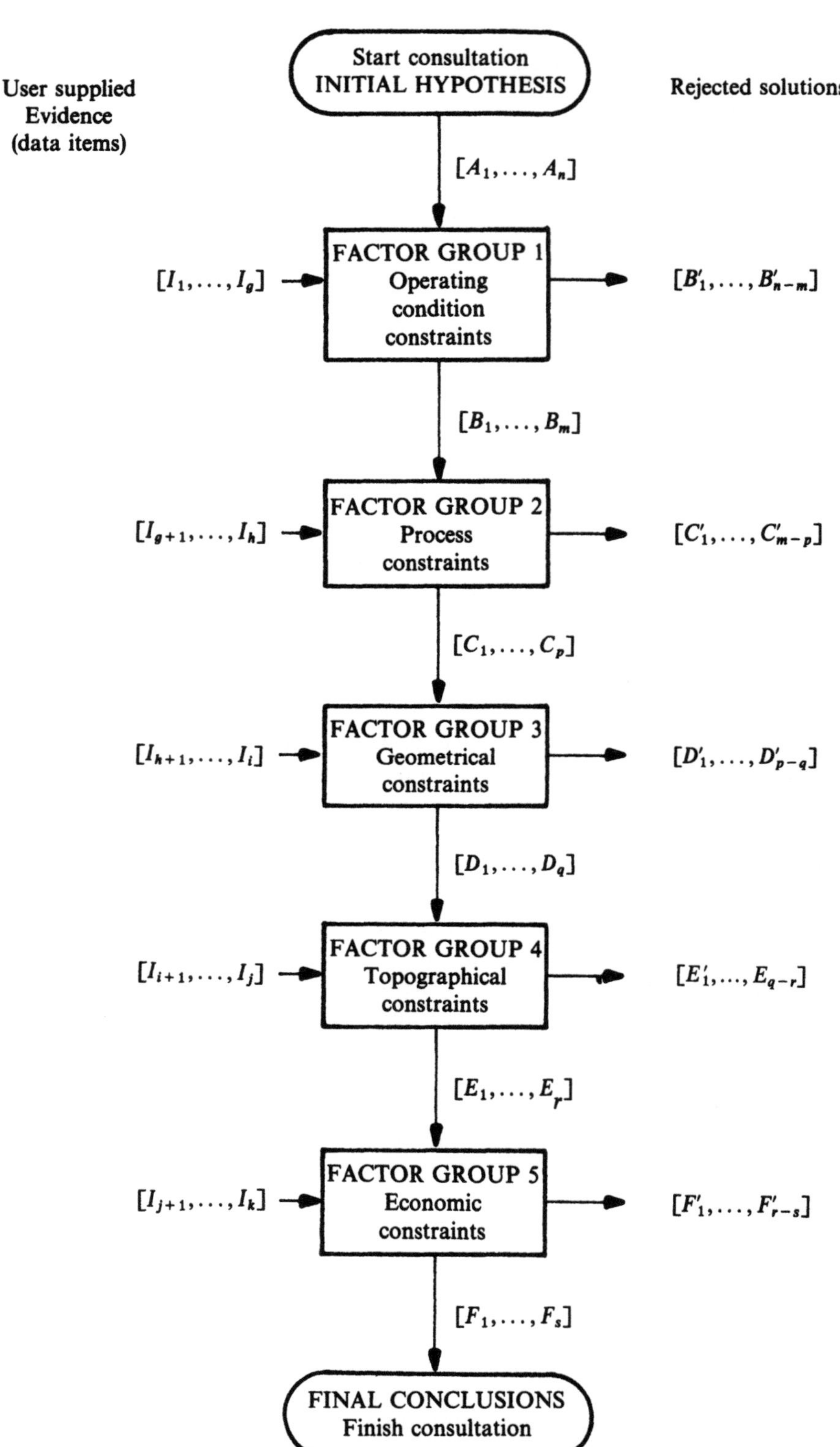

Fig. 8.4. Complete selection process.

or inferences suggested upon satisfaction of the antecedents. Figure 8.5 shows a rule in English format derived from the expert, the clausal representation replaces the IF A_1 & $A_2, \ldots, A_n$ THEN C format by C: $A_1, A_2, \ldots, A_n$ ($n>0$), where ‘,’=logical ‘and’, ‘;’=logical ‘or’ and ‘not’=logical ‘not’.

Both the heads and antecedents of the Horn clauses are Prolog atoms with English like format. This representation has been adopted because this syntax for rules makes the knowledge base easy to read and understand.

(i)	IF	Surface hardness range is 550–900 Hv
	and	maximum processing temperature allowed is <750°
	and	no restrictions on maximum component size
	and	T/C depth requirement is 0.5–5 mm
	and	substrate/T/C bond strength requirement is <yield strength of substrate
	THEN	process properties are satisfied for Induction Hardening.
(ii)	IF	Wear tolerance requirement is 0.5–5 mm
	OR	depth of maximum sub-surface stress is <5 mm
	THEN	T/C depth requirement is 0.5–5 mm.

Fig. 8.5. Example of processing constraints rules.

8.8 TESS – INFERENCE MECHANISM (IM)

The requirements of IM for the T/C consultation system are:

1. To apply the knowledge, expressed in the logic form, in effective and efficient ways, in order to solve domain problems. The aim being to emulate the reasoning skills of the domain expert.
2. To undertake the man/machine interaction in a friendly, helpful and efficient manner.
3. To assist the user to update the system knowledge base as new knowledge becomes available.

It has been widely accepted that the power of expert systems, for problem solving, is mainly derived from the knowledge and not from its inference procedure (Feigenbaum, 1981). Simple inference procedures have the virtue of clarity and build user confidence in the system reasoning.

Prolog offers a standard, built-in inference mechanism. This is a depth first, backward chaining search, incorporating multiple solutions capability. Hence this facility obviates the need to program the basic inference technique, which would need to be coded-in using any other language for building an expert system. However, there are many other fundamental facilities lacking in the basic Prolog core, which are needed for the T/C consultation system. Prolog is a powerful general purpose programming language and it offers the meta-level primitives, which lend themselves to implement these meta-logical facilities. These facilities include:

1. User interactions, e.g. requests for data, reporting of conclusions, etc.
2. Explanation facilities, both for questions and deductions.
3. Handling of uncertain data and knowledge.

8.9 USER/TESS INTERACTION

The TESS consultation system operates in an interactive manner. The user is questioned as the system applies the knowledge in the knowledge base. The user responses also control the path of the search through the knowledge base and hence no unnecessary line of questioning is followed, minimizing the total user inputs required for the consultation, i.e. inference generated requests for data. The 'maybe' and 'why' responses are available to the user at all stages of the consultation, providing explanation and further information regarding the query in order to help build up user confidence.

The sub-conclusions are reported by the system after each one of the five stages of the system knowledge has been applied, i.e. operating condition constraints, processing constraints, etc. The user has the option of rejecting any proposed sub-solution if he so wishes. This is done via the 'top-level interpreter' action, which offers facilities for carrying out 'what-if' simulations for a limited sensitivity analysis.

The final conclusions are reported to the user in two forms. The first group of recommendations are the 'certain' deductions. The second group are the 'possible' recommendations. These have numbers of 'maybe' responses allocated to them as an indication of possible uncertain factors that may rule out these conclusions. The system reports the certain conclusions and the user is allowed to interrogate further in order to evaluate 'maybe' conclusions.

A report is also produced by the consultation system, giving the part name, number, the data supplied by the user during consultation and the recommendations made by the system. Figure 8.6 shows an example report. This report is written to a file, which can be obtained as a hardcopy record of the consultation via printing.

8.10 EXPLANATION FACILITIES

One of the prime requirements of the consultation system is the ability to explain its line of reasoning. The ideal form of such a facility would be a 'natural language' interface, which can explain in 'English' its reasoning. This is an enormously difficult task, and is currently the topic being investigated by researchers (Li-Ming, 1986). An explanation facility has been developed, which offers advice in a much restricted fashion. This facility offers explanations of the questions being asked, the reasons for deductions made and trace back of the deductions through the chain of inferences used.

8.11 SYSTEM RULES, GOALS AND UNCERTAINTY HANDLING

A selection of rules from the TESS knowledge base is attached in Appendix A in order to give a better appreciation of the system structure. The whole subject of uncertainty handling is a controversial one and no coherent and agreed approach exists for this. Winston (1984) has stated that certainty handling in expert systems is at very best questionable.

Certainty factors are the most common type of approach used in this field (Shortliffe and Buchanan, 1975; Duda *et al.*, 1976). Hence TESS has simple and practicable approach, which reflects the expert practice in the surface T/C selection domain. The rules in the TESS knowledge base are fortunately mostly certainties. The rules that have uncertainty attached to their implications are represented as 'possible' recommendations. If more than one rule indicates 'possibility' of the current hypothesis during a line of reasoning, then the hypothesis is conclusively refuted. The uncertainties of user responses for the logical states of antecedents are treated as 'maybe' responses, these are then attached to the hypothesis being tested. They are reported with the final conclusion as a possible source of error. The system treats the 'maybe' responses as positive response in support of the current hypothesis. The system also affords the user to change the response to this request if it is encountered again in the ensuing line of investigation.

8.12 SYSTEM TERMINOLOGY DICTIONARY

It was found during the development of the consultation system that the user unfamiliar with the T/C domain would find the language used by experts difficult to understand. Since the knowledge base and hence the system questions are expressed using expert's terminology, design engineers

******************** CONSULTATION DATA ********************
********************   FOR   ********************

part name——————vane
part number—————1

****

the following are true:

component could have post treatment machining
component shape is sheet like
surfaces slide relative to each other
cface surface finish is better than 6.3 micro m ra
substrate is aluminium
substrate finish is better than 6.3 micro m ra
substrate is a metal
component is counterfacing another component

****

the following are the data supplied:

the quantity of parts rquired per year——————————7000000
substrate finish in micro m ra is——————————————0.8
post treatment tolerance in micro m is——————————100
post treatment surface finish in micro m ra is——————0.8
maximum dimension of the component in mm————————50
wear tolerance for the component in micro m——————5
max treatment processing temp allowable in deg c—————550
melting point of the substrate material in deg c—————850
counterface hv in vickers——————————————————150
substrate material hv in vickers hv——————————————150
yield strength of the component material in mn per sq m————250
maximum pressure at the surface concerned in mn per sq m——103
operating temperature of the component in deg c————————130

****

the following are the advised solutions by tess:

11 anodizing
the relative processing cost index is ic
component treatment cost is an economic cost
continuous production qty

12 electroless nickel
the relative processing cost index is ic
component treatment cost is an economic cost
continuous production qty

****

Fig. 8.6. Example of final report by TESS.

unfamiliar with the T/C domain would find difficulty in interpreting some of the questions. A dictionary of the terminology used can be called upon to explain the meaning of the technical language used for questioning the user and representing the knowledge at any time during the consultation.

There are two levels of possible terminology-related problems, they are:

1. Rule antecedents – these are requests for data where the whole question may not be obvious to the user.
2. The individual terms used in the antecedents.

For both the above cases, the request for terminology explanation of the antecedent or term is instantiated to the variable X for the dict(X) goal. The data base is searched for a clause with the head literal dict(X), the body of such a clause is the terminology explanation and is displayed to the user. Then the control is returned to the original calling rule and the consultation continues. If no explanation of that particular terminology exists, the default message, 'sorry, I cannot help any further', is output.

8.13 DISCUSSION AND FURTHER WORK

The consultation system developed can advise the design engineer in the surface T/C selection problem. The level of performance achieved is near to a consultant T/C specialist. This has been achieved using a knowledge base of around 530 rules and 30 possible conclusions. There are also 150 sub-conclusions, in total, that are possible within the knowledge base. It is implemented on Apollo DN3000 workstation but has recently been ported to IBM compatible PC. The Prolog programming system, used for encoding the consultation system, proved to lack certain facilities necessary for an interactive system. These include, for example, requests for data, and explanation and terminology facilities. These were, however, implemented by the use of Prolog itself to program the extended mechanisms required.

The linking of the TESS problem solver to a computer-aided draughting system, for automatic determination of properties, has been investigated. The work arose from the desire to integrate the DEM activities with the engineering drawing process. This was primarily to make the system more attractive for use by the design engineer by reducing the additional effort required for T/C analysis and to automate the selection process. Other benefits are also afforded by this approach, such as reduction of possible human errors in entering the geometrical data and reduction in total consultation time. The development of the system has contributed to the understanding of the T/C selection problem. The criteria used for selection can assist in the analysis of the component requirements and assess the T/C suitability without the need for exhaustive analysis of the component operating situation using expert knowledge encoded in the system.

It has been found, after a comprehensive testing programme of the system, that the TESS performance is at an expert level in all the cases where the system knowledge is complete. In cases where the knowledge is incomplete, the performance degrades substantially. Therefore, system performance has certainly been confirmed to be related to the quality and completeness of knowledge. Also it as found that the advice provided by the system would be more valuable if the data bases of the T/C suppliers, trade names and cost information could be supplied after the consultation. These areas of development pose challenges and are being looked at currently.

Further developments would be desirable, for example, in integrating the T/C selection with the materials selection process. This would provide a more comprehensive materials and surface engineering assistance to the designer. The interpreter developed for the TESS offers a useful tool for further investigations into similar selection domains as it has the relevant facilities necessary for such tasks already implemented.

REFERENCES

Andreason, M., Kahlen, S. and Lund, T. (1983) *Design for Assembly*, IFS (Publications) Limited and Springer-Verlag, New York.

Boothroyd, G. (1975) *Fundamentals of Metal Machining and Machine Tools*, McGraw-Hill, New York.

Burton, M.A., Shadbolt, N.R., Hedgecock, A.P. and Rugy, G. (1987) A formal evaluation of knowledge elicitation techniques for expert systems, in *Research and Development in Expert Systems* IV, (ed. S. Moralee), Cambridge University Press, Cambridge.

Chisholm, A.W.J. (1973) Design for economic manufacture. *Annals of CIRP*, **22**(2), 243–7.

Duda, R.O., Hart, P.E. and Nilson, N.J. (1976) Subjective Bayesian methods for rule-based inference systems. *Tech. Rep. 124*, SRI International, Project 4763.

Feigenbaum, E.A. (1981) *Innovation and Symbol Manipulation in Fifth Generation Computer Systems.* Proceedings of the International Conference on Fifth Generation Computer Systems, North Holland, Amsterdam, pp. 211–14.

Gladman, C.A. (1986) Design for production. *Annals of CIRP*, **16**, 3–10.

Hart, A. (1989) *Knowledge Acquisition for Expert Systems*, Chapman & Hall, London.

Li-Ming, M. (1986) *Natural Language Interface to Relational Data-base-Systems.* Proceedings of the 1st International Conference on Applications of AI in Engineering Problems, Vol. 1, pp. 215–26.

Schweickert, R., Burton, M.A., Taylor, N.K. *et al.* (1987) Comparing knowledge elicitation techniques: a case study. *Artificial Intelligence Review*, **1**, 245–53.

Sheldon, D.F. *et al.* (1990) Designing for whole life cycle costs. *Journal of Engineering Design*, **1**.

Shortliffe, E.H. and Buchanan, B.G. (1975) A model for inexact reasoning in medicine. *Mathematical Bio-Sciences*, **23**, 351–79.

Swift, K.G., Matthews, A. and Syan, C.S. (1985) *A Logic-based Approach to Problem Solving in Manufacturing Engineering.* Proceedings of the Workshop on Process Planning, Automation and Analysis, December 1985, SERC.

Syan, C.S. (1988) A computer-based surface coating selection methodology for advising the design engineer. *PhD Thesis*, University of Hull.

Syan, C.S., Matthews, A. and Swift, K.G. (1987) Knowledge-based expert system in surface coating and treatment selection for wear reduction. *Surface and Coating Technology*, **33** 105–15.

Wellbank, M. (1983) *A Review of Knowledge Acquisition Techniques for Expert Systems*, British Telecom/Martlesham Consultancy Services, UK.

Winston, P.H. (1984) *Artificial Intelligence*, 2nd edn, Addison-Wesley, Reading, MA.

CHAPTER 9

Expert system for casting design evaluation

I.C. You, C.N. Chu and R.L. Kashyap

9.1 INTRODUCTION

In the metal processing industry, casting is the most popular means for achieving a desired shape (net shape) in a single step. The casting process is heavily experience-oriented and casting design is an iterative task between casting designers and foundry experts. There is a lot of wastage of resources and time during the design–manufacturing cycle when a part is being engineered. Initial estimates of manufacturability are valuable to the designer as this avoids costly redesign efforts. With the emergence of abundant computing resources, computer-aided design (CAD)/computer-aided manufacture (CAM) of casting parts is poised for a big leap forward. Part descriptions of three-dimensional CAD models are in a form of basic geometry and topology to which manufacturing process and engineering analysis cannot be directly applied. Conventional solid models have been concentrated on representing physical information of objects, thus allowing the system to perform mathematical operations. However, recent developments in computer applications to manufacturing processes require more information for decision making in such areas as process planning and manufacturability analysis. Since domain knowledge is usually associated with manufacturing features rather than detailed geometric descriptions, obtaining manufacturing feature descriptions has been one of the most important aspects in CAD/CAM integration. A feature description varies depending on a particular manufacturing process because information required in one manufacturing process is different from that in other manufacturing processes. Different definitions of features are derived from different views of various application domains (Henderson, 1984; Joshi and Chang, 1988; Libardi, Dixon and Simmon, 1986).

Two major approaches, feature extraction and feature-based design, have been used by most researchers to obtain high level manufacturing feature information. Feature extraction is a process of identifying and classifying manufacturing features from low level details of topology and geometry. Feature extraction can also be regarded as a process of converting design feature descriptions into manufacturing feature descriptions. There are several methods for feature extraction: syntactic pattern recognition (Choi, Barash and Anderson, 1984; Staley, Henderson and Anderson, 1983) logic programming (Henderson, 1984), volume decomposition (Woo, 1983), two-dimensional projection (Yoshimura, Fujimura and Kunii, 1984) and graph-based heuristics (Joshi and Chang, 1988). Most feature extraction methods are carried out from boundary representation (B-rep) because of its uniqueness. However, in B-rep, an object is represented by topological and geometric information about its surfaces, hence it is very difficult and complex to build a general procedure to extract complete feature information. Some feature extraction has also been carried out from a constructive solid geometry (CSG) tree structure (Herbert, 1990; Lee and Fu, 1987). Since CSG provides a compact and concise data structure, it would seem useful for building a general feature extraction algorithm. However, it is difficult to extract general feature information because of the non-uniqueness of CSG. Furthermore, global shape information, which is essential in castability evaluation, cannot be extracted in a straightforward manner. Another method for feature extraction is skeleton extraction. This approach uses medial axis transformation (Lee, 1982; Montanari, 1969) to extract skeletons which capture the essence of shape. Due to the lower dimension of the skeleton as compared with the initial shape, geometric reasoning can be greatly simplified, but its application is mostly limited to two-dimensional objects. This approach has been attempted for three-dimensional models (Nackman, 1982), but no general algorithms are available to compute the skeleton of three-dimensional objects.

The other approach, designing-with-features (Dixon, 1988; Requicha and Vandenbrande, 1989; You, Chou and Kashyap, 1989), uses predefined manufacturing features as design elements, such as block, hole, pocket, etc., to create a desired part. The advantage of feature-based design is its ability to capture the manufacturing information of a design, since it uses predefined manufacturing features as design elements. Theoretically, any type of manufacturing feature can be a primitive for feature-based design. However, most of the work done to date can handle only a small application domain. Recently, an expert from feature modeling shell, which overcomes the limited geometric coverage of feature-based design, has been proposed. However, it is difficult to integrate additional features with the knowledge base of a particular application area. Improving geometric coverage is of little practical value in itself if the system must be modified along with the additional geometric coverage. In order to overcome these drawbacks, generic designing of primitives which have a large geometric coverage while conveying a valid and important set of geometric and topological entities to an application area are needed.

In this chapter, a formalism for symbolic representation of a three-dimensional object is presented to aid local and global shape analysis of casting design. A three-dimensional pattern model based on this formalism has been developed to provide a wide geometric coverage while conveying meaningful geometric and topological properties directly to the CAM application domain. Based on this model, three-dimensional net shape analysis is carried out for castability evaluation. Local shape is analyzed using symbolic representation and gobal shape analysis is performed by extracting the skeleton from the decomposed object. From the extracted skeleton, junction types are classified and used to aid geometry-based solidification simulation. The important aspects of this work can be summarized as follows:

1. A new symbolic representation based on pattern representation and sweep representation for describing three-dimensional objects is introduced. In this model, each generatrix is represented by concatenating pattern primitives. Five control parameters are used to control the sweeping motion. This model uses set operators to produce new objects.
2. A method to reason about local shape characteristics based on symbolic descriptions is developed. This method uses production rules to represent 256 cases of local shape variations. Binary primitive pairs are constructed based on 16 pattern primitives and form a basis for symbolic reasoning tables which are used to generate production rules. These production rules are used to reason about local shape characteristics.
3. A modified three-dimensional thinning algorithm is developed to aid the decision-making process in evaluating the global casting soundness. A skeleton is extracted from the discretized object and then a distance map is generated by calculating true euclidean distance from each node of the skeleton to the nearest boundary point. Also, all nodes of the skeleton are classified as several junction types and classified junction types are used to adjust values of the distance map. Since the cell decomposition is a unique representation, the extracted skeleton can be used to build a general algorithm for gobal shape analysis.

The paper is divided into five parts. Section 9.2 explains the feature-based design. Section 9.3 introduces the basic concepts of a three-dimensional pattern model and its use in the expert system. Section 9.4 explains local shape reasoning based on symbolic descriptions. In this section, we describe how the symbolic reasoning tables are mapped into the production rules. Section 9.5 discusses knowledge representation and control structure based on the three-dimensional pattern model. This section is devoted to present a knowledge refinement process based on symbolic descriptions. In section 9.6, skeleton extraction and distance mapping are explained and their applications for global shape analysis are described. Implementation is presented in Section 9.7, and the implementation of the algorithms and examples are provided for each section.

9.2 FEATURE-BASED DESIGN

In this section, an on-line evaluation expert system for designing complex axisymmetrical parts is explored by using the 'fixed-features based design' approach. A flywheel is used as an example because most other axisymmetrical parts can be represented by the features of a flywheel. Axisymmetrical parts are grouped into subclasses such as spacers, flanges, flywheels, splines, gears, pulleys, etc. The flywheel group, for example, has similar features which can be viewed from the cross-sectional drawing of each flywheel. More than 30 flywheels, currently manufactured at Cummins Engine Company, Columbus, IN, were examined in order to extract these feature representations. These similar features, which we call the 'fixed features', are cylindrical elements which are viewed as rectangles of different sizes in two-dimensional cross-sectional drawings. Figure 9.1 shows fixed-features of a flywheel.

A designer first selects the casting and pattern material. Casting materials currently used in EXCAST are gray, white or malleable cast iron, cast steel, aluminum alloy, and magnesium alloy. Pattern materials are wood, aluminum

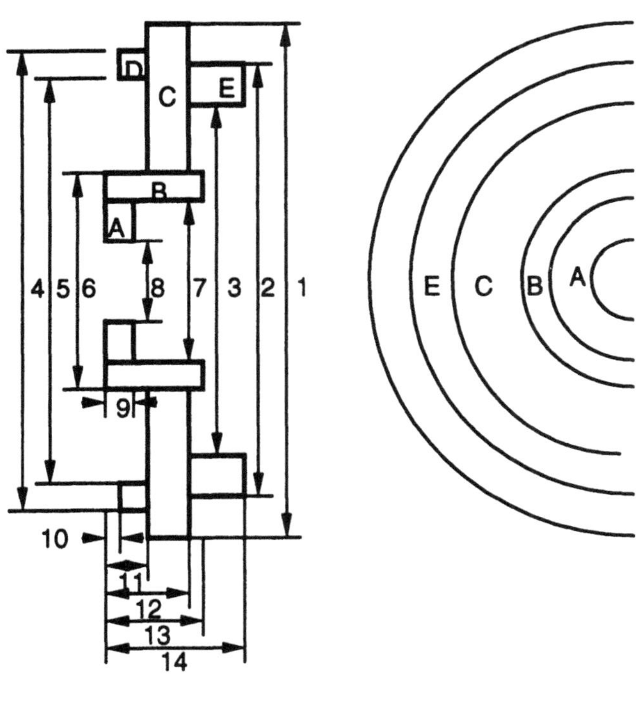

Fig. 9.1. Fixed-features of flywheels.

and steel. After materials are selected, EXCAST asks the designer for the dimensions of features 1–14 (Fig. 9.1) and then transfers them into memory storage corresponding to each feature. Pocket, hole, boss, fillet, rounding and chamfer can be added as desired (Fig. 9.2). EXCAST can retrieve the descriptions into intermediate frames between the memory and the knowledge base. A data element is a list of symbols that may represent any object including its own list processing functions. A symbol manipulation program uses symbolic expressions to work with data and procedures. An example data structure produced by EXCAST is shown below:

```
"(FLYWHEEL (800 780 740 480 400 320 250 150 50 20 70 110 100 300))"
"(FLYWHEEL-SECOND (10 0 0 0 0 0 0 RIGHT-IN EXTERNAL-ROUNDING B))"
"(FLYWHEEL-SECOND (30 20 0 0 0 0 0 LEFT-OUT EXTERNAL-CHAMFER B)"
"(FLYWHEEL-SECOND (30 20 0 0 0 0 0 LEFT-OUT EXTERNAL-CHAMFER C))"
"(FLYWHEEL-SECOND (10 5 0 0 0 0 0 LEFT-OUT EXTERNAL-CHAMFER D))"
"FLYWHEEL-CO (E 4 15 10 0.1 15 1 8 0 0 0 0 0 CIRCULAR-HOLE 4))"
"FLYWHEEL-CO (A 4 15 10 1 18 1 4 0 0 0 0 0 CIRCULAR-BOSS 0))"
"FLYWHEEL-ANGLE (0.0 T 6.2831855 8 0 0 0 0 0 CIRCULAR-HOLE E))"
"FLYWHEEL-ANGLE (0.0 T 6.2831855 4 0 0 0 0 0 CIRCULAR-BOSS A))"
"NIL"
```

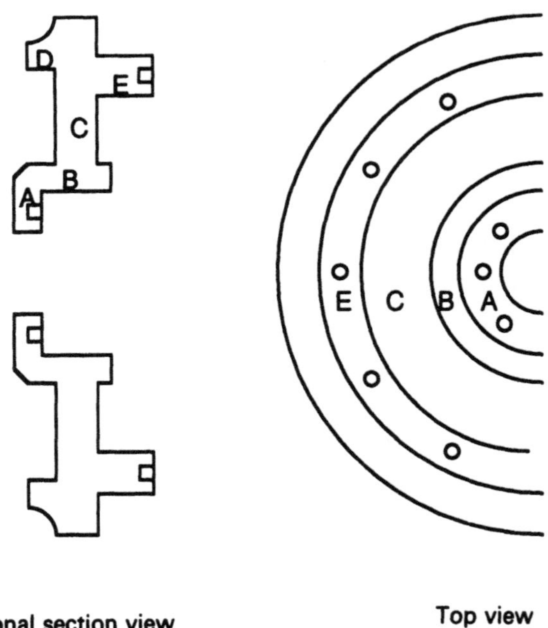

Fig. 9.2. Designed flywheel.

For example, the list of the first line (FLYWHEEL) represents the geometrical and topological information of fixed-features. Based on that list, the drawing is created as shown in Fig. 9.2. EXCAST provides error messages and recommendations and automatically generates parting lines. A few examples of error messages are shown below:

```
MODULUS FOR FEATURE A : 28.57143
MODULUS FOR FEATURE B : 20.533142
MODULUS FOR FEATURE C : 40.04975
MODULUS FOR FEATURE D : 20.0
MODULUS FOR FEATURE E : 20.0
--------ERROR.by-knowledge--------
Error is detected by : ($RULE 89)
|There is a potential hot spot at feature A.
+-----===>SUGGESTION <===-----+
|adjusting the dimensions of the feature B
 to increase its modulus.
|adding a chill to the surface of the feature A
 to lower the effective modulus.
|attaching a feeder to the feature A.

--------ERROR.by-knowledge--------
Error is detected by : ($RULE 95)
|Junction between feature C and E are not thermally neutral.
|Possible hot-spot.
+-----===>SUGGESTION<===-----+
|Change length of feature E.
|recommended length : less than 150.0
OR
|Add rounding.
|recommended radius : 15.0

--------ERROR.by-knowledge--------
Error is detected by : ($RULE 60)
|Incorrect Parting line for feature C.
| ==>Make combinational parting line with feature 'C_LEFT'

|------------------------------------------------------------------------------------------
|Suggested Parting Line ==> Stepped parting line
|Line 5.0 from (feature C_LEFT) and line 5.0 from (feature A_RIGHT)
|------------------------------------------------------------------------------------------
```

The advantage of feature-based design is its ability to capture the complete manufacturing feature information of a design. However, its geometric coverage is limited by the number of predefined features.

9.3 THREE-DIMENSIONAL PATTERN MODEL

Geometric reasoning requires extensive spatial reasoning and geometric manipulations. Any different geometric and topological combinations of the design can serve as a description of a particular situation for geometric reasoning. A set of all possible combinations or configurations is the space of problem states and is called *state description.* Since the state description is usually too large, it is difficult to apply conventional artificial intelligence techniques to design problems. One way to resolve this problem is to use high level symbolic representation with a topology-based design approach. By using a set of high level symbols for describing the design, the state description can be reduced and the extensive spatial reasoning and detailed geometric manipulation can be minimized. In the topology-based design, geometric entities are interconnected with each other and geometric information of each entity is associated with its neighboring entities. As a result, parameter variation of one entity will not affect its topological relationship with other entities. This property allows the system to carry out high level symbolic reasoning by mapping a particular situation to a member of the state description without going through detailed geometric manipulation. In order to represent the high level symbolic descriptions, a syntactic pattern representation (Fu, 1982) is adapted in this work. A pattern approach for geometry representation provides the capability of describing a large set of complex geometries by using a few sets of simple pattern primitives. Syntactic patterns, so called chain forms, have been used for a part description language for axisymmetrical parts (Jakubowski, 1982) and for a three-dimensional object representation (Lin and Fu, 1984). In Lin and Fu (1984), surfaces and grammatical production rules are used as primitives and structural relationship descriptors. Since the topological relationship among surfaces is explicitly specified in the grammatical production rules, it is useful for computer vision applications. However, each three-dimensional object must have its own structural descriptor, hence this scheme has the same drawbacks as the feature-based design.

The basic building blocks of a three-dimensional pattern model are the pattern representation scheme, the sweep representation scheme and CSG. The pattern representation and the sweep representation schemes are used to create design features; boolean operation is used as the input scheme and topological relationships embedded in the pattern representation are used to reason about local geometric shape. In order to model three-dimensional casting parts, sweeping schemes are used whenever possible. In the sweeping scheme, a set of geometric entities moving through space may produce a *volume* (a solid) that may be represented by the *set* of entities moving through a *path.* The set of entities is called the *generatrix.* For describing three-dimensional casting objects, generic pattern primitives (GPPs) are used to create path curves and generatrices which constitute three-dimensional objects. Figure 9.3 shows generic pattern primitives.

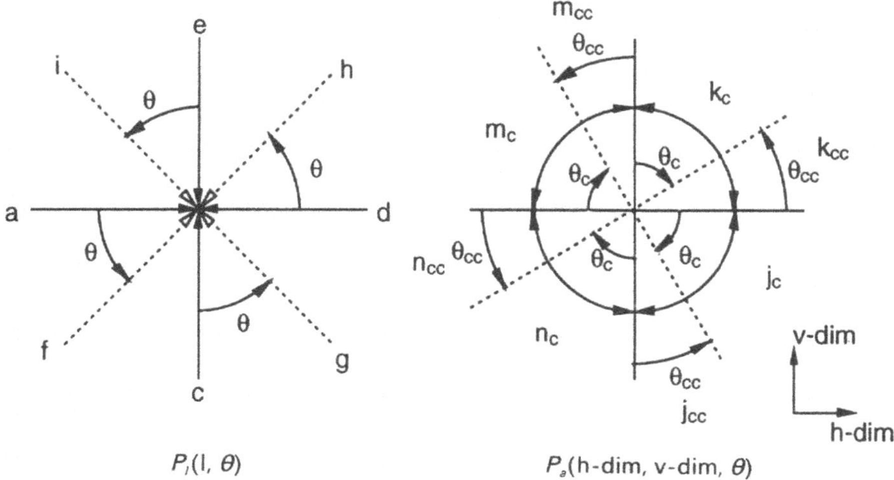

Fig. 9.3. GPPs. $P_1(1, q)$ and P_a(h-dim, v-dim, q) represent a line primitive and an arc primitive where 1, h-dim, v-dim and q represent length, horizontal-dimension, vertical-dimension, and angle. c (clockwise) and cc (counterclockwise) represent associated directions for arc primitives.

The most commonly used sweep operations are translational and rotational sweeping. Additionally, a pattern *skinning* technique is used to describe complex shapes. The basic idea of the pattern skinning technique is to vary the size ratio of each segmented cross-section while the generatrix moves through a path. These features are represented by generatrices (functional faces) with paths or axes. In order to create sweep based functional features, two basic steps must be carried out.

1. The designer must define a two-dimensional outline that describes the shape of the generatrix to be swept.
2. The designer must provide input data that describe the sweeping parameters. These parameters include the sweep type, such as linear and rotational, and size parameters, such as path length and rotational angles.

The design process is separated into two stages. In the first stage, functional features are created by sweeping functional faces along their associated paths. At this stage, local shape reasoning is carried out iteratively so as to obey manufacturing requirements. In the second stage, functional features are combined to make up the final designed part by applying boolean operations. At this stage, global shape analysis is carried out by using the extracted skeleton. The evaluation process following the second stage will result in parametric changes to the design. The revised design will be conveyed to the first stage to verify that the design still maintains functional requirements.

In order to facilitate the definition of contours, a sweep command interpreter has been developed and interfaced with an expert system. All information necessary to generate functional features is stored in a database so that the designer can select required functional features to create a part. These features are selected by using the menu choice, *design*, shown at the top of the CAD window as shown in Fig. 9.4. By applying boolean operation, a connecting rod is created and shown in Fig. 9.4, where connecting_rod = head U I_beam U (cylinder – hole). This is done by using the menu choice, *combine*. For boolean operation, a local coordinate system is used for each functional feature to carry out translational and rotational operations.

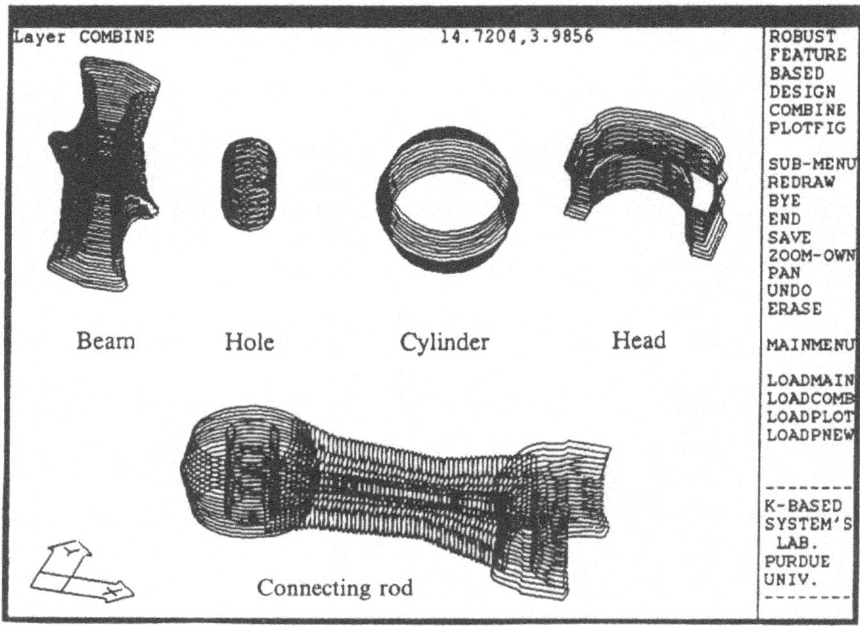

Fig. 9.4. Selected functional features and a connecting rod. Four functional features are shown where I-beam, hole, cylinder and head are positioned from the left to the right at the top of the CAD window. [Connecting_Rod = head U I_beam U (cylinder – hole)].

9.4 LOCAL SHAPE ANALYSIS BASED ON PATTERN PRIMITIVE PAIRS

The purpose of local geometric reasoning is to reason about local characteristics so as to alter the design to obey manufacturing requirements. For this purpose, binary pattern primitive pairs and symbolic reasoning tables are developed and used to aid extensive spatial reasoning and detailed geometric manipulations. The objectives here are to allow the system to (1) acquire heuristic knowledge and results of simulations, and (2) represent extracted knowledge in a hierarchical manner based on the high level symbolic descriptions. The number of cases which can occur for a *binary primitive pair* is 256 (16 × 16). Figure 9.5 shows some of possible binary pairs. Binary primitive pairs have been used as units for local shape geometric analysis. Based on these binary primitive pairs, symbolic reasoning tables, which are the basis for a knowledge base, have been constructed and some of them are shown in Table 9.1. In the symbolic reasoning table, '*A*' represents adjacency relationship for each binary primitive pair and ':' represents additional primitives which will lead to a new status. In the *constraints* column of Table 9.1, for example, *x A y* represents consecutive primitives x followed by *y*. Also, in the *new constraints* column of Table 9.1, *y:z* represents an additional primitive *z* which is added between primitives *x* and *y* to form *x A z A y*. In the columns, *Status* and *New Status* of Table 9.1, A, MPA, PA, MPNA and NA represent *acceptable* status, *most probably acceptable* status, *probably acceptable* status, *most probably not acceptable* status and *not acceptable* status, respectively. This symbolic reasoning table is formalized as production rules such as:

IF *a* adjacent to *g* THEN Most Probably Not Acceptable (Sharp edge formation)
Insert patterns (J_{cc}, K_{cc}) between *a* and *g* to blend smoothly.

A set of production rules is used to alter the unacceptable initial design into an acceptable design. There are two stages in the geometric reasoning process. At the first stage, only high level symbolic reasoning is carried out based on a *symbolic reasoning table*. It is noted that a symbolic reasoning table can be adjusted according to results of simulations or requirements of

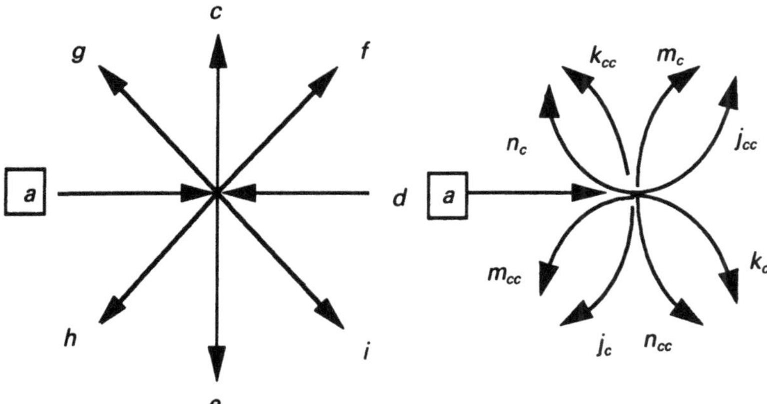

Fig. 9.5. Binary syntactic pattern primitive pairs. Each generic pattern primitive has 16 binary relations. For example, generic pattern primitive '*a*' has the following binary relations: (*a g*)(*a c*)(*a f*)(*a d*)(*a i*)(*a e*)(*a h*)(*a nc*)(*a kcc*)(*a mc*)(*a jcc*)(*a kc*)(*a ncc*)(*a jc*)(*a mcc*).

Table 9.1 Symbolic reasoning table for GPP a

Constraints	*Status*	*New constraints*	*New status*
$a\ A\ (j_{cc}$ or $k_c)$	A		
$a\ A\ (f$ or $i)$ $a\ A\ (m_c$ or $n_{cc})$	MPA	i: k_c f: j_{cc} n_{cc}: k_c m_c: j_{cc}	A
$a\ A\ (c$ or $e)$	PA	e: k_c c: j_{cc}	A
$a\ A\ (g$ or $h)$ $a\ A\ (k_{cc}$ or $j_c)$	MPNA	g: $(j_{cc} A\ k_{cc})$ h: $(k_c\ A\ j_c)$	MPA
		k_{cc}: j_{cc} j_c: k_c	A
$a\ A\ (nc$ or mcc or $d)$	NA		

A represents adjacency relationship for each binary primitive pair and ':' represents additional primitives which will lead to a new status. In the constraints column of the tables, for example, $x\ A\ y$ represents consecutive primitives x followed by y. Also, in the new constraints column of the tables, y:z represents an additional primitive z which is added between primitives x and y to form $x\ A\ z\ A\ y$. In the columns, Status and New status, A, MPA, PA, MPNA and NA represent acceptable status, most probably acceptable status, probably acceptable status, most probably not acceptable status and not acceptable status, respectively.

specific applications. The symbolic reasoning hierarchy is categorized into five levels, such as A, MPA, PA, MPNA and NA. These five levels of hierarchy are used for geometric reasoning. From the given constraints, the status is checked and then new constraints which will lead to a new status (improve status) will be generated. This process is continued until an acceptable status is reached. For each status change, pictorial interpretation will be given, thus allowing the designer to select the most properly matched shape for the functional requirements.

9.5 KNOWLEDGE REPRESENTATION AND CONTROL STRUCTURE FOR LOCAL SHAPE ANALYSIS

The context of manufacturability evaluation is different from others such as process planning and diagnosis. In process planning, for example, the system (knowledge base) addresses the problem of synthesizing a sequence of actions that will achieve a goal (task) with a given set of initial states. Therefore, a search tree for finding such a sequence is generated. In design for manufacturability evaluation, however, the system mainly deals with the problem of finding possible manufacturing difficulties so that the system can provide error messages and some recommendations to the designer. Therefore, the system in design (especially casting design) uses a control strategy (not search strategy) to find the sequence of rules which can be fired for manufacturability evaluation. The main issues in the development of an expert system in design are as follows:

1. Development of a multi-purpose data base which contains high level symbolic information.
2. Development of a suitable knowledge representation which can represent both the declarative and procedural knowledge of an application domain.
3. Development of a control strategy which can maintain and guide the consistency of the system to order a rule sequence.

This section describes a methodology of knowledge representation and a control structure based on the three-dimensional pattern model. A semantic data model is also used in the EXCAST system.

9.5.1 Knowledge representation

Both production rule systems and frames are used in the EXCAST system to represent the declarative and procedural knowledge sources. The production rule system alone does not seem suitable or convenient for design. However, when it is combined with meta-rules and frames, it is expected to be a useful knowledge representation method for design. The organization of this knowledge transfer is shown in Fig. 9.6.

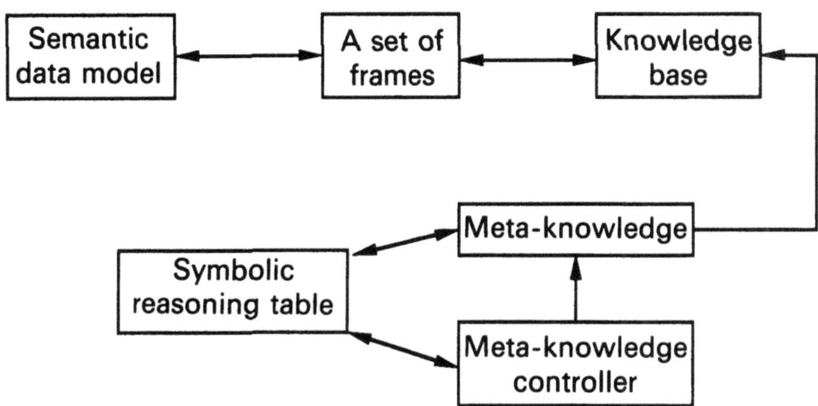

Fig. 9.6. Hierachical organization of knowledge.

The frames in EXCAST use association lists which are identified and extracted from the list using their keys. A frame is a network of modes and relations organized in a hierarchy, where the topmost nodes represent general concepts and the lower modes are more specific instances of those concepts. One way to represent a frame is as a nested association list. On the top level, the frame structure is:

(⟨Class_Name⟩(⟨slot name 1⟩...) (⟨slot name 2⟩...) ...)

In order to explain some frame structures as examples, a semantic data model is presented. Let *f*, *F*, *S* and *SS* be a set of *generatrices*, a set of *f*s, a set of *F*s and a set of *S*s, respectively. The data structure contains this information in a hierarchical manner as shown in Fig. 9.7. In Fig. 9.7, a *generalization* is an abstraction which enables a class of individual objects to be thought of generically as a single named object. For example, casting parts are defined as generalizations of SS_1, SS_2, SS_3,..., SS_n. An aggregation, on the other hand, is the process by which multiple lower level entities

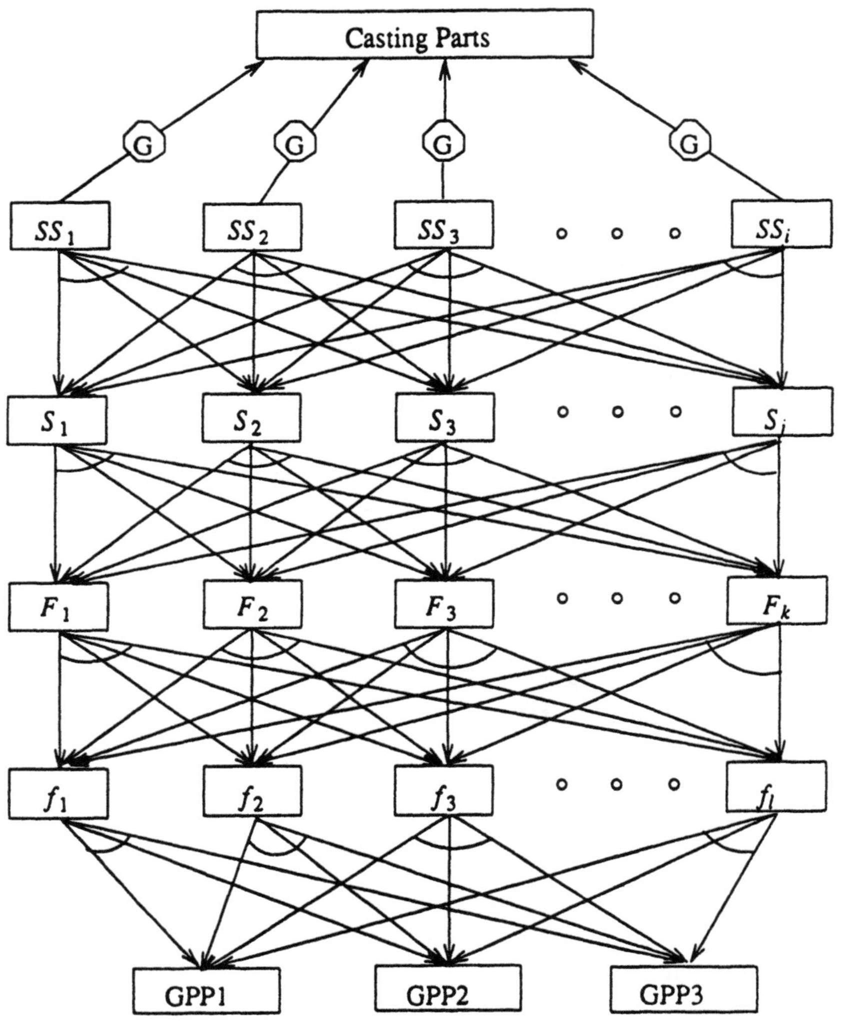

Fig. 9.7. A data structure of geometric and topological information. *t*, *F*, *S* and *SS* are a set of generatrices, a set of *f*s, a set of *F*s, and a set of *S*s respectively. G, generalization; and arc, aggregation.

combine to form a single higher level composite entity (Mohan and Kashyap, 1989). These methodologies seem particularly appropriate for CAD because the fundamental principles behind them are creation and manipulation of hierarchically defined objects (abstract data type) and these are exactly the activities done during the design process. Data abstractions provide several advantages for modeling CAD data, most importantly the data semantics introduced by the abstraction hierarchies. Each edge in these hierarchies represents an 'is a' or 'part of' relationship between two abstract data types. The abstract data types are therefore logically organized by these explicitly stated semantic relationships, rather than being a collection of independent data types (Spooner and Hardwick, 1988). A frame for SS_i, for example, possesses the following information.

1. An order of union operations associated with S_j.
2. Position and rotation parameters of each S_j.
3. Variational information.

This information is stored in a frame structure which is shown below:

```
(setq Part_List
  `(⟨Part_Name ( SS_i)⟩
    (⟨Order_Of_Union_Operation (Values)⟩
    ⟨Position_And_Rotation_Parameters (Values)⟩
    ⟨Variational_Information Symbolic_Representation⟩)))
```

A frame of S_j, as another example, contains a following information.

1. An order of set operations associated with F_k.
2. Position and Direction vectors of each F_k.
3. Variational Information.

This information is stored in a frame structure which is shown below:

```
(setq Discretized_Part_List
  `(⟨Discretized_Part_Name⟩
    (⟨Order_Of_Set_Operations (Values)⟩
    ⟨Position_And_Direction_Vectors (Values)⟩
    ⟨Variational_Information Symbolic_Representation⟩)))
```

When rules are invoked, frames are used as a 'medium' between semantic data model and production rules, and these frames are used as elements of production rules. This knowledge representation is used to bridge between a semantic data model and a knowledge base as shown in Figure 9.6. For example, binary relationship for pattern primitive *a* can be used to form a symbolic reasoning table which are formalized as production rules as shown below:

IF *a* adjacent to *g* THEN Most Probably Not Acceptable
Insert patterns (J_{cc}, K_{cc}) between *a* and *g*.

In this rule, the information about patterns *a* and *g* can be extracted by using frame *frame_get* such as:

- (frame_get 'Part_name 'Input_pattern 'neighbor_patterns *pattern1 pattern 2*)

where *pattern1* and *pattern2* are variables which are instantiated by *a* and *g*, respectively. The knowledge in EXCAST is represented by frames and production rules and is controlled by meta-rules. For instance, rule-50, shown below, checks if a particular section thickness causes manufacturability problems because of the fluidity of solidification of the molten metal.

```
RULE-50
IF
  Primary feature is 1
 AND
  Direction of hole is left OR right
 AND
  The distance from bottom of hole to side
  is less than minimum economical section thickness
THEN
  Give illegal dimension message and suggestion
  for economical minimum section thickness
```

The LISP version of this rule would be

```
(($ RULE 50)
(IF (and
  (equal primary-feature 1)
  (or (equal (frame-get 'rectangular-hole 'hole-face 'value 3)
     (equal (frame-get 'rectangular-hole 'hole-face 'value 4))
  (%lessp (- (frame-get 'cylinder 'length 'value)
          (frame-get 'rectangular-hole 'hole-axial 'value))
        (%minimum-economical-section
          (frame-get 'cylinder 'length 'value)))))
(THEN (%message-RH4)))))
```

This knowledge representation is explicit, machine-usable, modifiable and modular. In the EXCAST system, manufacturability knowledge concerning tolerancing, minimum thickness, draft angle, shrinkage, hot junction, parting line, rounding, boss, and solidification modulus are incorporated in the structure of the rules (see Appendix A).

9.5.2 Control structure

Since a rule sequence is determined according to the symbolic representation of the design, a heuristic control strategy is employed. This heuristic control strategy is embedded into the knowledge refinement process. The object level knowledge invocation is viewed in three steps (Davis, 1980): retrieval, refinement and execution. In retrieval, a set of rules (objective level) is grouped based on property of objective rules. During the refinement phase, rules in a set are pruned or possibly ordered based on a basic invocation criterion. The final phase is execution, in which one of the rules in the revised set is applied to the problem.

A reference by name technique is used in EXCAST for grouping a set of rules and is implemented by classifying consecutive GPPs and forming related rules into a set of lists. Some meta-rule lists that can be generated in EXCAST are shown below:

```
(META-RULE%I-BEAM1 (1 2 3 4 5 6)
(META-RULE%I-BEAM2 (15 16 17 18 19 20 21 22 23 24 25 26 27 28 29 30 31 32 33))
```

Manufacturability evaluation for I-BEAM2, for example, is carried out by rules 15–33 which represent binary pair rules from the symbolic reasoning table. These meta-rule lists are used to perform symbolic reasoning and their results are tested for further actions of detailed geometric reasoning. For example, if the designer wants to have an A (acceptable) status of the design, then these meta-rule lists are continuously updated until the A (acceptable) status is reached. Let meta-rule list MRL_1 be a set of five rules such that $MRL_1 = (2\ 5\ 11\ 23\ 37)$, and let us assume that the following result is generated after the first iteration:

Result After 1st Iteration = MPA(result of rule 2) A(result of rule 5)
A(result of rule 11)
MPNA(result of rule 23) A(result of rule 37)

then the system tries to improve all status where status is not A. In order to update their status, new constraints are added between the consecutive GPPs. Therefore, the result after the second iteration may be:

Result After 2nd Iteration = A A A MPA A

and the result of the third iteration can be:

Result After 3rd Iteration = A A A A A.

Notice that A status cannot be achieved if the knowledge base does not provide A status for all pattern primitive pairs (even after several iterations). The following algorithm explains the procedure. The first algorithm generates a set of meta-rule lists based on input. High level symbolic reasoning is carried out by the second algorithm based on the meta-rule lists generated by the first algorithm. The first algorithm is shown below:

ALGORITHM: *Forming_MetaRule_Lists*
INPUT: Sets of Functional Surfaces with Paths or axes
OUTPUT: Sets of Meta-Rule Lists
  WHILE there are functional surfaces DO
    •Get functional surface and initialize all variables

```
    WHILE there are binary pairs DO
      ••Extract a binary pair.
      ••Put its rule number into a list.
    END WHILE;
    •Save list.
  END WHILE;
END ALGORITHM;
```

The second algorithm is shown below:

```
ALGORITHM: Pattern_Based_Knowledge_Invocation
INPUT: Sets of Meta-Rule Lists (from the Algorithm:Forming_MetaRule_Lists)
OUTPUT: Manufacturability Evaluation
  WHILE there are meta-rule lists DO
    •Get a meta-rule list and initialize all variables.
    WHILE there are rules in the list DO
      ••Extract a rule from the list.
      ••Invoke the rule and save it in a list.
      ••Compare the elements of Result.
        ••If comparison is not satisfied, then get new status by calling the
      Algorithm:Forming_MetaRule_Lists, otherwise, list=f.
    END WHILE;
  •Do low level Geometric Reasoning if necessary.
  END WHILE;
END ALGORITHM;
```

9.6 GLOBAL SHAPE ANALYSIS

Symbolic reasoning tables associated with production rules are suitable for evaluation of local geometric shapes. However, global casting soundness should be evaluated by incorporating inter-shape interaction. Currently available knowledge of solidification is not in a suitable form to be incorporated in a rule base when the system does not possess predefined geometric primitives. In this work, a *skeleton* is extracted and the minimum distance from the skeleton to the boundary is calculated to generate a distance map. With this skeleton, global geometric characteristics are identified and solidification moduli are calculated to predict casting defects.

9.6.1 Skeleton extraction

The most important aspect of the skeletonization in global shape analysis is to allow the system to reason about the object without *a priori* knowledge about the object shape. In other words, the skeletonization should provide a distance map to build general purpose algorithms for global shape analysis. since inputs used in this work are in continuous space, they must be discretized first before extracting the skeleton. In order to discretize three-dimensional objects in continuous space, discontinuity must be carefully considered. Since digitization produces round off error, proper voxel size must be chosen in order to obtain required accuracy. The discretization process in this work is actually a filling process for each boundary contour. A filling algorithm fills the inside of boundary contour for each layer and this process is continued until all layers are filled. The main steps involved in the method are as follows:

(1) Boundary contours are digitized and stored in a three-dimensional array.
(2) Fill the rows until all voxels of the inside contours are filled.
(3) Continue step 2 for all layers.

After the decomposition, functional features are combined by applying digital set operators. In CSG (Requicha, 1980), regularized set operators are used to combine design features. In order to apply set operators to the digitized object, we define new operators for union, intersection, difference and complement. Let A be a three-dimensional array such that

$$A=\{x, y, z)|0\leqslant x, y, z\leqslant \text{array size}\}$$

and let $D_1(r)$ and $D_2(r)$ be two discretized objects, where $r=(x, y, z)$. Then there are two mapping functions for two discretized objects such that

$$D_1(r){:}A\rightarrow(f, b)$$
$$D_2(r){:}A\rightarrow(g, b)$$

where f and $g=1$ and $b=0$. The digitized set operators for the objects $D_1(r)$ and $D_2(r)$ are defined as follows:

- $D_1(r)\ U^d D_2(r)$ (union):
  If $D_1(r)=f$ or $D_2(r)=g$, THEN $(D_1\,U^d D_2)\ (r)=f$ or g, otherwise b.
- $D_1(r)\ll^d \mathrm{D}_2(\mathrm{r})$ (intersection):
  If $D_1(r)=f$ and $D_2(r)=g$, then $(D_1 \ll^d D_2)\ (r)=f$ or g, otherwise b.
- $D_1(r)-{}^d D_2(r)$ (difference):
  If $D_1(r)=f$ and $D_2(r)=\mathrm{g}$, then $(D_1-{}^d D_2)\ (r)=b$.
- $\mathrm{C}^d D_2(r)$ (complement):
  If $D_2(r)=g$, then $\mathrm{C}^d D_2(r)=b$, otherwise g.

These digitized set operations are carried out with transformation matrix. If an object is to be translated without rotation, and the vector displacement of each object point is to be $\mathbf{t}$, then the position vector $\mathbf{r}^*$ of the displaced point is related to its initial position $\mathbf{r}$ by the equation $\mathbf{r}^*=\mathbf{r}+\mathbf{t}$. When voxels of an object are rotated by an angle θ about the z-axis, we need to consider the relationship between the coordinate (x, y) of the original voxels and (x', y') of the rotated points. A two-dimensional rotation can be written in the matrix form:

$$r'=\begin{bmatrix} x' \\ y' \end{bmatrix}=\begin{bmatrix} \cos\theta & -\sin\theta \\ \sin\theta & \cos\theta \end{bmatrix}\begin{bmatrix} x \\ y \end{bmatrix}=Ar$$

The three-dimensional digitized object is generated by applying the digital set operators with translation and rotation of each separate functional feature. This digitized object is stored in a three-dimensional binary array having m rows, n columns and l layers which is denoted by an array $A=\{(i,j,k)|0\leqslant i<m,\ 0\leqslant j<n,\ 0\leqslant k<l\}$ and the value of each voxel either 0 or 1. The objects of interest in the image are represented by the set S which consists of a voxels with value 1 and the complement $\bar{S}$ which consists of all voxels with value 0. Every voxel $P=(i,j,k)$ has two types of neighbors:

1. 6-neighbor: $N_6(P=(i,j,k))=\{(r,s,t)|\ |r-i|+|s-j|+|t-k|=1\}$;
2. 26-neighbor: $N_{26}(P=(i,j,k))=\{(u,v,w)|\ \max(|u-i|,|v-j|,|w-k|)=1\}$,

and is 6− (26−) adjacent to its 6− (26−) neighbors. In order to avoid ambiguity, opposite types of connectedness are always used for S and $\bar{S}$. Since objects are more important than background, we use 26 connectedness and 6 connectedness for objects and background, respectively. This three-dimensional array is used as an input for thinning algorithm. In the discrete space, the boundary voxels of S are iteratively deleted until no more voxel can be deleted. The remaining voxels constitute a skeleton. While the object is being thinned, the topological properties of the object must be preserved. Depending on the property being preserved, different approaches have been reported (Gong and Bertrand, 1990; Morgenthaler, 1980; Tsao and Fu, 1981). In general three conditions must be satisfied in order to delete a boundary voxel P:

- Connectivity must be preserved after the deletion of a voxel P.
- Isomorphism must be preserved after the deletion of a voxel P.
- A voxel P must satisfy non-end voxel conditions.

In order to ensure connectedness, extended Euler's formula can be used:

$$V-E+F=2-2G$$

where V, E, F and G denote the number of vertices, edges, faces of the polyhedron and genus, respectively. The genus number of a three-dimensional digital image is defined to be the number of object minus number of holes plus the number of cavities (Morgenthaler, 1980). The number, $2-2G$, is called the Euler characteristic of the surface and is a topologically invariant property (Meserve, 1955). Then, the connectivity number N of the objects in a three-dimensional digital picture is defined as the sum of the Euler characteristics

$$N=\Sigma_i(2-2G_i).$$

A voxel V in S is called simple if its removal is connectivity preserving. If the difference of the genus numbers before and after the deletion of the center voxel V is equal to zero, then the voxel V is simple and removal of V is allowed. However, simplicity of the voxel does not guarantee isomorphism.

A separate routine is required to prevent hole generation and to preserve isomorphism. In this work, the following routines were employed:

- The number of objects after deletion of the center voxel is counted instead of using checking windows. If the number of objects after the deletion of the center voxel is one, then isomorphism is preserved.
- Since the simultaneous removal of simple points separates the original objects or even eliminates the original objects completely, a sequential rechecking connectivity is carried out after labeling all deletable voxels.

The three-dimensional thinning algorithm used in this work is shown in Appendix B. Figure 9.8 shows the skeleton of a connecting rod. The connecting rod in Fig. 9.8 is used as an imput. For discretization, $128 \times 128 \times 128$ digital space was used.

9.6.2 Distance mapping

Euclidean distance between two voxels $p(x, y, z)$ and $q(x', y', z')$ is defined as follows:

$$d_e(p, q) = \sqrt{(x'-x)^2 + (y'-y)^2 + (z'-z)^2}$$

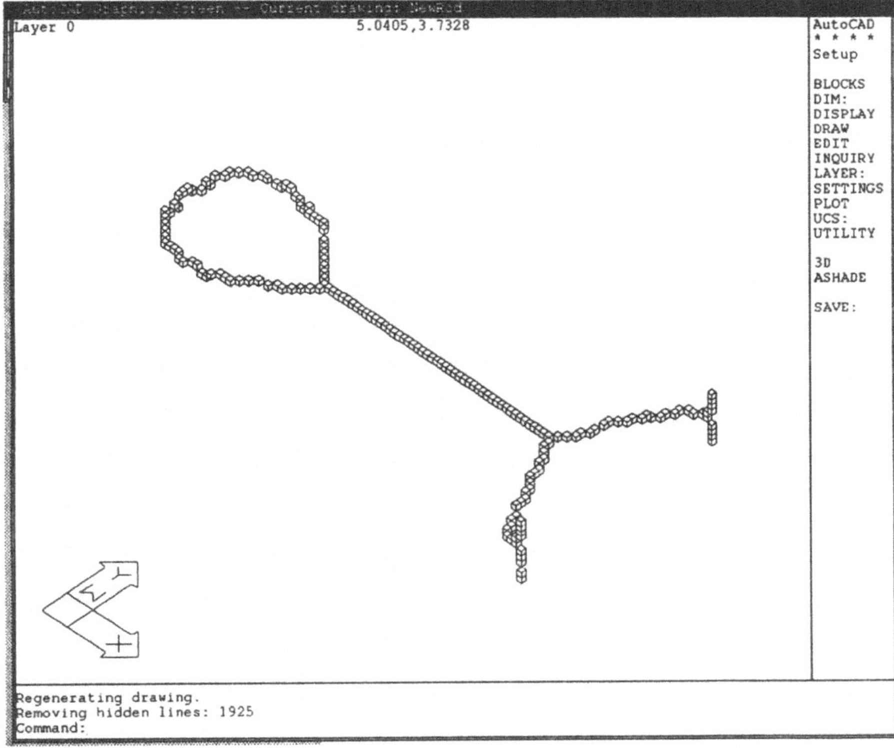

Fig. 9.8. Extracted skeleton of the connecting rod.

Given a three-dimensional binary array with two sets of voxels.

$$S = \text{set of 1s, the objects}$$
$$\bar{S} = \text{set of 0s, the background}$$

a distance map **D** is a three-dimensional array such that for each voxel

$$p = (i, j, k) \in S$$

there is a corresponding point $L(p)$ where

$$L(p) = \min_{q \in \bar{S}} d_e(p, q) = \text{minimum distance of } p \text{ to boundary}$$

A distance map of character L is shown as an example in Fig. 9.9. The center layer is shown as an example where the minimum euclidean distance is labeled at the skeleton and at the boundary. Table 9.2 shows coordinate, minimum distance to the boundary and junction type for each voxel of the skeleton. Four different types of node are classified based on the number of neighbors as follows:

1. *end node*: a voxel which has only one neighbor.
2. *M node*: a voxel which has two neighbors.
3. *T node*: a voxel which has three neighbors.
4. *X node*: a voxel which has four neighbors.

These junction types are used to adjust labeled distances at the boundary contour for proper cooling factor assignment. In general, the cooling factors for external corners are much higher than the sand thickness, since they are in contact with a large area of the mold, and the heat flow diverges out into the sand. At internal corners, the cooling factors are small, since the heat

```
layer = 4
[3.8][2.8][2.4][2.0][2.4][2.8][3.8][0.0][0.0][0.0][0.0][0.0][0.0][0.0]
[3.4][*.*][*.*][*.*][*.*][*.*][3.4][0.0][0.0][0.0][0.0][0.0][0.0][0.0]
[3.0][*.*][*.*][2.0][*.*][*.*][3.0][0.0][0.0][0.0][0.0][0.0][0.0][0.0]
[3.0][*.*][*.*][3.0][*.*][*.*][3.0][0.0][0.0][0.0][0.0][0.0][0.0][0.0]
[3.0][*.*][*.*][3.0][*.*][*.*][3.0][0.0][0.0][0.0][0.0][0.0][0.0][0.0]
[3.0][*.*][*.*][3.0][*.*][*.*][3.0][0.0][0.0][0.0][0.0][0.0][0.0][0.0]
[3.0][*.*][*.*][3.0][*.*][*.*][3.0][0.0][0.0][0.0][0.0][0.0][0.0][0.0]
[3.0][*.*][*.*][3.0][*.*][*.*][*.*][3.0][3.0][3.0][3.0][3.4][3.8][4.2]
[3.0][*.*][*.*][3.0][*.*][*.*][*.*][*.*][*.*][*.*][*.*][*.*][*.*][3.8]
[3.0][*.*][*.*][3.0][*.*][*.*][*.*][*.*][*.*][*.*][*.*][*.*][*.*][3.4]
[3.0][*.*][*.*][3.0][3.0][3.0][3.0][3.0][3.0][3.0][3.0][*.*][*.*][3.0]
[3.4][*.*][*.*][*.*][*.*][*.*][*.*][*.*][*.*][*.*][*.*][*.*][*.*][3.4]
[3.8][*.*][*.*][*.*][*.*][*.*][*.*][*.*][*.*][*.*][*.*][*.*][*.*][3.8]
[4.2][3.8][3.4][3.0][3.0][3.0][3.0][3.0][3.0][3.0][3.0][3.4][3.8][4.2]
```

Fig. 9.9. A distance map $D(S)$ for a character L. Center layer is shown as an example where the minimum euclidean distance is labeled at the skeleton and at the boundary.

Table 9.2 Distance mapping table which includes coordinate, minimum distance to the boundary and junction type for each voxel of the skeleton

Node number	*Coordinate*	*Distance*	*Node type*
1	(5,4,4)	2	end node
2	(5,5,4)	3	M node
3	(5,6,4)	3	M node
4	(5,7,4)	3	M node
5	(5,8,4)	3	M node
6	(5,9,4)	3	M node
7	(5,10,4)	3	M node
8	(5,11,4)	3	T node
9	(5,12,4)	3	M node
10	(6,12,4)	3	T node
11	(7,12,4)	3	M node
12	(8,12,4)	3	M node
13	(9,12,4)	3	M node
14	(10,12,4)	3	M node
15	(11,12,4)	3	M node
16	(12,12,4)	3	end node

flow is convergent and the mold is prone to getting hot. These external and internal corner effects are considered for initial cooling factor assignment based on junction types. For example, corners near end nodes are considered as external corners and a corner near T nodes with small angle side is considered as an internal corner.

9.6.3 Solidification modulus

In order to carry out global casting analysis, Chvorinov's rule is used to determine the relative freezing time among various segments of the casting. The solidification modulus is the ratio of the casting volume to the surface area (V/A) of the casting (Wlodawer, 1966). According to Chvorinov, solidification time, t_s, is:

$$t_s = \text{constant (solidification modulus)}^2.$$

The order of solidification within a casting is decided by comparing the modulus and then the sections freezing last would serve as feeding paths for the riser or, if feed metal was lacking, then would contain shrinkage porosity. Using this concept, global casting analysis can be carried out to decide riser positions and predict possible casting defects. The modulus calculation is based on the subdivision of a complex part into simple basic components. Simple geometries which are pieced together to produce most of familiar complex casting designs are T-sections, X-sections and L-sections. Since the most common casting defects occur at the junction, a complex part is subdivided into those sections using the extracted skeleton. Then the ratio modulus is calculated for each component and compared

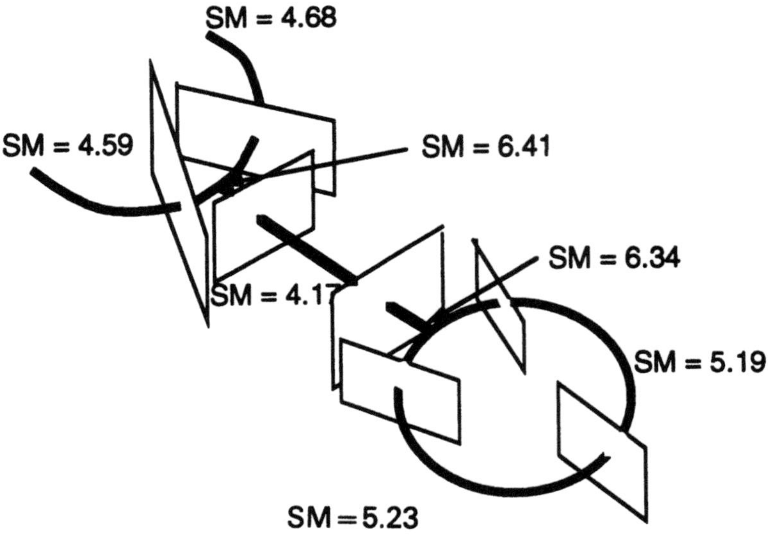

Fig. 9.10. Solidification modulus calculation from the skeleton model.

with the sectional modulus of directly neighboring components. Figure 9.10 shows the automatic solidification modulus calculation using the extracted skeleton of the connecting rod. It can be seen that the two T junctions can possibly cause a solidification problem.

The effects of section thickness and round corners have been studied by Kotschi and Loper using Chvorinov's rule (Kotshi and Loper, 1974). Figure 9.11 shows a representative T-shaped casting cross-section composed of a supporting arm of thickness T2 and a cross arm of thickness T1, joined with fillets of radius R. Region CA represents the cross arm away from the T-section junction, region J represents the T-section junction and region SA represents the supporting arm away from the T-section junction. The junction region was taken to extend an additional length equal to the cross arm and supporting arm thickness from the geometric edge of the junction. Finite element simulation of solidification in sand casting is carried out in order to validate solidification modulus approach. The problem is modeled as a transient phase change over the range of solidification of the metal. The enthalpy method was used to model the latent heat of fusion during solidification. The above procedure was implemented using the ANSYS Engineering Analysis System on a SUN 4/280 workstation. The simulation results were validated for a theoretical solution of an isothermal freezing problem and experimental results of a cylinder casting. Details of the simulation procedure and the effects of interface resistance, boundary condition and pouring temperature on the solidification behavior of T-shaped castings were reported in Cadarso, Chy and Kashyap (1991).

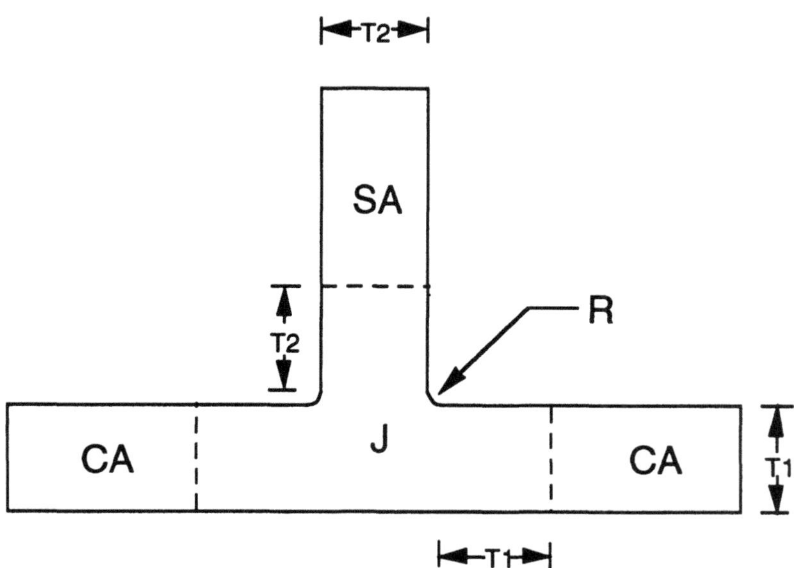

Fig. 9.11. Illustration of a T-shaped casting cross-section.

Figure 9.12(a–d) shows the isochronal 70% solid fraction profiles when the cross arms are 10.2 cm (4 in) thick and infinitely long, and the supporting arms are infinitely long and (a) 2.54 cm (1 in) thick, (b) 5.08 cm (2 in) thick, (c) 7.62 cm (3 in) thick and (d) 10.2 cm (4 in) thick, respectively. Only half of the T-section is modeled because of symmetry. When the supporting arm section is 2.54 cm (1 in) thick, the supporting arm section solidifies first, acting as a chill, next the junction solidifies and, last, the cross arm section where the critical solid fraction, f_{cr}, loop disappearance point is located. As the supporting arm section thickness increases, the f_{cr}, loop disapparance point moves toward the middle of the junction. Obviously the solidification time is longer when the supporting arm section is 10.2 cm (4 in) thick rather than 2.54 cm (1 in), due to the higher modulus. The results of the finite element method simulations agree very well with the results of Kotschi and Loper based on solidification modulus calculations using the section definition shown in Fig. 9.11. By changing the thickness of the supporting arm section, it can act as a chill or as a riser over the cross arm, and the f_{cr} loop disappearance point can be moved from the junction to the cross arm section. Some other casting knowledge base is shown in Appendix A.

9.7 IMPLEMENTATION

The EXCAST system was implemented with an interactive CAD environment on a Sun workstation 3/60 UNIX using Sun Common-LISP and the AutoCad graphics package. Both production rule systems and frames were used for

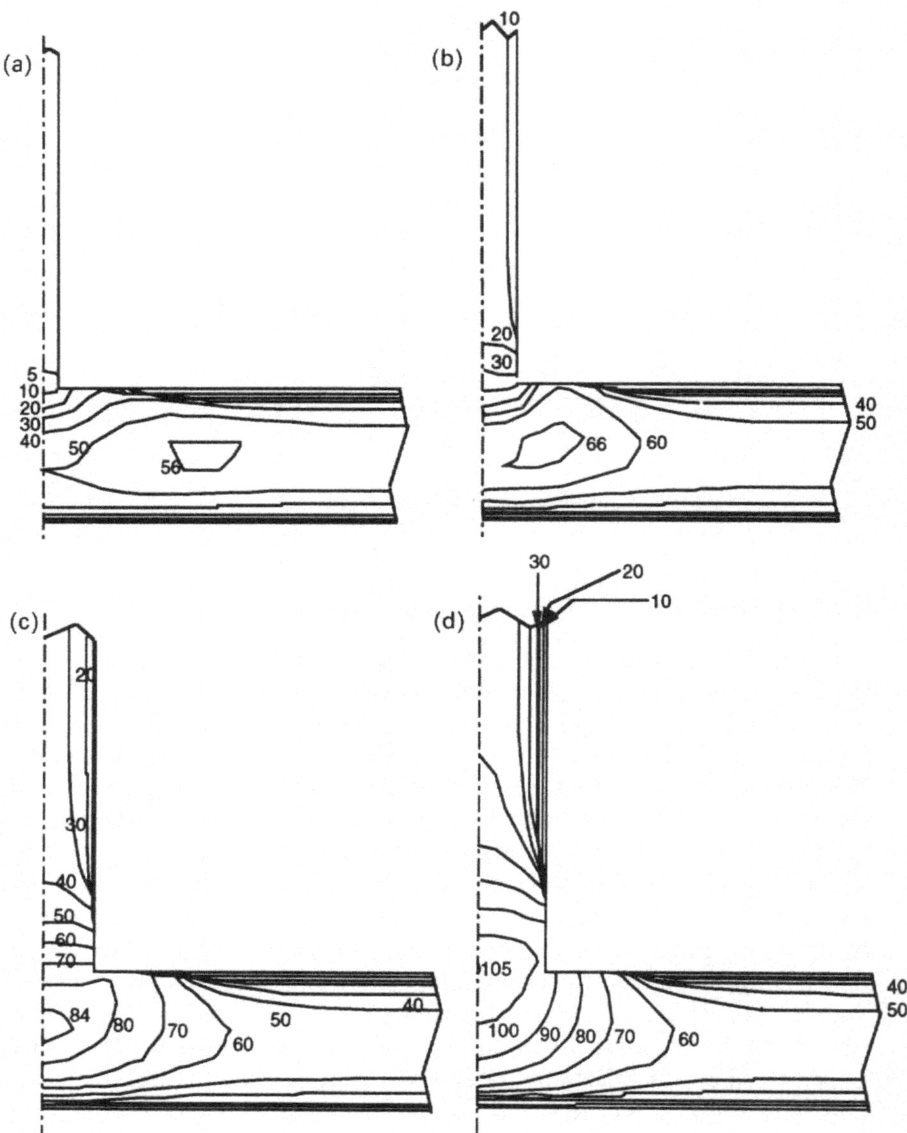

Fig. 9.12. Isochronal 70% solid fraction profile of a 0.38% C eq. steel T-Section solidification time in minutes. Cross arm: 4 in thickness and infinite length. Supporting arm: (a) 1 in thickness and infinite length, (b) 2 in thickness and infinite length, (c) 3 in thickness and infinite length, (d) 4 in thickness and infinite length.

representing the declarative and procedural knowledge sources. Low level geometry manipulations are implemented with C language and high level symbolic manipulations are implemented with Sun Common-LISP for the new EXCAST system. For a boolean operation, a local coordinate system is used to translate and rotate an object. Three 2 MB three-dimensional arrays are used to carry out the thinning algorithm. For the thinning algorithm, 26 connectedness and 6 connectedness are used for objects and background, respectively. There are about 250 rules for local shape analysis. This system has been successfully implemented at Cummins Engine Company, Columbus, IN.

9.8 CONCLUSIONS

A formalism for three-dimensional geometric analysis is developed to aid local and global feature extraction and casting design. Local feature analysis is carried out based on binary syntactic pattern primitives to reason about local geometric characteristics. Usually, this local feature analysis can be easily carried out in a feature-based design approach of a specific domain. However, casting components generally have complicated geometry and local feature analysis cannot properly address the complicated solidification behavior arising from interfeature interactions. A three-dimensional parallel thinning algorithm is developed in order to extract the global topology of the object. In this scheme, a geometry is digitized and the skeleton line of the geometry is extracted. The extracted skeleton allows the expert system to mimic the decision making of a casting expert. The solidification modulus approach is used to evaluate global casting soundness so as to avoid shrinkage porosity or to decide the feeding path for a riser. The modulus calculation is based on the subdivision of a complex part into simple basic components, and then the solidification modulus is calculated for each component and compared with the sectional modulus of neighboring components. The subdivision is carried out by classifying and identifying the junction types of each element of the extracted skeleton.

The framework enables the system to reason about geometric shape based on pattern primitives and skeletons providing meaningful geometric and topological properties for castability valuation. Both production rule systems and frames were used to represent the declarative and procedural knowledge sources in the EXCAST system. Automatic meta-rule formation based on symbolic representation is a part of the knowledge refinement process and a semantic data model is also used to aid various CAM applications. The integration of the procedural and rule-based approaches makes the EXCAST system efficient and easy to used by novices, especially in a small job shop. With this system, we can expect a smooth transition from design to evaluation, allowing automated castability evaluation at the design stage so as to minimize repeated trial and error in the casting process.

ACKNOWLEDGMENTS

This work was supported by the National Science Foundation under grant CDR 8803017 to the Engineering Research Center for Intelligent

Manufacturing Systems. The authors also acknowledge the encouragement and support of the Cummins Engine Company.

REFERENCES

Bralla, J.G. (1986) *Handbook of Product Design for Manufacturing*, McGraw-Hill, New York.

Cadarso, L., Chu, C.N. and Kashyap, R.L. (1991) Finite element simulation of T-shaped castings. *AFS Transactions*, **99**, 369–77.

Choi, B.K, Barash, M.M. and Anderson, D.C. (1984) Automatic recognition of machined surfaces from a 3D solid model. *CAD*, **16**, 81–6.

Davis, R. (1980) Content reference: reasoning about rules. *Artificial Intelligence*, **15**, 223–39.

Dixon, J.R. (1988) *Designing with Features: Building Manufacturing Knowledge into more Intelligent CAD systems*. Proceedings of ASME Manufacturing International, pp. 51–7.

Fu, K.S. (1982) *Syntactic Pattern Recognition*, Prentice-Hall, Englewood Cliffs, N.J. 1982.

Gong, X. and Bertrand, G. (1990) *A Simple Parallel 3D Thinning Algorithm*. 10th IEEE International Conference of Pattern Recognition, pp. 188–90.

Henderson, M.R. (1984) *Feature Recognition in Geometric Modeling*. CAM–I's 13th Annual Meeting and Technical Conference, pp. 5.1–5.12.

Herbert, P.J., Hinde, C.J., Lauders, V.A. *et al.* (1990) Feature recognition within a truth maintained process planning system. *International Journal of Computer Integrated Manufacturing*, **3**, 121–32.

Jakubowski, R. (1982) Syntactic characterization of machine parts shapes. *International Journal of Cybernetics and Systems*, **13**, 1–24.

Joshi, S. and Chang, T.C. (1988) Graph-based heuristics for recognition of machined features from a 3-D solid model, *CAD*, **20**, 58–66.

Kotschi, R.M. and Loper, C.R. (1986) Design of T and X sections of castings, *AFS Transactions*, 535–42.

Lee, D.T. (1982) Medial axis transformation of a planar shape. *IEEE Transactions on Pattern Analysis and Machine Intelligence*, **PAMI-4**, 363–9.

Lee, Y.C. and Fu, K.S. (1987) Machine understanding of CSG: extraction and unification of manufacturing features. *IEEE Computer Graphics and Applications*, 20–32.

Libardi, E.C., Dixon, J.R. and Simmon, M.K. (1986) *Designing with Features: Design and Analysis of Extrusions as an Example*. Proceedings of Mechanical Design Conference, pp. 1–8.

Lin, W.C. and Fu, K.S. (1984) A syntactic approach to 3-D object representation. *IEEE Transactions in Pattern Analysis and Machine Intelligence*, **PAMI-6**, 351–64.

Meserve, B.E. (1955) *Fundamental Concepts of Geometry*, Addison Wesley, Reading, MA.

Mohan, L. and Kashyap, R.L. (1989) *Abstractions in Object-Oriented Data Models: A Formalized Representation Scheme*. Proceedings of the First International Conference on Software Engineering and Knowledge Engineering, pp. 79–84.

Montanari, U. (1969) Continuous skeletons from digitized images. *Journal of the Association for Computing Machinery*, **16**, 534–49.

Morgenthaler, D.G. Three dimensional digital topology: the genus. *TR-980*, Computer-vision Laboratory, Computer-Science Center, University of Maryland, College Park.

Nackman, L.R. (1982) Three-dimensional shape description using the symmetric axis. *PhD Thesis*, University of North Carolina.

Requicha, A.A.G. (1980) Representation for rigid solids: theory, methods, and systems. *The Association for Computing Machinery Computing Surveys*, **12**, 437–64.

Requicha, A.A.G. and Vandenbrande, J.H. (1989) *Form Features for Mechanical Design and Manufacturing*. Proceedings of ASME Computers in Engineering Conference, pp. 47–52.

Spooner, D.L. and Hardwick, M. (1988) A conceptual framework for data management in mechanical CAD, in *Geometric Modeling for CAD Applications*, (ed. M.J.Wozny, H.W. McLaughlin and J.E. Encarnacao), North-Holland, Amsterdam, pp. 317–29.

Staley, S.M., Henderson, M.R. and Anderson, D.C. (1983) Using syntactic pattern recognition to extract feature information from a solid geometric data base. *Computers in Mechanical Engineering*, 61–6.

Tsao, Y.F. and Fu, K.S. (1981) A parallel thinning algorithm for 3-D pictures. *Computer Graphics and Image Processing*, **17**, 315–31.

Wieser, P.F. (1980) *Steel Castings Handbook*, Steel Founders' Society of America.

Wlodawer, R. (1966) *Directional solidification of steel castings*, Pergamon Press, Oxford.

Woo, T.C. (1983) Interfacing solid modeling to CAD and CAM: data structures and algorithms for decomposing a solid, in *Computer Integrated Manufacturing*, (ed. M. L. Martinez and M.C. Leu), New York, pp. 39–46.

Yoshiura, H., Fujimura, K. and Kunii, T.L. (1984) Top-down construction of 3-D mechanical object shapes from engineering drawings. *IEEE Computer*, 32–9.

You, I.C., Chu, C.N. and Kashyap, R.L. (1989) Expert system for castability evaluation: using a fixed-features based design approach. *Robotics and Computer-Integrated Manufacturing*, **6**, 181–89.

CHAPTER 10

Expert system approaches to the selection of materials handling and transfer equipment

Jacob Rubinovitz and Reuven Karni

10.1 INTRODUCTION

Materials handling and transfer (MHT) is defined as the movement of physical objects – raw materials, components parts subassemblies, assemblies and finished goods – along the factory floor from receiving through shipping (Eastman, 1987, p. 1). The purpose of moving the material should be to increase its value. However, 'the handling, transporting, housing and controlling of materials and goods adds nothing but cost to the system' (Sims, 1991, p. xiii). Thus MHT is usually regarded as a burden; and MHT design is, therefore, often carried out as a final step after product, process and layout design has been completed, and in isolation from the overall design process.

Only recently, with the trend towards cooperative design in the form of concurrent engineering (CE), has it been recognized that MHT has an overwhelming effect on successful plant operation, and that the MHT function has to be an integral part of engineering design.

Our chapter contributes to this aim by presenting an artificial intelligence (AI)-based qualitative approach to one phase of MHT design – selection of a specific equipment type – which allows a large number of product, process and production factors to be taken into account. The expert system format allows decisions regarding MHT to be taken in conjunction with other design phases, and provides a mutual interface and feedback mechanism necessary to support a CE design process.

10.2 MHT AND EXPERT SYSTEMS

Several prototype expert systems for MHT equipment selection have been developed during the last decade (Gabbert and Brown, 1988; Hosni, 1989; Malmborg *et al.*, 1986, 1987; Matson, Swaminathan and Mellichamp, 1990). Most of these systems assist in an initial selection of equipment category or type by using a structured evaluation of the handling task and general equipment characteristics.

Malmborg *et al.* (1986) suggest three phases in the development of a knowledge-based system for specifying MHT equipment: knowledge base construction, expert system implementation, and system refinement and validation. They outline a three-level hierarchical structure for organizing the knowledge base. The top level should contain rules and principles for selection of a general equipment category (such as conveyor or truck), the middle level should focus on knowledge (codified as relations and rules) for selection of a specific equipment type (such as belt conveyor or hand truck) and the bottom level should incorporate facts about individual equipment alternatives at the level of commercially available products.

A complete implementation of EXIT, an expert system for industrial truck selection, is described by Malmborg *et al.* (1987). The system deals with the intermediate level of truck type selection and is implemented on a PC using Prolog. Based on a taxonomy for industrial trucks, it uses a pattern-directed inference engine to select a truck type from a simple MHT problem description. The knowledge base representation incorporates semantic nets, production rules and frames. Some important suggestions for further development of such expert systems are made, such as explanation modules, learning capabilities, and a mechanism for parallel certainty inferences.

Other expert systems target the top two levels – the selection of the category and/or type of MHT equipment. These include MHES (Hosni, 1989), MAHDE (Gabbert and Brown, 1988) and EXCITE (Matson, Swaminathan and Mellichamp, 1990). MHES (Hosni, 1989) is based on the material handling equation of MATERIAL + MOVE = METHOD (Apple, 1977) and on tree-based branching as the core of an inference engine for a prototype expert system which selects 'first cut' MHT equipment and methods. The system uses a frame-based knowledge representation, and incorporates forward and backward chaining and confidence factors in a rule-based inference mechanism. The MAHDE (Gabbert and Brown, 1988) knowledge base has a hierarchical frame structure. An important component of MAHDE is a preferential mechanism, which utilizes weights for the calculation of a joint utility function for candidate MHT equipment types. This mechanism relies on input from the user regarding the relative importance of equipment attributes. EXCITE (Matson, Swaminathan and Mellichamp, 1990) has a knowledge base composed of production rules with a forward chaining inference mechanism. The system is implemented on a PC using OPS53. It contains 340 rules for the selection of 35 equipment options. In a manner similar to MAHDE, user preferences for equipment attributes are used to score system recommendations.

10.3 MHT AND ENGINEERING DESIGN

As a result of the success of Japanese manufacturing and intense competition in international markets, it has been realized that the 'design bottleneck' has to be removed. Concepts such as concurrent engineering (CE) have been mooted as showing the way to bringing products to the market more rapidly and reliably; however, no 'CE design science' has yet emerged (Suh and Sharon, 1992). One of the aims of this chapter is to demonstrate the incorporation of AI-based principles into aspects of engineering design, with emphasis on the integration of MHT with other design activities.

Engineering design can be defined as a decision-making process which searches for and develops physical, operational or organizational artifacts or schemes that achieve certain goals and abide by certain constraints. The outcome of the design process is a plan or specification for producing devices, systems or procedures.

Bedworth, Henderson and Wolfe (1991, p. 135) describe current design practice as follows: 'The traditional design process is a serial process in which the design is passed through the various modules ... marketing experts give their needs to the designers, who determine product specifications, and, in turn, send their product design to the manufacturing experts, who specify the production system to make the design ... and make decisions on purchasing new equipment ... then the production system (including fabrication and assembly together with quality control) is designed ...'. Even though their book has only recently been published, materials handling and transfer are nowhere mentioned, except for the cryptic comment that '... a manufacturing axiom for many years has called attention to the fact that materials handling in manufacturing is non-productive and should be eliminated as much as possible ...' (Bedworth, Henderson and Wolfe, 1991, p. 505). Such sentiments are also echoed by Eastman (1987) and Sims (1991). This attitude, and the compartmentalized approach to engineering design, probably account for the manner in which MHT design itself is presented: MATERIAL + MOVE = METHOD (Apple, 1977), where little regarding the manufacturing environment or the contribution of MHT to this environment is taken into consideration.

Only recently has a wider approach been advocated for MHT design (Eastman, 1987; Sims, 1991), which recognizes that 'effective materials handling planning and engineering has a major impact on every component of process flow planning, package design, plant layout, manufacturing and physical distribution system design, industrial building design and many other business activities. It cuts across the whole enterprise and must be recognized as a basic business function' (Sims, 1991, p. 5). Moreover, the trend towards concurrent engineering, which 'has as its purpose to detail the [product] design while simultaneously developing production capability, field-support capability, and quality' (Bedworth, Henderson and Wolfe, 1991, p. 141), requires a more comprehensive view of engineering design. This view encompasses (cf. Halevi, 1992):

1. Product design – functions, structure, materials, quality.
2. Design for production – for fabrication, assembly, handling and transfer.
3. Process design – fabrication, assembly, machines, routing.
4. Material flow design – layout, unit loads, handling, transfer, facilities.
5. Operations design – scheduling, inventory, distribution.

This total approach enables MHT to be integrated correctly into the CE design process, not only by accepting it as a basic component of design, but, in turn, requiring other design modules to recognize the ramifications of MHT-related decisions.

10.4 AI-BASED ADVISORY DESIGN TOOLS

AI-based tools provide a new and systematic format for carrying out design processes whilst taking a larger number of factors into account. MHT design can take advantage of such methodologies, as it requires the collection and organization of a large amount of data – often unclear or imprecise – in order to come to a decision. Once these data have been accumulated into an 'active' form in a knowledge base, the resultant expert systems can fulfill several functions within the integrated procedures of concurrent design:

1. They can serve as 'virtual team members' (Krause and Ochs, 1992) and advisers during both preliminary and advanced stages of design. Whilst product and process aspects are being developed, questions can be directed to an expert system to check for negative impacts on other design modules.
2. They can suggest ways of offsetting these impacts, and thus provide feedback between design decisions in other modules and those relating to the expert system domain.
3. They can serve as expert design critics whose function it is to 'provide a critique of the user's judgment and knowledge in terms of what the program thinks is wrong with the user-proposed solution' (Silverman and Mezher, 1992).
4. They can aid in making final decisions regarding the related design module.

In order to develop engineering design expert systems we require the following basic components: a taxonomy of the domain, prioritization of the taxonomy (when required), a conceptual framework for knowledge representation and specific paradigms for the design problems involved.

10.5 A TAXONOMY OF MHT

Within the framework described above, a design domain taxonomy must incorporate two aspects – the functional environment, and the artifact to be designed.

Based on the MHT equation MATERIAL + MOVE = METHOD, Apple (1977) presents a basic taxonomy which describes eleven factors for consideration in MHT problem analysis:

1. MATERIAL	(i) type; (ii) characteristics; (iii) quantity.
2. MOVE	(iv) source and destination; (v) logistics; (iv) characteristics; (vii) type.
3. METHOD	(viii) handling unit; (ix) equipment; (x) manpower; (xi) physical restrictions.

Eastman (1987) divides his analysis into seven groups of questions:

1. What is to be moved?
2. Where does the product move from and go to?
3. Why is the material being moved?
4. Who will move the material?
5. When should the material be moved?
6. How much material must be moved?
7. What are the present layout and [production] equipment?

In order to incorporate CE thinking, and differentiate between function and artifact, we propose a rearranged and extended version of the MHT equation, and a consequent taxonomy: MATERIAL + MOVE + MANUFACTURE + MEANS = METHOD

1. MATERIAL	(i) transferability; (ii) problematics.
2. MOVE	(iii) source and destination; (iv) smoothness of flow; (v) nature of handling.
3. MANUFACTURE	(vi) continuity; (vii) capability.
4. MEANS	(viii) building and utilities; (ix) manpower; (x) maintenance.
5. METHOD (Apple, 1977; Eastman, 1987)	(i) manual handling; (ii) conveyors; (iii) hoists, cranes and monorails; (iv) industrial trucks; (v) automatic storage and retrieval.

This taxonomy provides a framework for organizing the multitude of aspects to be considered in selecting MHT equipment. In this chapter we have delineated some 50 aspects (Table 10.1), gleaned from various sources (such as Muther and Haganas, 1969; Apple, 1977; Tompkins and White, 1984; Eastman, 1987). These have been incorporated into the various expert systems detailed in the following sections. Obviously, a much fuller list can be developed. For each attribute, we have listed a series of qualitative values that can be assigned. The values are ordered such that, in our estimation,

Table 10.1 Materials handling and transfer taxonomy

(a) MATERIAL – characteristics of the material to be handled

(1) Stability of the load during handling and transfer

BATCH	Quantity per transfer batch
SHAPE	Shape of transferred item or load
VARY	Changes in shape of items in product mix
SIZE	Size of objects to be handled and transferred
CHANGE	Changes in sizes of objects in flow mix
WEIGHT	Weight of transferred item or load
STACK	Ability to nest or stack individual items
EVEN	Evenness of weight of item or load
CARRY	Load support method

(2) Problematic material properties

HAZARD	Hazardous properties of material
BREAK	Vulnerability to damage in transit
STATE	Surface condition of material

(b) MOVES – nature of the moves through the shop floor

(3) Sources and destinations

LENGTH	Distance through which item or load to be moved
PATH	Nature of source and target points
ROUTE	Nature of sources and targets
RIGID	Permanency of sources and targets

(4) Smoothness of flow during transfer

CROSS	Cross traffic
SMOOTH	Smoothness of flow between points
LINEAR	Direction of flow

(5) Nature of material handling

MANUAL	Method of loading and unloading
ENTRY	Accessibility of load and unload points
ACCESS	Accessibility of material at load/unload point

(c) MANUFACTURE – nature of the underlying production process

(6) Continuity of production flow

FLOW	Continuity of process flow
LINK	Synchronization of process and material handling
WAIT	Proportion of handling in overall flow activity
WORK	Existence of intermediate work-in-process inventory
FOLLOW	Requirement for operator to accompany load

(7) Capability required to maintain production flow

SPEED	Production throughput rate
VOLUME	Quantity of material flowing through the system
PLAN	Ability to plan production schedules and routings
LOCATE	Requirement to locate and identify load

(d) MEANS – resources available and required for handling and transfer

(8) Nature of the building and shop floor

SLOPE	Differences in height along handling route
FLOOR	Quality of floor surface

Table 10.1—*Contd.*

SPLIT	Partitioning of shop floor
STOPS	Obstacles existing along handling paths
SPACE	Availability of floor space for handling and transfer
HEIGHT	Clearance height available
AISLES	Availability of demarcated aisles for transfer
CROWD	Congestion in open spaces and aisles
COVER	Path protection against elements
ROOF	Strength of factory roof for material handling and transfer
POWER	Availability of material transfer power sources
(9) Nature of transfer equipment maintenance	
GOOD	Reliability of handling and transfer equipment
SERVE	Requirement for availability of rapid service on failure
(10) Nature of manpower for handling and transfer	
AVAIL	Availability of manpower for material handling and transfer
SKILL	Skill level of material handling and transfer personnel
SIMPLE	Standard of maintenance personnel available

they range from the more restrictive to the more comprehensive; if a solution can satisfy an attribute at a specified value, it can satisfy it at all lower values as well. This provides a 'benefit' scale for each attribute, which is utilized (1) to create a more restricted set of values by grouping levels into a 'yes' (Y – restrictive) and 'no' (N – comprehensive) dichotomy; and (2) to enable deviations to be classified as positive or negative. The full set of attributes and values is given in Table 10.A1(a) and the dichotomized values in Table 10.A3. The types of MHT equipment covered in this chapter are set out in Table 10.A1(b).

10.6 A CONCEPTUAL FRAMEWORK FOR CONSTRUCTING EXPERT SYSTEMS FOR SELECTION

Figure 10.1 illustrates a generalized framework for constructing expert systems for equipment selection. The goal of the expert system is to compare a set of attributes of the intended operating environment with a set of attributes of MHT equipment, and thereby select the most appropriate equipment type and model (Fig. 10.1a). However, MHT design requires consideration of a large number of factors; and equipment comes in a wide variety of types, models, makes and sizes. This requires an extensive set of environmental and equipment characteristics, and a highly complex pattern matching mechanism to best fit equipment to requirements. Thus it may be preferable to categorize environmental conditions and equipment capabilities (Fig. 10.1b), and combine this with several approaches to matching attributes, in order to be able to develop manageable expert systems.

(a)

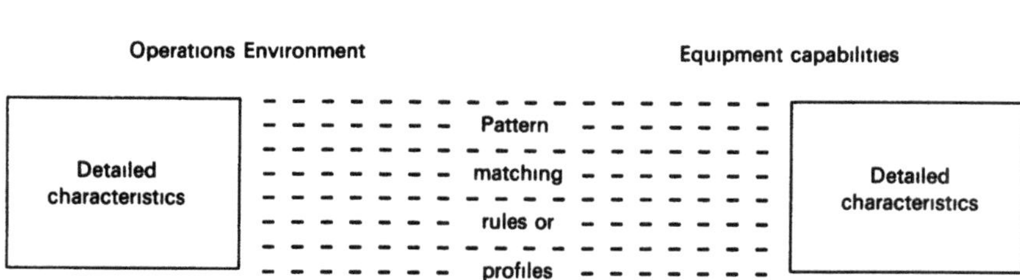

(b)

Operations Environment
Equipment capabilities
Detailed characteristics
Pattern matching
General categories
General categories
rules or profiles
Detailed characteristics

(c)

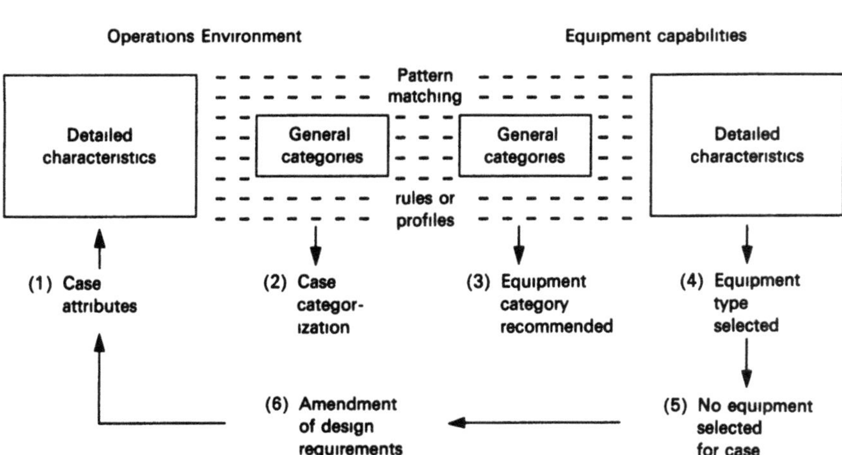

Fig. 10.1. Expert system framework. (a) Knowledge base structure – single level; (b) knowledge base structure – decomposed levels; (c) expert system consultation.

1. The outcome can be decomposed into several levels (Malmborg *et al.*, 1986). First, we determine the recommended category ('METHOD' – conveyor, truck, crane, etc.). Second, we decide upon a suitable equipment type (e.g. belt conveyor, fork-lift truck, bridge crane, etc.). Third, we select a specific model and make. An appropriate expert system can be developed and utilized at each level. In this chapter, the expert systems deal with the first two (category and type) levels.
2. Relationships between environmental and equipment factors can be expressed qualitatively and implicitly, rather than quantitatively or explicitly. For example, factors can be measured on qualitative or ordinal scales and the contribution of each equipment characteristic to supporting each environmental requirement can be expressed on an intensity scale. In this way we avoid the necessity to develop a complex quantitative model with many variables or an expert system with a complex set of decision rules.
3. The decision-making structure, as the overall design process proceeds, usually progresses from generalized and preliminary concepts and decisions to more specific ones. We may, therefore, wish to determine initially whether a particular category or type has a distinct advantage over other categories or types without having to compile and enumerate an exhaustive list of environmental attributes before coming to a decision. We can then determine how each candidate equipment type matches up to the envisaged production environment. Eventually, we can come to a definite selection decision.
4. Finally, as design is an iterative process, preliminary selection decisions may be acceptable or may conflict with designers' requirements. The expert system must provide a means to locate and resolve these conflicts, so that an ultimate decision can be reached (Fig. 10.1c).

10.7 PARADIGMS FOR EXPERT SYSTEMS FOR SELECTION

Karni and Gal-Tzur (1990, 1992) propose 10 expert system paradigms for solving a wide variety of problems in manufacturing. Three of these are appropriate for our purposes:

1. A confidence-building expert system (CBES), which asks a series of yes/no questions – in decreasing order of relevance or importance – regarding environmental factors, and develops a cumulative 'contribution' of each alternative to coping with that environment.
2. A profile-matching expert system (PMES), which asks a series of questions regarding the environment, and then calculates a weighted 'contribution' of each alternative to coping with that environment.
3. A production-rule expert system (PRES), which asks a series of questions regarding the environment, and uses the answers to guide a search – and

ask further questions – through a rule base in order to select the most appropriate alternative. The structure of these systems is described in Table 10.2.

Table 10.2 Expert system paradigms

(a) Confidence building expert system

(1) Selection environment

The environment is described by a set of characteristics with dichotomized values, such as: restricted/comprehensive or present/absent in a given situation. An item is evaluated on the basis of its cumulative compatibility with, or support it contributes to, these characteristics.

(2) Knowledge representation – facts

(i) Attributes – those characteristics of the environment by which the suitability of the candidate is determined (usually listed in order of descending importance); (ii) candidates – the possible selections or choices available to the user.

(3) Knowledge representation – rules

(i) Profiles – for each candidate, a score indicating the degree (0–5) to which the candidate is compatible with each attribute, should it have the first of the two values; and a second score indicating the degree (0–5) to which the candidate is compatible with each attribute, should it have the second of the two values; (ii) weights – the relative impact or importance of each attribute on the overall candidate evaluation.

(4) Inference

The user is asked to indicate which of the two dichotomized values best represents the environment ('yes' or 'no'). The corresponding profile score, multiplied by the normalized attribute weight, is added to an accumulated score for each candidate to give an aggregate confidence level for each candidate, on the same scale of 0–5. The candidate with the highest aggregate score, on reaching a set number of attributes or a given cumulative weight cover, is selected. Alternatively, candidates may be ranked by their aggregate scores.

(b) A profile matching expert system

(1) Selection environment

The environment is described by a set of characteristics which exist at one of several discrete levels in a given situation. An item is selected if it is closest to the preferred values of these levels.

(2) Knowledge representation – facts

(i) Attributes – those characteristics or properties of the candidates by which their suitability or relevance can be measured. The measure of each attribute is expressed by a discrete number of levels associated with the attribute; (ii) candidates – the possible selections or choices available to the user.

(3) Knowledge representation – rules

(i) Profiles – the level achievable by each candidate on each attribute; (ii) weights – the relative impact or importance of each attribute on the overall candidate evaluation.

(4) Inference

The user is asked for his/her preferred level for each attribute. The closeness score of each candidate to the required profile is measured by the sum of weighted normalized absolute deviations between the candidate and preferred profile levels.

Table 10.2—*Contd.*

This sum ranges between 0 (far off) to 1 (close match). In addition, the sum of weighted algebraic deviations is also measured. It ranges between -1 (far off, from below) through 0 (approximate match) to 1 (far, from above), and provides an indication as to whether closeness is from below or from above, and – for high negative deviations – which attributes are poorly supported by the alternative. The candidate with the highest aggregate score is selected. Alternatively, candidates may be ranked by their aggregate scores.

(c) A production-rule expert system

(1) Selection environment

The environment is described by a set of multi-valued characteristics which take on one of these values in a given situation. An item is selected on the basis of specific values of these characteristics.

(2) Knowledge representation – facts

(i) Findings – attributes or characteristics of the environment that have a specific value in a given situation or state; (ii) hypotheses – intermediate categorizations that can be inferred by the system from findings or other hypotheses, and receive a specific inferred value; (iii) classes – intermediate categorizations that are inferred by the system from findings or hypotheses, and can receive one or more specific inferred values; (iv) conclusions – the final choices or selections to be made by the system (more than one may be possible).

(3) Knowledge representation – rules

IF–THEN relations between facts such that: a set of specific values of findings and/or hypotheses serves to attribute a specific value to further hypotheses; a set of specific values of findings and hypotheses serves to assign one or more specific values to classes; and a set of specific values of findings, hypotheses and classes serves to assign one or more specific values to further conclusions.

(4) Inference

The user is requested to indicate appropriate values for the set of findings. Rules are tested against each finding or derived hypothesis or class value. For a rule with a hypothesis as consequent: if it is refuted, it is eliminated, if confirmed, the hypothesis is given a value and the rule is eliminated together with any further rules regarding the same hypothesis. For a rule with a class as consequent: if it is refuted, it is eliminated; if confirmed, the class is given a value and the rule is eliminated. Further rules regarding the same class remain active, as a class may have more than one value. This also holds for rules with a conclusion as consequent. The process continues until all conclusions have been either confirmed or rejected.

Using these three paradigms, the taxonomy detailed in Tables 10.1 and 10.A1, and attribute–candidate profiles set out in Table 10.A2, three complete MHT equipment selection knowledge bases have been developed (see Tables 10.A3–10.A5). These knowledge bases are structured as follows:

1. CBES

   Attributes — list of attributes measurable at two levels: the situation is restrictive (Y) or non-restrictive (N); and the accompanying priority weights (Table 10.A3).

Profiles	for each attribute level and each alternative: a value indicating the degree (0 – not at all; through 5 – very much so) to which each alternative is compatible with the attribute level (Table 10.A2).
Choices	list of alternative candidates (Table 10.A1a).

2. PMES

Attributes	list of attributes and their levels (Table 10.A1a).
Profiles	for each attribute and each alternative: a value indicating the level at which each alternative is compatible with the attribute (Table 10.A2).
Weights	list of attributes and their priority weights (Table 10.A7).
Candidates	list of alternative candidates (Table 10.A1b).

3. PRES

Findings	list of attributes and their levels (Table 10.A1a).
Hypotheses	intermediate categorizations of groups of attributes (Table 10.A5).
Classes	intermediate categorizations of groups of alternatives (Table 10.A5).
Selections	list of alternative candidates (Table 10.A5).
Rules	production rules for determining categorizations and selections.

Once the basic taxonomy has been built up, the expert's function is to develop attribute–candidate profiles and production rules. Profiles are created by taking each attribute–candidate pair and determining which attribute level best represents the relationship, support or contribution of the pair to providing a suitable solution. Production rules are far more difficult to formulate, as each candidate has to run the gauntlet of a large number of assertions (the IF part of the rule) in order to be selected and each attribute level should presumably appear in at least one assertion. This leads to an underlying 'combinatorial' problem regarding AND or OR combinations of assertions. Until more systematic methods enable systematic complex rule bases to be developed, their correctness and comprehensiveness has to be based on experience in using the expert system. In our case, we have first formulated rules to establish hypotheses, then classes (equipment categories), and finally candidates (equipment types). We have also checked that assertions cover all attributes – although not necessarily all attribute values. Very often the NOT assertion implicitly covers such 'missing' values.

10.8 ATTRIBUTE PRIORITY WEIGHTS

Within a design framework (as opposed, say, to a diagnostic framework), attributes represent those parameters and constraints to be taken into account both when formulating solutions and evaluating them. They can be of three kinds: restrictions that cannot be changed or vital functions that must be supported; situations that the designer feels are important, but is prepared to reconsider or modify if necessary; and requirements which he/she is exploring and has no particular attitude towards them at a given stage in the design process. Therefore, in addition to enumerating all relevant environmental attributes and relating them to the alternatives to be evaluated, it is crucial to ascribe a degree of importance of achieving or supporting the desired value of each attribute.

We therefore need to be able to rank attributes in order of importance, and incorporate this ranking both into the paradigm of the expert system and the analysis of the results provided by the system.

The most common and generally acceptable manner for ranking attributes is to ascribe a cardinal (quantitative) weight to each one. Eckenrode (1965), Saaty (1980) and Hwang and Yoon (1980) propose several methods for determining such weights:

1. Relative rating: a relative numerical weight is allocated directly to each attribute, such that the total sums to an agreed ('normalized') value.
2. Absolute rating: an absolute numerical level of importance is allocated directly to each attribute, based on a scale ranging from 'unimportant' to 'important'. The values are then normalized.
3. Comparative rating: attributes are first sequenced in descending order of importance. The most important attribute is rated at, say 100. The remaining attributes are then rated relative to this. The ratings are then normalized.
4. Qualitative paired comparisons: each pair of attributes is compared, and the more important one is noted. Each attribute is rated by the total number of times it is preferred over the second attribute. The ratings are then normalized.
5. Quantitative paired comparisons: each pair of attributes is compared, and a ratio describing the relative importance of the one to the other is noted. A complex algorithm (see below) is then used to derive normalized numerical weights.

In view of the significance of attribute weights, the large number of attributes usually associated with expert systems and the advisability of being able to show how weights have been obtained, the quantitative paired comparison method – also known as the analytical hierarchy process (AHP) (Saaty, 1980, 1982) – is probably the recommended approach for use with expert systems. The procedure is illustrated by determining weights (relevant to the case study described below) for the seven basic MHT categories. Calculations are displayed in Table 10.3. In order to facilitate computation and avoid mathematical problems associated with the AHP (Karni, Sanchez and Rao Tummala, 1990), we have used the

Table 10.3 Priority weight matrices (Saaty method); attribute CATEGORY weights (total = 1000)

Attribute	*1*	*2*	*3*	*4*	*5*	*6*	*7*	*Product*	*Root*	*Saaty*	*Final*
1 MATERIAL	1	$\frac{3}{2}$	2	$\frac{3}{2}$	7	6	0	1701	2.894	300	300
2 MOVES	$\frac{2}{3}$	1	1	1	5	4	7	$93\frac{1}{3}$	1.912	199	200
3 HANDLING	$\frac{1}{2}$	1	1	$\frac{2}{3}$	4	3	5	20	1.534	159	160
4 PROCESS	$\frac{2}{3}$	1	$\frac{3}{2}$	1	5	5	8	200	2.132	221	220
5 BUILDING	$\frac{1}{7}$	$\frac{1}{5}$	$\frac{1}{4}$	$\frac{1}{5}$	1	1	1	$\frac{1}{700}$	0.392	41	40
6 MAINTAIN	$\frac{1}{6}$	$\frac{1}{4}$	$\frac{1}{3}$	$\frac{1}{5}$	1	1	$\frac{3}{2}$	$\frac{1}{240}$	0.457	48	50
7 MANPOWER	$\frac{1}{9}$	$\frac{1}{7}$	$\frac{1}{5}$	$\frac{1}{8}$	1	$\frac{2}{3}$	1	$\frac{1}{3780}$	0.308	32	30
									9.629		1000

logarithmic least squares approach (see Karni, Feigin and Breiner, 1992) for estimating the weights.

1. Pairwise comparisons (21 in all) are made between the seven categories, using a nine-point scale with the following interpretation (Saaty, 1980): 1 – equal importance; 3 – weak preference; 5 – strong preference; 7 – demonstrable preference; 9 – absolute preference; 2–8 – intermediate preferences. In addition, we have utilized the ratio 3/2 to represent a marginal preference. These comparisons are organized into a preference matrix (Table 10.3). For example: MATERIAL is considered as being marginally more important than MOVES; 3/2 is entered in the MATERIAL–MOVE cell and 2/3 (the reciprocal) is entered in the MOVE–MATERIAL cell. BUILDING is considered to be of equal importance as MANPOWER; so 1 is entered into both cells.
2. The product of all entries along each row in the matrix is computed.
3. The *n*th root of each product is calculated, where *n* is the number of columns (and rows) in the matrix, to provide an unnormalized weight.
4. The values are normalized by the required sum-of-weights to provide the Saaty weights.
5. Finally, we have made rounding adjustments – usually to the nearest 5 or 10 – to obtain the final weights used in this chapter.

Eight Saaty matrices have been developed, one for all categories and one for each category (Tables 10.3 and 10.A6). In two cases, PROCESS and BUILDING, several attributes were considered to be very marginal (exceeding a ratio of 9) relative to the most important attribute. They were, therefore, lumped together for comparison purposes and the aggregate weight obtained was then evenly divided amongst them. These Saaty matrices provide an explicit picture as to how the designer derived his/her weights. The final values are listed in Table 10.A7.

Weights play an essential role in each of the three paradigms. In confidence building, we order the attributes in descending order of importance and decide how many are needed to cover a meaningful proportion of the total weight (say, 80 or 90%). This can reduce the amount of information required at various stages of design. Moreover, a normalized weighted score (between 0 and 5, with the same interpretation as for individual attributes) for each candidate enables us to both rank and evaluate each alternative. In profile matching, the closeness of each candidate to the desired profile is given by the sum of weighted distances from that profile and significant deviations from the desired profile can be highlighted. Finally, in production rules, the importance of an assertion leading to rejection of a rule can help make a decision as to whether that assertion should be relaxed. This subject is taken up further in Section 10.11.

10.9 CASE STUDIES

The case described in this chapter is based upon a student term project carried out under the supervision of the authors. A large maintenance facility has been established for rebuilding engines for military, industrial and agricultural vehicles. The engines are brought into the facility and disassembled at a disassembly station. Individual parts are sorted and separated into three groups for handling and transfer. They are then transferred to various processing stations in the plant. The three groups are:

1. Engine blocks. These blocks measure about 80 × 50 × 60 cm on the average and weigh about 150 kg. They are moved from the disassembly area to cleaning baths and from there to the rebuild section. Storage before and after disassembly is limited – transfer to cleaning should be as soon as possible. Both manual and automated equipment is used to handle the blocks.
2. Rework parts. These parts include pistons, shafts, cranks, valves, covers, etc. They are sorted and stacked on pallets, which may accumulate in the sorting area. Pallets are then sent to three different work areas, according to part type. After rework, parts are transferred to the rebuild section.
3. Scrap parts. These parts include bolts, nuts, spacers, springs, gaskets, etc. They are placed in containers, for delivery to a salvage area.

We created a materials handling and transfer specification in the form of a questionnaire listing the interface design attributes and their possible values (as in Table 10.A1). In addition, we filled out the Saaty matrices (Tables 10.3 and 10.A6) in accordance with our evaluation of the rebuilding operation. Environmental attribute weights and values relevant to each of the three part groups are detailed in Tables 10.4 and 10.A7. These tables constitute the design specifications for three case studies.

10.10 EXPERT SYSTEM OUTPUTS – INITIAL EQUIPMENT SELECTION

The results of consultations in all three cases using the three expert systems are detailed in Table 10.5.

10.10.1 Confidence building (Table 10.5a)

Only 17 out of the 48 attributes – one-third of the total information – are needed to cover 80% of the weight assigned to all attributes. This is quite a significant reduction in information requirements. In the first case, the best candidate, JIB, has a clear advantage at 80%, although somewhat

Table 10.4. Environmental conditions for case studies

(a) Handling of engine blocks – responses for full and restricted levels

BATCH	(1)	(Y)	Individual items
SHAPE	(2)	(Y)	Uniform shape
VARY	(2)	(Y)	Uniform shapes
SIZE	(4)	(N)	Large
CHANGE	(1)	(Y)	Fixed sizes
WEIGHT	(3)	(N)	Heavy weight
STACK	(1)	(Y)	Items cannot be nested or stacked
HAZARD	(1)	(Y)	No particular hazard problem
EVEN	(1)	(Y)	Even distribution of weight
BREAK	(1)	(Y)	Robust
STATE	(2)	(N)	Some surface problems (dirty, hot, rough edges)
CARRY	(4)	(N)	No support provided for load
LENGTH	(2)	(Y)	Moderate (usually short)
PATH	(1)	(Y)	Only point-to-point
ROUTE	(2)	(N)	Work center to work center
CROSS	(2)	(Y)	Some cross traffic to be bypassed
SMOOTH	(3)	(Y)	Variable flow tolerated
LINEAR	(1)	(Y)	Single linear flow direction
RIGID	(1)	(Y)	Permanent sources and targets
MANUAL	(2)	(Y)	Manual and automatic loading and unloading
ENTRY	(1)	(Y)	Free access to machines
ACCESS	(4)	(N)	Material requires specialized maneuvering
VOLUME	(2)	(Y)	Medium volume
FLOW	(2)	(Y)	Intermittent (but regular) flow
LINK	(3)	(N)	Material handling must be synchronized with process
WAIT	(1)	(Y)	Negligible handling activity
WORK	(3)	(N)	No work-in-process inventory tolerated
SPEED	(3)	(N)	Some sensitivity to fluctuations in material flow
PLAN	(2)	(Y)	Production schedule has some uncertainties
FOLLOW	(3)	(N)	No operator required to accompany load
LOCATE	(3)	(N)	Load must be locatable and identifiable
SLOPE	(1)	(Y)	All routes on one plane
FLOOR	(2)	(N)	Semi-smooth floors
SPLIT	(2)	(N)	Shop floor is partitioned into separate halls
STOPS	(2)	(Y)	Known and fixed obstacles along handling paths
SPACE	(2)	(Y)	Some floor space available
HEIGHT	(2)	(Y)	Medium-high ceilings
AISLES	(2)	(Y)	Demarcated wide aisles
CROWD	(2)	(N)	Somewhat congested spaces and aisles
COVER	(2)	(Y)	Most routes indoors
POWER	(2)	(N)	Power is available at restricted points
ROOF	(2)	(N)	Roof has enough strength to allow overhead operations
GOOD	(2)	(Y)	Some regular maintenance tolerable
SERVE	(3)	(N)	Rapid or in-house service required
AVAIL	(3)	(N)	Manpower is somewhat limited
SKILL	(1)	(Y)	Untrained personnel available
SIMPLE	(2)	(N)	Trained personnel available

(b) Handling of rework parts – responses for full and restricted levels

BATCH	(2)	(Y)	Unit loads
SHAPE	(3)	(N)	Irregular shape

Table 10.4.—*Contd.*

VARY	(3)	(N)	Changeable shapes
SIZE	(3)	(Y)	Medium
CHANGE	(2)	(N)	Uniform sizes
WEIGHT	(2)	(Y)	Medium weight
STACK	(1)	(Y)	Items cannot be nested or stacked
HAZARD	(1)	(Y)	No particular hazard problem
EVEN	(1)	(Y)	Even distribution of weight
BREAK	(2)	(Y)	Fairly robust
STATE	(2)	(N)	Some surface problems (dirty, hot, rough edges)
CARRY	(2)	(Y)	Load placed on wooden (light) pallet or skid
LENGTH	(2)	(Y)	Moderate (usually short)
PATH	(1)	(Y)	Only point-to-point
ROUTE	(2)	(N)	Work center to work center
CROSS	(2)	(Y)	Some cross traffic to be bypassed
SMOOTH	(3)	(Y)	Variable flow tolerated
LINEAR	(1)	(Y)	Single linear flow direction
RIGID	(1)	(Y)	Permanent sources and targets
MANUAL	(3)	(N)	Manual loading and unloading
ENTRY	(1)	(Y)	Free access to machines
ACCESS	(2)	(N)	Material requirs some maneuvering to handle
VOLUME	(3)	(N)	Relatively high volume
FLOW	(2)	(Y)	Intermittent (but regular) flow
LINK	(2)	(Y)	Material handling and process are buffered
WAIT	(2)	(Y)	Some handling activity
WORK	(2)	(N)	Short-term work-in-process inventory tolerated
SPEED	(3)	(N)	Some sensitivity to fluctuations in material flow
PLAN	(2)	(Y)	Production schedule has some uncertainties
FOLLOW	(3)	(N)	No operator required to accompany load
LOCATE	(3)	(N)	Load must be locatable and identifiable
SLOPE	(1)	(Y)	All routes on one plane
FLOOR	(2)	(N)	Semi-smooth floors
SPLIT	(2)	(N)	Shop floor is partitioned into separate halls
STOPS	(2)	(Y)	Known and fixed obstacles along handling paths
SPACE	(2)	(Y)	Some floor space available
HEIGHT	(2)	(Y)	Medium-high ceilings
AISLES	(2)	(Y)	Demarcated wide aisles
CROWD	(2)	(N)	Somewhat congested spaces and aisles
COVER	(2)	(N)	Most routes indoors
POWER	(2)	(N)	Power is available at restricted points
ROOF	(2)	(N)	Roof has enough strength to allow overhead operations
GOOD	(2)	(Y)	some regular maintenance tolerable
SERVE	(2)	(Y)	Service agency in vicinity satisfactory
AVAIL	(3)	(N)	Manpower is somewhat limited
SKILL	(1)	(Y)	Untrained personnel available
SIMPLE	(2)	(N)	Trained personnel available

(c) Handling of scrap parts – responses for full and restricted levels

BATCH	(3)	(N)	Bulk loads
SHAPE	(3)	(N)	Irregular shape
VARY	(3)	(N)	Changeable shapes
SIZE	(2)	(Y)	Small
CHANGE	(3)	(N)	Variable sizes are uniform

Table 10.4.—*Contd.*

WEIGHT	(1)	(Y)	Light weight
STACK	(2)	(N)	Items can be nested or stacked
HAZARD	(1)	(Y)	No particular hazard problem
EVEN	(2)	(N)	Uneven distribution of weight
BREAK	(1)	(Y)	Robust
STATE	(3)	(N)	Major surface problems (sticky, corrosive, sharp edges)
CARRY	(1)	(Y)	Load in containers or in suspension mechanism
LENGTH	(3)	(N)	Relatively unlimited (usually short)
PATH	(3)	(N)	Multiple source and target points
ROUTE	(3)	(N)	Area to area
CROSS	(2)	(Y)	Some cross traffic to be bypassed
SMOOTH	(3)	(Y)	Variable flow tolerated
LINEAR	(4)	(N)	Random flow directions
RIGID	(2)	(N)	Some changes in sources and targets
MANUAL	(3)	(N)	Manual loading and unloading
ENTRY	(1)	(Y)	Free access to machines
ACCESS	(1)	(Y)	Material can be handled easily
VOLUME	(3)	(N)	Relatively high volume
FLOW	(2)	(Y)	Intermittent (but regular) flow
LINK	(2)	(Y)	Material handling and process are buffered
WAIT	(3)	(N)	Significant handling activity
WORK	(2)	(Y)	Short-term work-in-process inventory tolerated
SPEED	(1)	(Y)	Material flow not a factor in determining throughput rate
PLAN	(2)	(Y)	Production schedule has some uncertainties
FOLLOW	(3)	(N)	No operator required to accompany load
LOCATE	(1)	(Y)	No requirement to locate and identify load
SLOPE	(1)	(Y)	All routes on one plane
FLOOR	(2)	(N)	Semi-smooth floors
SPLIT	(2)	(N)	Shop floor is partitioned into separate halls
STOPS	(2)	(Y)	Known and fixed obstacles along handling paths
SPACE	(2)	(Y)	Some floor space available
HEIGHT	(2)	(Y)	Medium-high ceilings
AISLES	(2)	(Y)	Demarcated wide aisles
CROWD	(2)	(N)	Somewhat congested spaces and aisles
COVER	(2)	(Y)	Most routes indoors
POWER	(2)	(N)	Power is available at restricted points
ROOF	(2)	(N)	Roof has enough strength to allow overhead operations
GOOD	(2)	(Y)	Some regular maintenance tolerable
SERVE	(2)	(Y)	Service agency in vicinity satisfactory
AVAIL	(3)	(N)	Manpower is somewhat limited
SKILL	(1)	(Y)	Untrained personnel available
SIMPLE	(2)	(N)	Trained personnel available

high (3.8 out of 5); this falls to 3.6 when all attributes are taken into account. In the second case HAND has a clear advantage at the beginning; but POWER becomes a second preferred alternative, both scoring around 3.7. In the last case both BELT and POWER are indicated at all coverages, scoring 3.9 overall. All preferred candidates exhibit high scores (around 3.8); other alternatives score far less (down to 2.3).

Table 10.5 Expert system recommendations: (a) CBES

Stage	*Belt*	*Roller*	*Power*	*Jib*	*Mono*	*Bridge*	*Hand*	*Fork*	*AGVS*
(i) MHT equipment scores for engine blocks									
After 80%	2.47	2.40	3.01	3.81	3.54	2.96	3.28	3.02	3.26
After 90%	2.72	2.66	3.11	3.64	3.49	2.86	3.29	3.02	3.36
After 100%	2.83	2.76	3.19	3.56	3.46	2.89	3.21	3.00	3.40
(ii) MHT equipment scores for rework parts									
After 80%	3.30	2.94	3.73	3.44	3.69	2.94	3.93	3.64	2.43
After 90%	3.40	3.09	3.73	3.35	3.64	2.88	3.80	3.54	2.58
After 100%	3.48	3.12	3.75	3.27	3.57	2.89	3.71	3.50	2.69
(iii) MHT equipment scores for scrap parts									
After 80%	4.11	4.03	4.13	2.09	2.62	3.06	3.47	3.50	2.69
After 90%	3.94	3.87	3.97	2.29	2.69	3.11	3.41	3.41	2.63
After 100%	3.88	3.81	3.86	2.31	2.68	3.08	3.38	3.42	2.62

Table 10.5 Expert system recommendations: (b) PMES

	Candidate	*Closeness*	*Conformance*
(i) MHT equipment ranking for engine blocks			
(BELT)	Belt conveyor	0.497	−0.067
(ROLLER)	Roller conveyor	0.536	−0.074
(POWER)	Power-and-free conveyor	0.562	0.162
(JIB)	Jib crane (fixed location)	0.350	0.012
(MONO)	Monorail	0.463	−0.037
(BRIDGE)	Bridge crane (movable)	0.333	0.179
(HAND)	Hand truck or cart	0.303	−0.058
(FORK)	Powered fork-lift truck	0.459	0.139
(AGVS)	Automated guided vehicle (floor guidance)	0.685	−0.056
(ii) MHT equipment ranking for rework parts			
(BELT)	Belt conveyor	0.493	−0.097
(ROLLER)	Roller conveyor	0.601	−0.089
(POWER)	Power-and-free conveyor	0.659	0.105
(JIB)	Jib crane (fixed location)	0.398	−0.038
(MONO)	Monorail	0.683	−0.130
(BRIDGE)	Bridge crane (movable)	0.468	0.119
(HAND)	Hand truck or cart	0.373	−0.123
(FORK)	Powered fork-lift truck	0.468	0.131
(AGVS)	Automated guided vehicle (floor guidance)	0.630	−0.030
(iii) MHT equipment ranking for scrap parts			
(BELT)	Belt conveyor	0.557	−0.063
(ROLLER)	Roller conveyor	0.512	−0.100
(POWER)	Power-and-free conveyor	0.637	0.099
(JIB)	Jib crane (fixed location)	0.410	−0.114
(MONO)	Monorail	0.585	−0.149
(BRIDGE)	Bridge crane (movable)	0.517	0.087
(HAND)	Hand truck or cart	0.514	−0.118
(FORK)	Powered fork-lift truck	0.514	0.090
(AGVS)	Automated guided vehicle (floor guidance)	0.452	−0.035

Table 10.5 Expert system recommendation summary: (c) PRES

MHT equipment selected for engine blocks	MONO or BRIDGE or FORK
MHT equipment selected for rework parts	POWER or ROLLER or HAND or FORK
MHT equipment selected for scrap parts	FORK

Table 10.5 Expert system recommendation summary: (d) Overall rating summary

	Belt	*Roller*	*Power*	*Jib*	*Mono*	*Bridge*	*Hand*	*Fork*	*AGVS*
(i) Confidence building expert system – equipment scores									
Engines	2.83	2.76	3.19	**3.56**	3.46	2.89	3.21	3.00	3.40
Rework	3.48	3.12	**3.75**	3.27	3.57	2.89	**3.71**	3.50	2.69
Scrap	**3.88**	3.81	**3.86**	2.31	2.68	3.08	3.38	3.42	2.62
(ii) Profile matching expert system – equipment ranks									
Engines	0.497	0.536	0.562	0.350	0.463	0.333	0.303	0.459	**0.685**
Rework	0.493	0.601	0.659	0.398	**0.683**	0.468	0.373	0.468	0.630
Scrap	0.557	0.512	**0.637**	0.410	0.585	0.517	0.514	0.514	0.452
(iii) Production rule expert system – equipment selected									
Engines					**	**		**	
Rework		**	**				**	**	
Scrap								**	

10.10.2 Profile matching (Table 10.5b)

In the first case, AGVS is clearly closest to the desired profile, achieving an average of 69% of requirements – from below. In the second case, MONO reaches 68% of requirements – from below – whilst POWER (66% – from above) could also be considered in this case. In the last case POWER (64% – from above) is that alternative most preferred.

10.10.3 Production rules (Table 10.5c)

The execution trace for the production rule expert system is listed in Table 10.A8. Only the first page of the trace is included, with the full trace being placed on a diskette for the interested reader. From this trace we see that, in the first case, all alternatives have been rejected. The reasons for rejection appear in Table 10.A9, which is further discussed below. In order to provide a recommendation, some constraints have to be relaxed. Three possibilities exist, provided only one assertion has led to rejection: if we allow a monorail to pass through several halls, it can be accepted; and if we are prepared to hire and train special MHT staff then a bridge crane or powered fork-lift truck can be recommended. In the second case, a power-and-free conveyor has been accepted unconditionally. If, however, we are prepared to hire and train special MHT staff, then a powered fork-lift truck can be recommended; if we are prepared to increase staff available for MHT then a hand truck or cart is suitable; and if we can solve the problem of identifying and locating loads then a roller conveyor comes into consideration. In the third case, a powered fork-lift truck is again a possibility if we are prepared to hire and train special MHT staff.

In general, then, improving the MHT staff situation can lead to greater freedom in selecting suitable MHT equipment.

10.11 USING THE EXPERT SYSTEM IN DESIGN

Integrating the expert system into the design process and coming to final recommendations is done via the following procedure:

1. Design specifications	the list of attributes and assigned values
2. Candidate performance	consulting the expert system
3. Candidate selection	determining a set of favorable candidates
4. Critique	analyzing conformance to design specifications
5. Final recommendations	design specifications and candidates

10.11.1 Input – design specifications

We regard the set of attributes and their desired values and weights as representing design specifications passed to the MHT designer. They include knowledge about materials, moves, manufacture and means, as provided by product, process and production designers. The aims of the MHT designer are then to: (1) critique the specifications and determine their acceptability to MHT design, and the ability of MHT design to conform to them; (2) indicate negative conformance-to-specification of proposed MHT equipment, and how conformance may be improved; (3) recommend changes in the specification and (4) recommend suitable MHT equipment.

10.11.2 Output – results from the expert system

The expert system provides three measures of performance for each candidate: (1) how well each rates compared to the maximum rating attainable; (2) how well each rates compared to other alternatives and (3) what factors contribute to lowering ratings. These are based on two measures: performance and non-conformance.

(a) *Confidence building*

The candidate rating (0–5), i.e. the normalized weighted profile score over all attributes 1 through j, is given by the relation

$$r_i = \sum_j w_j x_{ij} \Big/ \sum_j w_j$$

where r_i is the rating of candidate i, w_j is the weighting of attribute j and x_{ij} is the yes/no profile value of candidate i on attribute j.

The non-conformance level (0–1) of a candidate to any attribute, i.e. the weighted degree to which the profile score deviates negatively from the rated score (r_i), is given by the relation

$$d_{ij} = w_j \Big/ \sum_j w_j - w_j x_{ij} \Big/ \sum_j w_j x_{ij} \quad (r_i > x_{ij})$$

$$= \left(w_j \Big/ \sum_j w_j \right) * (1 - x_{ij}/r_i)$$

where d_{ij} is the non-conformance level of candidate i on attribute j and w_j is the weighted degree to which x_{ij}, being less than r_i, leads to a lowering of the average score.

We look for those attributes which lead to a significant level of non-conformance for a given candidate.

(b) Profile matching

The candidate rating (0–1), i.e. the weighted absolute deviation of the profile level from the desired level over all attributes, is given by the relation

$$r_i = 1 - \sum_j [w_j * |x_{ij} - x_j^*| / \max(l_j - x_j^*, x_j^* - 1)]$$

where x_j^* is the desired (design) level of attribute j and l_j is the number of levels in attribute j and the normalizing denominator expresses the maximum interval between the desired level and the first or last level for the attribute.

The non-conformance level (0–1) of a candidate to any attribute, i.e. the weighted degree to which the profile level deviates negatively from the rated level, is given by the relation

$$d_{ij} = w_j(x_{ij} - x_j^*)/\max(l_j - x_j^*, x_j^* - 1) \quad \text{for} \quad x_{ij} < x_j^*$$

We look for those attributes which lead to a significant level of non-conformance for a given candidate.

(c) Production rules

Candidates are rated by being confirmed, or the degree to which the specifications must be modified (or constraints ignored) in order to accept the candidate. The degree of non-conformance is then expressed by those constraints which are ignored in order to accept the candidate.

10.11.3 Determining favorable candidates

(a) Confidence building

Candidates are ranked by their overall scores. Those having an acceptable score (say, 3.5 or more) are considered further.

(b) Profile matching

Candidates are ranked by their overall rating. Those having an acceptable score (say, 0.65 or more) are considered further.

(c) Production rules

All accepted candidates or those with acceptable category levels (conveyor, crane, truck) or conformance violation levels are considered further.

10.11.4 Critique – analyzing conformance to design specifications

Two types of critique can be carried out: for those candidates proposed by the expert systems and for those candidates proposed *a priori* by the designer. We illustrate both approaches for the cases studied.

For each candidate to be considered we make a list of attributes with a significant level of non-conformity. We next analyze the reasons for non-conformance. Finally, we decide whether to suggest modifications to the design specification or to find other means of overcoming the disadvantages associated with poor conformance.

10.11.5 Final recommendations – design specification and candidates

Finally, we propose recommendations in accordance with the aims set out above in Section 10.11.2.

10.12 SELECTING MHT EQUIPMENT

For the three case studies, the initial design specifications are given in Table 10.4 (stage a). Output results from the expert systems are summarized in Tables 10.5 and 10.A9–A11 (stage b). From these results, initial favorable candidates are as follows (stage c):

1. Engine blocks

	BELT	ROLLER	POWER	JIB	MONO	BRIDGE	HAND	FORK	AGVS
CBES				(1)	(2)				
PMES									(1)
PRES					(2)	(1)		(1)	

2. Rework parts

	BELT	ROLLER	POWER	JIB	MONO	BRIDGE	HAND	FORK	AGVS
CBES	(2)		(1)				(1)	(2)	
PMES			(2)		(1)				
PRES		(2)	(1)				(2)	(2)	

3. Scrap parts

	BELT	ROLLER	POWER	JIB	MONO	BRIDGE	HAND	FORK	AGVS
CBES	(1)	(1)	(1)						
PMES			(1)		(1)				
PRES								(1)	

A power-and-free conveyor seems to be that most favored for rework and scrap parts; there is less consistency between the expert systems in selecting equipment for handling engine blocks.

Stage (d): critique – analyzing conformance to design specifications. Deviations from specification for all equipment types recommended by the three expert systems are detailed in Table 10.A9. For each possible candidate we need to make a list of the reasons for low conformance to specification, and decide, in conjunction with the other phases of design, how to deal with these deviations. We demonstrate the approach in two ways.

(1) We discuss the adoption of a power-and-free conveyor system for all cases, even though this alternative has not been suggested for engine blocks by any of the expert systems. This is equivalent to the designer wishing to find out whether making an *a priori* selection of a power-and-free conveyor for engine blocks – and thus for all three cases – is advisable. In the other two instances, he/she wishes to study the interaction between the recommendation and the design specifications.

To this end, as a first step, low-conformance attributes detected for power-and-free conveyors by the three expert systems, for all cases, have been concentrated in Table 10.A10. For confidence building and profile matching, negative deviations above 0.015 have been considered significant; for production rules, assertions which have led to rejection of power-and-free conveyors – a traceback through the inference process – have been listed. As a second step, these attributes have been analyzed in Table 10.6, which compares power-and-free capabilities to specifications, and suggests corresponding recommendations or reactions. These are then fed back into the design process for consideration.

As a third step, from the observations in Table 10.6, we can summarize our comments as follows. Regarding engine blocks, it is clear that transporting heavy items would require far stronger equipment than that needed for rework or scrap parts; and that added specialized equipment would be necessary to integrate loading and unloading with the processing of these blocks. This would probably preclude the use of this type of equipment. Regarding rework and scrap parts, appropriate design of the material carriers to both accommodate the kinds of parts stripped from the engines, and facilitate swift loading and unloading, would enable a power-and-free conveyor system to be a successful solution for MHT.

(2) Returning to engine blocks: the production rule expert system has come out in favor of a bridge crane. This may appeal to the plant designers, as it has the advantage of facilitating immersion and withdrawal of the blocks from the cleaning baths. Here, then, we wish to investigate the desirability of proposing a bridge crane for MHT operations on engine blocks. Low conformance is analyzed in Table 10.A11, and recommendations are outlined in Table 10.7. The two main drawbacks seem to be the necessity for an appropriate carrier design to handle the blocks rapidly and efficiently, and the employment of suitable bridge crane operators.

Stage (e): final recommendations – design specification and candidates. We summarize our final recommendations as follows:

1. A bridge crane is recommended for handling engine blocks and a power-and-free conveyor for handling rework and scrap parts.
2. Suitable carriers need to be designed for both equipment types.
3. Trained operating personnel are required for the bridge crane.

Table 10.6 Analysis of non-conformance to specifications (power-and-free)

Aspect	*Specifications*	*Power-and-free capability*	*Recommendations (descending order of severity)*
(i) Environmental conditions for handling of engine blocks			
Loading access	Special maneuver	Easy access	Probably demands sophisticated load-unload equipment
Load support	None provided	Pallet or skid	Probably demands sophisticated load-unload equipment
Loading method	Automated	Manual	Probably demands sophisticated load-unload equipment
Part weight	Heavy parts	Medium weight	Design appropriate carriers to carry heavy parts
Part size	Large parts	Medium parts	Design appropriate carriers to carry large parts
Move distance	Short distances	Long distances	Check economics of a short-distance configuration
Move routing	Point-to-point	Multiple points	Can provide flexibility for location of work stations
Flow volume	Medium volume	Medium volume	Constraint can be ignored for this type of conveyor
(ii) Environmental conditions for handling of rework parts			
Product shape	Irregular shape	Regular shape	Design appropriate carriers to support irregular shapes
Flow volume	Relatively high	Medium volume	Design appropriate carriers for fast loading and unloading
Loading access	Some problems	Easy access	Load-unload procedures should be studied and improved
Move distance	Short distances	Long distances	Check economics of a short-distance configuration
Move routing	Point-to-point	Multiple points	Can provide flexibility for location of work stations
(iii) Environmental conditions for handling of scrap parts			
Flow volume	Relatively high	Medium volume	Design appropriate carriers for fast loading and unloading
Load unit	Bulk loads	Unit loads	Design appropriate carriers to support bulk loads
Product shape	Irregular shape	Regular shape	Design appropriate carriers to support irregular shapes
Part size	Small parts	Medium parts	Design appropriate carriers to carry small parts
Part surface	Major problems	Some problems	Design appropriate carriers to protect against hazards
Load location	No requirement	Tagging needed	Load tagging required
Move routing	Multiple points	Point-to-point	Constraint can be ignored for this type of conveyor

Table 10.7 Analysis of non-conformance to specifications (bridge conveyor) (environmental conditions for handling of engine blocks)

Aspect	*Specifications*	*Bridge Conveyor capability*	*Recommendations (descending order of severity)*
Loading method	Automated	Manual	Probably demands sophisticated load-unload equipment
Part weight	Heavy parts	Medium weight	Design appropriate carriers to carriers to carry heavy parts
Part size	Large parts	Medium parts	Design appropriate carriers to carry large parts
Flow volume	Medium volume	Low volume	Design appropriate carriers for fast loading and unloading
Operating skill	Untrained personnel	Highly trained personnel	Requires suitable operators to be hired and/or trained
Flow regularity	Intermittent flow pattern	Irregular flow pattern	Design appropriate scheduling procedures for block MHT
Load location	Tagging needed	Not usually	Load tagging required
Move distance	Short distances	Long distances	Check economics of a short-distance configuration
Load handling	Special maneuver	Difficult maneuver	Constraint can be ignored for this type of conveyor
Loading access to machines	Free access	Impeded access	Constrain can be ignored for this type of conveyor
Product shape	Regular shape	Irregular shape	Constraint can be ignored for this type of conveyor
Accompanying the load	None required	Move monitored by operator	Constraint can be ignored for this type of conveyor

4. Efficient scheduling procedures must be developed to maintain the desired process flow.

10.13 DISCUSSION

This chapter has presented a detailed description of the use of expert systems for the selection of MHT equipment types within a concurrent design framework. It has also set out a design procedure, based upon expert systems, which provides both an appropriate design and feedback regarding difficulties in conforming to design specifications. This procedure may be summarized in the following steps:

1. Define the design domain.
2. Develop a taxonomy of the design domain.
3. Enumerate the attributes of the design domain.
4. Enumerate the possible value levels for the attributes.
5. Define the alternative designs to be selected.
6. Describe the actual design environment as a specific set of attribute values.
7. Assign a specific set of priority weights for the attributes.
8. Run the expert system.
9. Determine which design alternatives are acceptable.
10. Determine low-conformance attributes for each alternative.
11. Decide whether low conformance or conflicting constraints may be modified or ignored.
12. Provide a recommendation of one (or more) preferred design alternative.
13. Provide recommended changes to the design specification.

REFERENCES

Apple, J.M. (1977) *Plant layout and material handling*, Wiley & Sons, New York.

Bedworth, D.D., Henderson, M.R. and Wolfe, P.M. (1991), *Computer-integrated design and manufacturing*, McGraw-Hill, New York.

Eastman, R.M. (1987) *Materials handling*, Marcel Dekker, New York.

Eckenrode, R.T. (1965) Weighting multiple criteria. *Management Science (Series A)*, **12**(3), 180–92.

Gabbert, P.S. and Brown, D.E. (1988) *General Electric: A Case Study in Materials Handling System Design*. 1988 International Industrial Engineering Conference Proceedings, pp. 153–8.

Halevi, G. (1992) *Lecture Notes*. IFIP Working Groups 5.2, 5.3 and CATE/TF Workshop Manufacturing in the Era of Concurrent Engineering, Herzlia, Israel.

Hosni, Y.A. (1989) Inference engine for materials handling selection. *Computers and Industrial Engineering*, **17**(1–4), 79–84.

Hwang, C.L. and Yoon, K. (1980) *Multiattribute Decision Making—Methods and Applications, Lecture Notes in Economics and Mathematical Systems 186*, Springer Verlag, Berlin.

Karni, R. and Gal-Tzur, A. (1990) Paradigms for knowledge-based systems in industrial engineering. *International Journal of Artificial Intelligence in Engineering*, **5**(3), 126–41.

Karni, R. and Gal-Tzur, A. (1992) Frame-based architectures for manufacturing planning and control. *International Journal of Artificial Intelligence in Engineering*, to appear.

Karni, R., Sanchez, P. and Rao Tummala, V.M. (1990) A comparative study of multiattribute decision making methodologies. *Theory and Decision*, **29**, 203–22.

Karni, R., Feigin, P. and Breiner, A. (1992) Multicriterion issues in energy policy-making. *European Journal of Operational Research*, **56**(1), 30–40.

Krause, F.-L. and Ochs, B. (1992) *Potential and Advanced Concurrent Engineering Methods*. IFIP Working Groups 5.2, 5.3 and CATE/TF Workshop Manufacturing in the Era of Concurrent Engineering, Herzlia, Israel.

Malmborg, C.J., Simons, G.R. and Agee, M.H. (1986) *Knowledge Engineering Approaches to Material Handling Equipment Specification*. 1986 Fall Industrial Engineering Conference Proceedings, pp. 148–51.

Malmborg, C.J., Krishnakumar, B., Simons, G.R. and Agee, M.H. (1987) EXIT: a PC-based expert system for industrial truck selection. *International Journal of Production Research*, **27**(6), 927–41.

Matson, J.O., Swaminathan, S.R. and Mellichamp, J.M. (1990) *Knowledge-Based Material Handling Equipment Selection*. 1990 International Industrial Engineering Conference Proceedings, pp. 212–17.

Muther, R. and Haganas, K. (1969) *Systematic Handling Analysis*, Management and Industrial Research Publications, Kansas City, MO.

Saaty, T.L. (1980) *The Analytical Hierarchy Process in Planning, Priority-setting and Resource Allocation*, McGraw-Hill, New York.

Saaty, T.L. (1982) *Decision Making for Leaders*, Lifetime Learning Institute, Belmont CA.

Silverman, B.G. and Mezher, T.M. (1992) Expert critics in engineering design: lessons learned and research needs. *AI Magazine*, **13**(1), 45–62.

Sims, R.E. (1991) *Planning and Managing Industrial Logistics Systems*, Elsevier, Amsterdam.

Suh, N.P. and Sharon, A. (1992) *The Biggest Bottleneck in Concurrent Engineering*. IFIP Working Groups 5.2, 5.3 and CATE/TF Workshop Manufacturing in the Era of Concurrent Engineering, Herzlia, Israel.

Tompkins, J.A. and White, J.A. (1984) *Facilities Planning*, John Wiley, New York.

CHAPTER 11

A knowledge-based system for scheduling in a flexible manufacturing system

Suranjan De and Anita Lee

11.1 INTRODUCTION

The production flexibility and efficiency of a flexible manufacturing system (FMS) is made possible by having (1) a set of multi-purpose workstations capable of performing many different operations with minimum set-up time between successive operations; (2) an automated transport/material handling system that enables parts to move between any pair of workstations; and (3) a network of supervisory computers to direct and coordinate simultaneous part processing. These features account for additional complexity in managing an FMS production as opposed to that of flow shop or job shop. This chapter describes a knowledge-based approach to modeling the essential features and interactions within a highly complex system such as FMS. The resulting knowledge-based system (KBS) allows us to investigate the use of an 'intelligent' search strategy based on filtered beam search to solve scheduling problems in FMSs.

This chapter is organized as follows: Section 11.2 defines the scheduling problems addressed by our KBS; Section 11.3 describes the building blocks of our KBS: (1) the knowledge base, (2) the problem solver and (3) the user interface; Section 11.4 illustrates in detail how the different types of knowledge in the knowledge base is represented; Section 11.5 describes how the stored knowledge is accessed and utilized in production scheduling with the use of an example; finally, the implementation aspects and versatility of our KBS in modeling various different FMS environments are discussed in Section 11.6.

11.2 FMS SCHEDULING PROBLEM

Given an FMS with (1) m workstations, each of which can perform a set of non-assembly-type and/or assembly-type operations at a different rate; (2) one or more products to be produced, where each product consists of a certain number of parts and/or subassemblies that give rise to j jobs; and (3) each job has a linear precedence constraint defined among its operation requirements, the FMS scheduling problem is to sequence the jobs to enter the FMS and route them from one workstation to another to complete the operation requirements of the jobs such that the precedence constraints are satisfied while achieving certain objectives such as minimum makespan, minimum lateness and minimum tardiness.

We consider only the deterministic (the quantities needed to define the problem are known and fixed) and static (number of jobs and their ready times are known and fixed) aspects of the FMS scheduling problem by observing the following assumptions: (1) the number of workstations and their capabilities are known and fixed, (2) all workstations are available at the same time and remain available throughout the scheduling period, (3) all processing times are sequence-independent, (4) no pre-emption is allowed, i.e. each operation once started must be completed without interruption, (5) the number of jobs to be scheduled is known and fixed, (6) each job has a known and fixed set of operation requirements and (7) no job cancellation is allowed. Furthermore, depending on whether assembly-type operations are allowed or not, two types of scheduling problems can be defined. If no assembly operation is required, as in a flexible machining system, the jobs can be assumed to be independent and available at the same release time. This assumption allows us to model the problem as a *single-stage* scheduling problem with general resource structure and *linear* precedence constraints. On the other hand, if assembly operations are required, as in a flexible assembly system, the jobs are related to one another by means of a general precedence structure according to the assembly requirements. The resulting problem can be characterized as a *multi-stage* scheduling problem with general resource structure and *general* precedence constraints. In solving the latter scheduling problem, the additional complexity introduced by the job precedence constraints is reduced by decomposing the original multi-stage problem into a number of single-stage problems using an integrated hierarchical approach, as described later on in Section 11.5.

11.3 KBS ARCHITECTURE

Our KBS is made up of three components: (1) a knowledge base that captures both factual and procedural knowledge about an FMS and is represented in such a way as to permit storage and retrieval efficiencies; (2) a problem solver that stores various types of solution strategies and/or algorithms to make use of the knowledge in the knowledge base in problem solving; and (3) a user interface that facilitates the use of the system to solve problems, explain system behavior and improve or increase the repository of knowledge accumulated in the knowledge base. The architecture of our KBS is shown in Fig. 11.1.

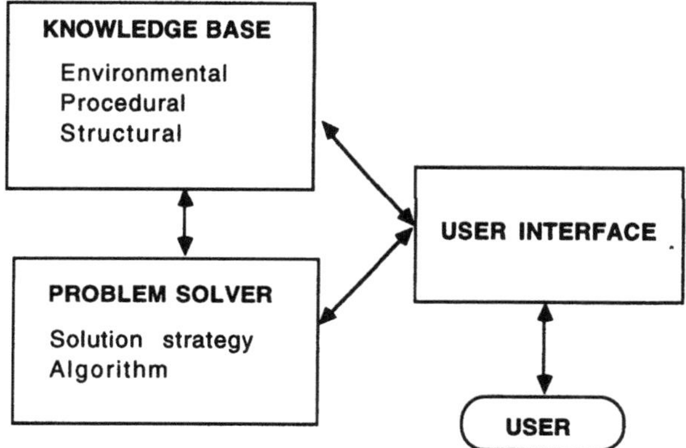

Fig. 11.1. Architecture of our KBS.

Three categories of knowledge are represented in the knowledge base:

1. Environmental knowledge. This includes knowledge about products to be scheduled, operation requirements for each product, operations performed on each machine and structure of the products. It represents the factual information about the various aspects of an FMS.
2. Procedural knowledge. This includes algorithms for node expansion, heuristic used in node generation and node evaluation, and dispatching rules used in conflict resolution. It represents the users' strategies in production scheduling.
3. Structural knowledge. This includes search nodes and the Gantt chart. It represents information generated during the solution process, which allows the user to examine the intermediate stages of the solution process and to fine tune or revise an existing schedule.

The problem solver stores a library of solution strategies that will take advantage of the knowledge of the problem domain to construct a production schedule selectively and efficiently. It contains (1) a search strategy called filtered beam search; (2) a two-level hierarchical solution approach in which a rough cut schedule based on critical path analysis is generated to reduce the problem complexity in the first phase (this rough cut schedule is then fed into a detailed scheduling phase giving rise to a complete schedule); and (3) a post analysis to seek schedule improvements.

The user interface is in the form of a menu. Users can maintain the knowledge base, initiate and manipulate the scheduling process, and examine system generated schedules through different choices provided by the

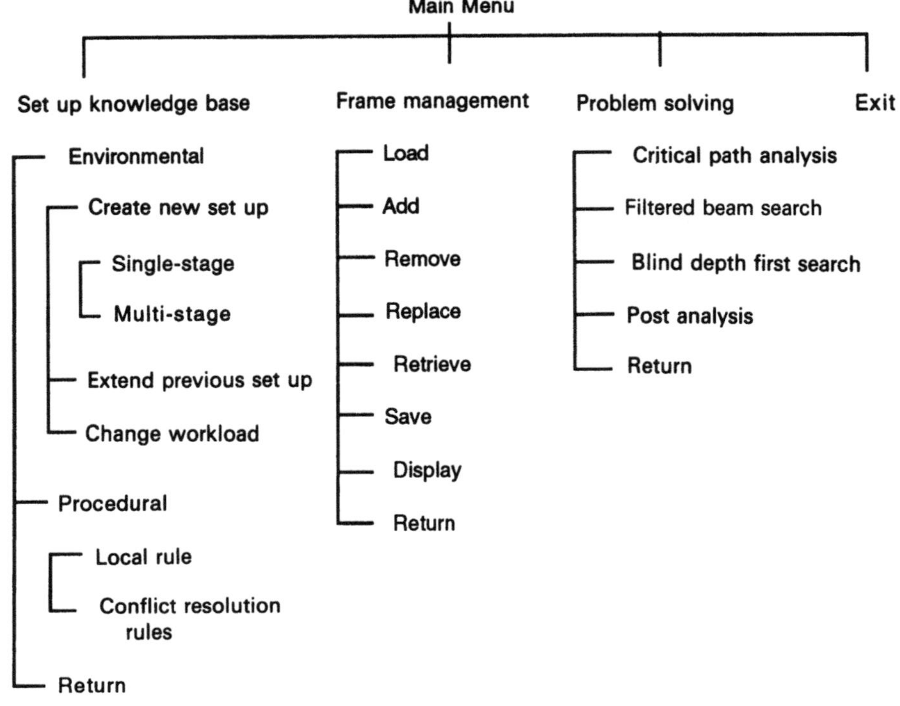

Fig. 11.2. System functions available to users.

menu. The system functions available to users through the user interface is displayed in Fig. 11.2.

11.4 KNOWLEDGE REPRESENTATION

Frames are chosen as the representation scheme in our KBS. Frames are put forward by Minsky (1975) as a formal language for expressing all forms of knowledge as a nested association list with up to five levels of embedding:

```
(frame
        (slot (facet (datum (label message...))
                      (datum...)...)
              (facet...)...)
                          ...)
        (slot...)
...)
```

A *slot* is essentially an attribute of a frame. It can have multiple values and a set of properties called *facets*. The value of a slot is a *datum* under a

$VALUE facet. The data in slots may be function calls which may be evaluated automatically – this is known as *procedural attachment*. Nested substructures, such as the (*label message*) list after a datum, are optional. Four powerful features are associated with such a scheme, i.e. inheritance, demon, default and perspective. Inheritance provides a mechanism to guide description movement from class descriptions to individual descriptions. Demons are procedures for computing descriptions; this feature is also known as procedural attachment. Default allows one to determine descriptions in the absence of specific knowledge. Perspective enables one to derive descriptions from context. Several frame-based programming and representational languages have been developed, e.g. KRL (Bobrow and Winograd, 1977), FRL (Goldstein and Roberts, 1977) and KLONE (Brachman, 1978).

Generally speaking, knowledge representation techniques provide a structure in which knowledge can be stored. They allow the system to understand how pieces of knowledge are related so as to manipulate those relationships. The choice of a representation technique depends on how the designer thinks about the knowledge in question and which representation method leads itself most efficiently to retrieving and deducing the information. Frames are chosen here as the representation scheme because they store a large amount of inter-related information in a hierarchical manner. Information common to a class of objects can be stored with the 'parent' frame to avoid unnecessary duplication of information in the 'child' frame, thereby saving storage space and alleviating the update anomalies problem. Moreover, the hierarchy of object types is itself a piece of important information for certain purposes, such as generating the search tree in scheduling. The following sections illustrates how frames might be used to represent environmental, procedural and structural knowledge in our knowledge base.

11.4.1 Environmental knowledge

Environmental knowledge captures the factual aspects of an FMS such as the products to be scheduled, operation requirements for each part, and operations performed at each workstation. We introduce the notion of ORDER as a request for the production of a given number of a known product; the notion of PRODUCT to store the information of the product structure, i.e. what parts are needed and the order in which they are assembled; the notions of PARTS and SUBASSEMBLY to store the operation sequence required to produce the parts and subassemblies, respectively; finally, the notion of JOB to represent each part or subassembly to capture its unique routing sequence. In addition to the information above, the knowledge base also keeps track of the resources and their capabilities in two entities called RESOURCE and ACTIVITY. The scheme of the environmental knowledge is shown in Fig. 11.3.

Each of the entities such as ORDER, PRODUCT, PARTS, JOBS, RESOURCE, etc., is represented by a corresponding frame with the same

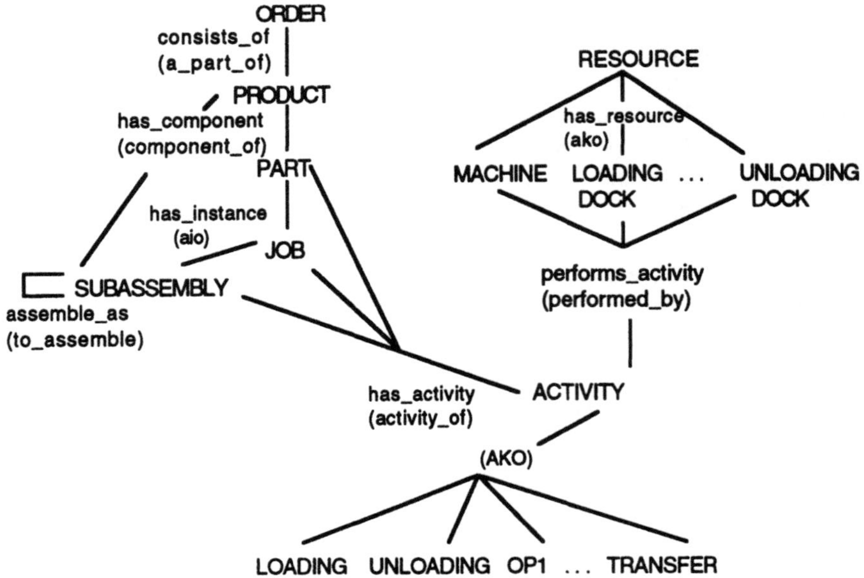

Fig. 11.3. Schema of the environmental knowledge.

name. An arc represents the relationship between a pair of frames. The name of the relationship is given next to the arc. Hence PRODUCT1 (an instantiation of PRODUCT) 'has_component'(s) PART1, PART2, etc. The inverse relationship between a pair of frames is given in parentheses. Hence, PART1 is a 'component_of' PRODUCT1. Examples of the generic frame structures for these entities and their instantiations are given below:

Generic frame structure	Instantiation
ORDER	
(*Order_Name*	(*ORDER1*
(AKO ($VALUE (*ORDER*)))	(AKO ($VALUE (*ORDER*)))
(CONSISTS_OF	(CONSISTS_OF
($VALUE (*product-name order_size*))	($VALUE (*(PRODUCT1 10) (PRODUCT2 5)*))
($IF_ADDED (*procedure to create product frame*))	($IF_ADDED (*ADD_CHILD*))
($IF_REMOVED (*procedure to remove product frame*)))	($IF_REMOVED (*REMOVE_ CHILD*)))
(PRIORITY ($IF_NEEDED (procedure to retrieve priority information))))	(PRIORITY ($IF_NEEDED (ASK (TYPE:REQUEST)))))

PRODUCT

```
(Product_Name
  (AKO ($VALUE (PRODUCT)))
  (A_PART_OF
    ($VALUE (Order_Name))
    ($IF_ADDED (procedure to add
                order frame))
    ($IF_REMOVED (procedure to
                  update order
                  frame)))
  (HAS_COMPONENT
    ($VALUE (Part/Subassembly_
             Name))
    ($IF_ADDED (procedure to add
                part/subassembly
                frame))
    ($IF_REMOVED (procedure to
                  remove part/
                  subassembly
                  frame))))
```

```
(PRODUCT1
  (AKO ($VALUE (PRODUCT)))
  (A_PART_OF
    ($VALUE (ORDER1))
    ($IF_ADDED (ADD_PARENT))
    ($IF_REMOVED
      (UPDATE_PARENT)))
  (HAS_COMPONENT
    ($VALUE (PART1 PART2 A1))
    ($IF_ADDED (ADD_CHILD))
    ($IF_REMOVED
      (REMOVE_CHILD))))
```

PART

```
Part_Name
  (AKO ($VALUE (PART)))
  (COMPONENT_OF
    ($VALUE (Product_Name))
    ($IF_ADDED (Procedure to add
                product frame))
    ($IF_REMOVED (Procedure to
                  update pro-
                  duct frame)))
  (ASSEMBLE_AS
    ($VALUE (Subassembly_Name))
    ($IF_ADDED (Procedure to add sub-
                assembly frame))
    ($IF_REMOVED (Procedure to
                  remove
                  subassembly
                  frame)))
  (HAS_INSTANCE
    (VALUE (Job_Name))
    ($IF_ADDED (Procedure to
                add job frame))
    ($IF_REMOVED (Proc. to
                  remove job
                  frame)))
  (HAS_ACTIVITY
    ($VALUE (activity precedence))))
```

```
(PART1
  (AKO ($VALUE (PART)))
  (COMPONENT_OF
    ($VALUE (PRODUCT1))
    ($IF_ADDED (ADD_PARENT))
    ($IF_REMOVED
      (UPDATE_PARENT)))
  (ASSEMBLE_AS
    ($VALUE (A1))
    ($IF_ADDED (ADD_CHILD))
    ($IF_REMOVED
      (REMOVE_CHILD)))
  (HAS_INSTANCE
    ($VALUE (JOB1))
    ($IF_ADDED (ADD_CHILD))
    ($IF_REMOVED
      (REMOVE_CHILD))
  (HAS_ACTIVITY
    ($VALUE ((LOADING 1)
      (OP2 2)))))
```

SUBASSEMBLY

```
(Subassembly_Name
  (AKO ($VALUE
    (SUBASSEMBLY)))
  (TO_ASSEMBLE
    ($VALUE (Part/Subassembly_
              Name))
    ($IF_ADDED (Procedure to add
                 product frame))
    ($IF_REMOVED (Procedure to
                   remove
                   product
                   frame)))
  (HAS_INSTANCE
    ($VALUE (Job_Name))
    ($IF_ADDED (Procedure to
                 add job frame))
    ($IF_REMOVED (Proc. to
                   remove job
                   frame)))
  (HAS_ACTIVITY
    ($VALUE (activity precedence))))
```

```
(A1
  (AKO ($VALUE
    (SUBASSEMBLY)))
  (TO_ASSEMBLE
    ($VALUE (PART1 PART2))

    ($IF_ADDED (ADD_PARENT))

  ($IF_REMOVED
    (UPDATE_PARENT)))

  (HAS_INSTANCE
    ($VALUE (JOB1))
    ($IF_ADDED (ADD_CHILD)))

    ($IF_REMOVED
      (REMOVE_CHILD)))

  (HAS_ACTIVITY
    ($VALUE ((OPa1 1) (OPa3 2)))))
```

JOB

```
(Job_Name
  (AIO ($VALUE (Part/Subassembly_
                 Name)))
    ($IF_ADDED (Procedure to add
                 part/subassembly
                 frame))
    ($IF_REMOVED (Procedure to
                   update part/
                   subassembly
                   frame)))
```

```
(JOB1
  (AIO ($VALUE (PART1)))

    ($IF_ADDED (ADD_PARENT))

    ($IF_REMOVED
      (UPDATE_PARENT)))
```

ACTIVITY

```
(Activity_Name
  (AKO ($VALUE (ACTIVITY)))
  (REQUIRE_RESOURCE
    ($IF_NEED (Procedure to
                retrieve resource
                frame)))
  (PRECONDITION ($VALUE
    (condition)))

  (CONSEQUENT ($VALUE
    (condition))))
```

```
(LOADING
  (AKO ($VALUE (ACTIVITY)))
  (REQUIRE_RESOURCE
    ($IF_NEEDED
      (GET_RESOURCE)))

  (PRECONDITION
    ($VALUE (LOADING_DOCK:
      IDLE)))
  (CONSEQUENT ($VALUE
    (LOADING:DONE))))
```

```
RESOURCE
'Resource_Name
  (AKO ($VALUE (RESOURCE)))
  (PERFORMS_ACTIVITY
    ($VALUE ((Activity_Name
      Processing_time)))))
```

```
(M1
  (AKO ($VALUE (RESOURCE)))
  (PERFORMS_ACTIVITY
    ($VALUE ((OP1 10 (:UNIT MIN))
             (OP2 20 (:UNIT
               MIN)))))
```

Frames can be organized into taxonomies by using two constructs that represent relationships between frames: AKO (short for 'A Kind Of') or subclass links, representing class containment or specialization, and AIO (short for 'An Instance Of') or member links, representing class membership. AKO is a relationship between generic objects or between categories of objects, whereas AIO is a relationship between an individual and its category. An individual can inherit properties from the more generic category frame. JOB1, for example, could inherit information such as activity and product reference from its generalization, the PART1 frame. Frames can, therefore, be related hierarchically for more economical storage of information.

Frames provide various ways of attaching procedural information expressed in some other language (e.g. LISP) to frames. This procedural capability enables one to model how objects might behave in an application domain. As an illustration of procedural attachment, in the first case, consider the $IF_ADDED and $IF_REMOVED facets in the ORDER frame. The datum in the $IF_ADDED and $IF_REMOVED facets is a procedure which will be triggered when the frame is created and deleted respectively. The procedure creates/removes an appropriate number of different types of PRODUCT frames as determined by its order size. In other words, this is a mechanism for triggering procedures that supports the operation of creating/removing an instance of a frame and filling in values for its slots. In the second case, consider the $IF-NEEDED facet in the ACTIVITY and ORDER frames. The $IF_NEEDED facet allows access to procedures for supplying values through value inheritance in the ACTIVITY frame and through user interaction in the ORDER frame.

There are many features in our existing implementation which are not explicitly demonstrated in this small example. Consider, for example, the DEFAULT feature. One could specify that priority associated with a product, unless otherwise stated, could be characterized as 'REGULAR'. This information could be stored in the generic ORDER frame (e.g. as in ORDER1). In that case we do not need to store that information for each product that must be produced to complete the ORDER. If only one of the products in ORDER1 (say, PRODUCT1) has a different priority (say, 'URGENT'), then that information could be explicitly stored as the value of the PRIORITY slot of PRODUCT1. This would then override the default value.

11.4.2 Procedural knowledge

It is characteristic of any heuristic search strategy to use information about the problem domain (heuristic information) to reduce the search effort. In our search mechanism, the heuristic information used can be classified into two categories: (1) local rules and (2) conflict resolution rules. The two sets of rules together govern how the search process progresses. The following are typical examples of local rules:

Rule 1.1. IF a job is ready
THEN start first operation

Rule 1.2 IF an operation of a job is complete
THEN start next operation

Rule 1.3 IF the last operation of a job is complete
THEN the job is done

These rules are local, since decisions are made independent of the status of other jobs. However, since conflicts are inevitable, we use conflict resolution rules to resolve them. The following are typical examples of conflict resolution rules:

Rule 2.1 IF more than one job is waiting for a resource
THEN choose the job with the least slack

Rule 2.2 IF more than one resource is capable of performing an operation
THEN assign the resource that would finish the operation earliest

In our knowledge base, procedural knowledge in the form of the rules above is also stored as frames as shown below:

Generic frame structure	Instantiation
LOCAL RULE (*Rule_Name* (AKO ($VALUE (*LOCAL_RULE*))) (PRECONDITION ($VALUE (*if portion*))) (ACTION ($VALUE (*then portion*)))	 (*RULE1.1* (AKO ($VALUE (*LOCAL_RULE*))) (PRECONDITION ($VALUE (*JOB:READY*))) (ACTION ($VALUE (*START_FIRST_OP*))))
CONFLICT RESOLUTION RULE (*Rule_Name* (AKO ($VALUE (*CONFLICT_RESOLUTION_RULE*))) (PRECONDITION ($VALUE (*if portion*))) (ACTION ($VALUE (*then portion*))))	 (*RULE2.1* (AKO ($VALUE (*CONFLICT_RESOLUTION_RULE*))) (PRECONDITION ($VALUE (>*JOB_WAITING 1*))) (ACTION ($VALUE (*SLACK (:BEST LEAST)*)))))

There are several advantages to expressing knowledge declaratively as rules: (1) a rule's use in a specific circumstance can be automatically explained and defended to the system's users; (2) developers and users can realistically expect to make changes to a few rules without affecting the entire system; and (3) new knowledge can be added to the system simply by adding new rules – this is essential to a system that is designed to learn from new knowledge or from past experiences.

11.4.3 Structural knowledge

In generating a feasible schedule, our solution mechanism uses both the environmental knowledge and the procedural knowledge described above. There is a third category of knowledge, which we refer to as structural knowledge, generated during the solution process that we find useful to represent explicitly using frames. This category of knowledge differs from the other two in that it is related to solution generation process and can be discarded once the solution is found. It includes search nodes, resource assignment and the Gantt chart. Search nodes provide information about the state of the system at a given node. The Gantt chart is a collection of resource assignment frames representing the assignment of machines and other resources as dictated by the scheduling algorithm. The frame representations of these knowledge are:

Generic frame structure	Instantiation
SEARCH NODE	
(*Node_Name*	(*NODE1*
(AKO ($VALUE (*SEARCH_NODE*)))	(AKO ($VALUE (*SEARCH_NODE*)))
(LEVEL_NUMBER ($VALUE (*depth of node*)))	(LEVEL_NUMBER ($VALUE (*1*)))
(BRANCH_NUMBER ($VALUE (*spread of node*))	(BRANCH_NUMBER ($VALUE (*1*)))
(PARENT_NODE ($VALUE (*ancestor*))	(PARENT_NODE ($VALUE (*NODE0*)))
(STATUS_*Job_Name* ($VALUE (*new_state resource completion_ time*))	(STATUS_*JOB1* ($VALUE (*LOADING :DONE LOADING_ DOCK 1*))) (STATUS_*JOB2* ($VALUE (*JOB :READY 1*)))
(EST_TF_*Job_Name* ($VALUE (*estimated finish time*))	(EST_TF_*JOB1* ($VALUE (*74.5*))) (EST_TF_*JOB2* ($VALUE (*84.7*)))
(*Resource_Name* ($VALUE (*:idle* (*start_interval end_interval*))))	(LOADING_DOCK ($VALUE (*:IDLE (1*)*)))
(OPERATOR_TO_APPLY ($VALUE (*Job_Name Operation Resource Start Finish*)))	(OPERATOR_TO_APPLY ($VALUE (*(JOB1 OP1 M3 3 29)* *(JOB2 LOADING LOADING_ DOCK 1 2)*))

```
  NODES_CREATED ($VALUE
    (successor))))
```

```
  NODES_CREATED ($VALUE
    (NODE3))))
```

GANTT CHART

```
(GANTT_CHART1
  (AKO ($VALUE
    (GANTT_CHART)))
  (HAS_ASSIGNMENT ($VALUE
    (Resource_Assignment_Name))))
```

```
(GANTT_CHART1
  (AKO ($VALUE
    (GANTT_CHART)))
  (HAS_ASSIGNMENT ($VALUE
    ((LOADING_DOCK_ASSIGN1)
    (M3_ASSIGN1))))
```

RESOURCE ASSIGNMENT

```
(Resource_Assignment_Name
  (AKO ($VALUE
    (RESOURCE_ASSIGNMENT)))
  (RESOURCE_REFERRED
    ($VALUE (Resource_Name)))
  (ACTION_SEQUENCE ($VALUE
    (Job_Name Operation Start Finish
      (:Node_Name))
  (RESERVED_FOR ($VALUE
    (Gantt_Chart_Name)))
```

```
(M3_ASSIGN1
  (AKO ($VALUE
    (RESOURCE_ASSIGNMENT)))
  (RESOURCE_REFERRED
    ($VALUE (M3)))
  (ACTION_SEQUENCE ($VALUE
    ((JOB1 OP1 3 29 (:NODE 3))
    (JOB3 OP3 29 47 (:NODE 4)))))
  (RESERVED_FOR ($VALUE
    (GANTT_CHART1))))
```

Structural knowledge allows the user to examine the intermediate stages of the solution process and use the information for subsequent fine-tuning or revision of the schedule generated thus far.

11.5 PRODUCTION SCHEDULE GENERATION

A three-phase solution strategy is used to generate the schedule. Phase 1 determines a rough-cut schedule by critical path analysis (CPA) with the use of environmental knowledge. The rough-cut schedule contains information about the sequence in which the jobs ought to be scheduled. Phase 2 uses the job sequence information from phase 1 as well as procedural and structural knowledge to produce a detailed production schedule, which involves the assignment of time and resources to each job. The solution mechanism in phase 2 uses a heuristic search technique called filtered beam search. Phase 3 performs a post-analysis of phase 2's detailed schedule by seeking local improvements in the schedule. Structural knowledge, in particular, the Gantt chart representation of phase 3's detailed schedule is employed in this phase. The overall solution strategy can be summarized as follows:

Phase 1. Rough-cut scheduling using CPA

- Use the average completion time of each job to perform a critical path analysis, and determine, for each job i, the earliest start time $Es(i)$, earliest finish time $Ef(i)$, latest start time $Ls(i)$, latest finish time $Lf(i)$ and the slack $Ls(i)-Es(i)$.

- Based on the slack information determine a critical path. Multiple paths are resolved arbitrarily. Jobs that lie on the critical path are classified as critical. The rest are classified as non-critical.
- Generate a rough-cut schedule that makes use of the critical path information and the product structure to group jobs into different scheduling stages. The stages are set up in such a way that the job precedence constraints are preserved and proper priority is given to the critical jobs.

Phase 2. Detailed scheduling using filtered beam search
- Schedule jobs in the sequence as determined in phase 1.
- The actual start and finish times of each job on certain resources are determined by using filtered beam search stage by stage.

Phase 3. Post-analysis
- Analyze the resource assignments in phase 2.
- Structural knowledge is used to detect in-process idle periods.
- The number of in-process idle periods are reduced by a series of left shifts and operation switchings.

The above solution strategy can be summarized in pseudocode form as shown in Fig. 11.4. The pseudocodes for phases 1, 2 and 3 are given in Appendix A, B and C, respectively. A detailed description of each of these phases can be found in De and Lee (1991).

```
Given:  ni units of Pi products are to be scheduled, i=1,..., N

1.      For each Pi with i=1 to N
            DO PHASE-ONE
        Next i

2.      Pi' ← { Pi in descending order of Ef(Pi) }

3.      For each Pi' with i'=1 to N
            For each jth unit of Pi' with j=1 to ni'
                DO PHASE-TWO
            Next j
        Next i'

4.      DO PHASE-THREE
```

Fig. 11.4. Pseudo-code for the three-phase solution strategy.

11.5.1 Illustrative example

We have three workstations W1, W2 and W3, one loading dock, and one unloading dock. The operation time (in minutes) for each workstation to perform its set of operations is given in Table 11.1. Transfer time between any pair of workstations or between a workstation and the loading/unloading dock is 2 min. Loading and unloading time is 1 min. The problem here is to find a minimum makespan schedule for scheduling a unit of PRODUCT1, PRODUCT2, PRODUCT3 and PRODUCT4.

Table 11.1 Workstation operation times

Workstation	*OP1*	*OP2*	*OP3*	*OPa1*	*OPa2*
W1	28	20	15	–	–
W2	19	25	17	–	–
W3	–	–	–	15	20
Average processing time	23.5	22.5	16	15	20

PRODUCT1 has a product structure as shown in Fig. 11.5. PRODUCT2 to PRODUCT4 do not need to be assembled, i.e. they are single parts. The operation requirements for these products are:

PRODUCT1 (PART1 or J1);	Loading OP1 OP3
PRODUCT1 (PART2 or J2):	Loading OP3
PRODUCT1 (PART3 or J3):	Loading OP1 OP2 OP3
PRODUCT1 (A1 or J4):	OPa1
PRODUCT1 (A2 or J5):	OPa2 Unloading
PRODUCT2 (PART4 or J6):	OP3 OP1
PRODUCT3 (PART5 or J7):	OP1
PRODUCT4 (PART6 or J8):	OP1

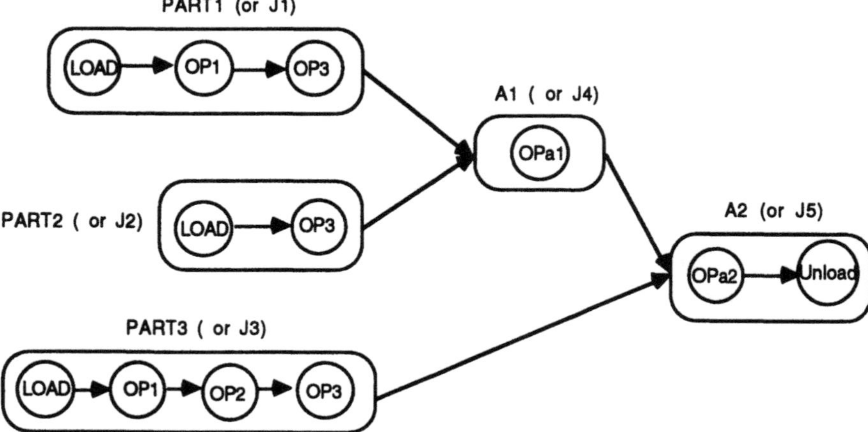

Fig. 11.5. Product structure of PRODUCT 1.

In phase 1, a rough-cut schedule is generated that specifies the sequence in which the various jobs should be scheduled. CPA is applied to each product to determine its estimated finishing time (average completion time of jobs along the critical path) as well as the sequence of the jobs within that product, giving priority to critical jobs. Job sequences from different products are then concatenated in descending order of the estimated finishing time of their corresponding product. In our example, the critical path for PRODUCT1 is J3 → J5. Mapping the critical path information with the knowledge of the product structure shows that the J3 should be scheduled first. Then before J5 can be scheduled, J4, which is the immediate predecessor of J5, has to be scheduled; and in turn J1 and J2 being the immediate predecessor of J4 have to be scheduled. Hence, the job sequence for PRODUCT1 is (J3) → (J1 and J2) → (J4) → (J5). Since PRODUCT1 has the smallest estimated finishing time (84 min), followed by PRODUCT2 (39.5 min), PRODUCT3 (23.5 min) and PRODUCT4 (23.5 min), the job sequence for the four products is (J3) → (J1 and J2) → (J4) → (J5) → (J6) → (J7) → (J8). Figure 11.6 depicts the rough-cut schedule produced in phase 1.

In phase 2, a detailed schedule is generated stage by stage using the filtered beam search strategy according to the job sequence determined in phase one; this means that the sequence of procedure calls to the filtered beam search algorithm is: (J3) → (J1 and J2) → (J4) → (J5) → (J6) → (J7) → (J8). Filtered beam search resembles a breadth-first search because it expands nodes in the search tree level by level without backtracking. However, it does not expand all the nodes at each level of the search tree. It selects a certain number of nodes (called beam width) for expansion. The choice of the nodes is made on the basis of an evaluation function f^* which is a natural measure of goal distance. For each node selected, it generates a specified number of successors (called filter width). The procedure for finding the successors of a node is called node expansion strategy (De and Lee, 1990). Both the beam and filter width can be specified by the user. A search with beam width two and filter width two is used here. The detailed schedule from phase 2 is shown in Fig. 11.7.

Phase 1 did not take into account any resource requirements of the jobs in generating the rough-cut schedule. Jobs are sequenced according to their precedence relationship and work content in relation to the length of the product's completion time. When resources are assigned in phase two, priority is given to jobs in earlier stages. The resource requirements of jobs in subsequent stages are ignored. If jobs are grouped in different stages for reasons other than to maintain their precedence relationship (as in the example here, J7 is sequenced after J6 according to the EFT rule), jobs in later stages may be blocked from making full use of a given resource because of an earlier resource assignment. Notice in Fig. 11.8, J8 is blocked by J6 in using W2. Although J8 is ready at time 0, it is not scheduled until J6 has completed its OP1. Hence, J8 cannot be scheduled until time 95. W2 is

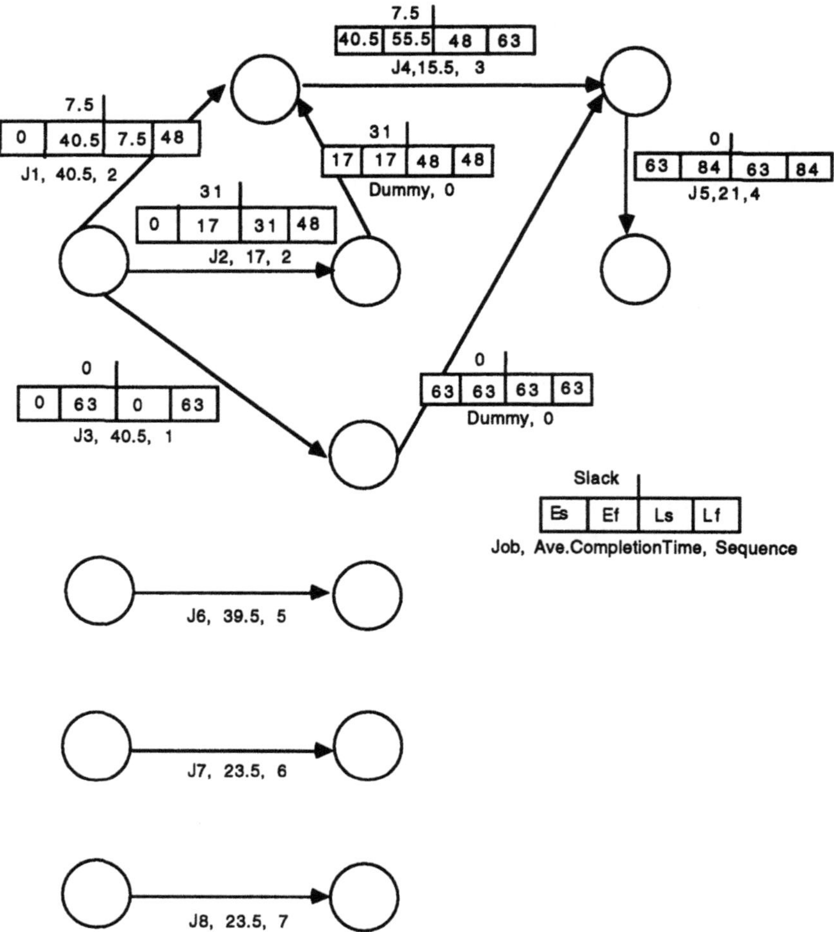

Fig. 11.6. Phase 1's rough-cut schedule.

available from time 58 to time 76, but it is not long enough for J8 to perform its OP1 which requires 19 time units. Phase 3 quickly recognizes the flaw in phase 2's schedule (W2 has an in-process idle period of 18 time units) and proceeds to amend the schedule by first, switch J6 with J8 and move J8 forward to the beginning of the idle-period. The resulting schedule has a makespan of 102, showing an improvement of 10.53% as compared to a makespan of 114. Figure 11.8 shows the result of phase 3. The interaction between the user and the system in setting up the knowledge base, retrieving information from the knowledge base, and solving this sample scheduling problem are recorded in Appendix D, E and F, respectively.

Our KBS has been implemented and tested using Common-LISP on a MAC SE/30 with 4 MB main memory. The test environment consists of five workstations: two workstations in the machining area and three work-

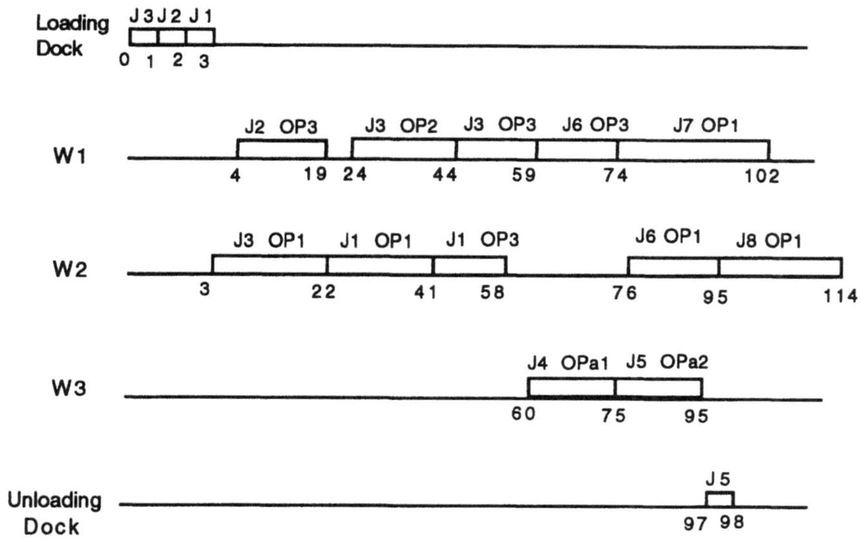

Fig. 11.7. Phase 2's detailed schedule.

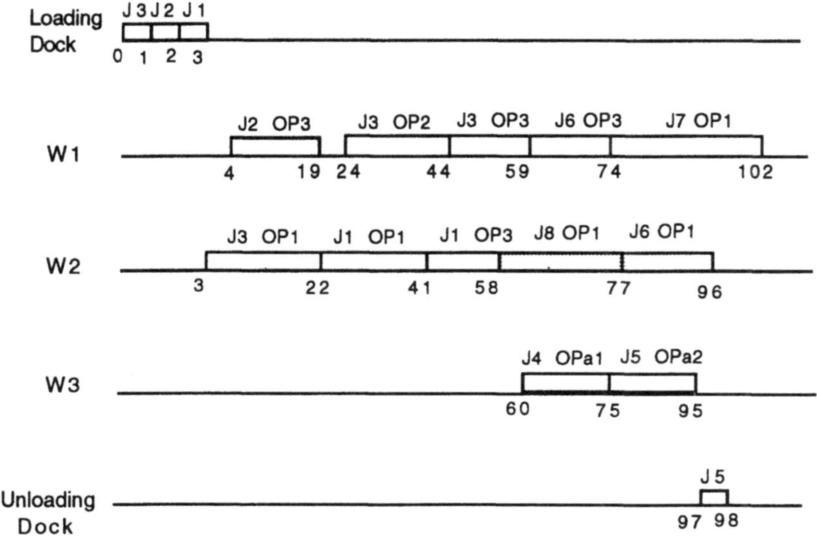

Fig. 11.8. Phase 3's improved schedule. □ blocking operation in phase 2; ■ blocked operation in phase 2.

stations in the assembly area. The loading/unloading area has one loading/unloading dock. Automated guided vehicles are used for material handling. The workstations can perform a set of ten operations. Multiple batches of three assembled products with number of operations ranges from

14 to 243 are tested. Our experimental experience indicates that the application of phase 1 followed by phase 2 yields better schedules in shorter time than using phase 2 alone (0.12–17.28% improvement in performance measure; 27.63–93.79% improvement in run time). In addition, phase 3 is especially useful in multi-product scheduling (0.05–0.59% improvement in performance measure). For further details of the experimental design and results, see De and Lee (1991).

11.6 CONCLUSION

Our KBS is designed to model a wide range of (1) system configurations, (2) product structures and (3) scheduling problems. In terms of system configuration, our KBS is not limited to a particular plant layout, number of workstations, capabilities of each workstations or method of material handling. It considers explicitly the transfer time between workstations, loading/unloading operations and routing flexibilities. In terms of product structure, our KBS is designed to handle structures of different complexities, i.e. the partial ordering of jobs are not restricted in a tree of a forest (for algorithms that worked well by assuming a tree or a forest structure, see Hu, 1961; Maxwell and Mehra, 1968; Kusiak, 1989). Also, the jobs are not restricted to be strictly assembly jobs or strictly non-assembly jobs; in other words, a job can have both assembly and non-assembly operations. In terms of solving the scheduling problem, our KBS can be used to solve (1) a single product problem, (2) a single-batch problem, (3) a N-product problem and (4) a N-batch problem as defined in Kusiak (1989). Expert knowledge as well as knowledge about the solution structure are used by a heuristic search strategy called filtered beam search to limit the search space to a manageable size so that a good schedule can be produced in a reasonable time. The scheduling strategy works well for any objective function, such as makespan or various tardiness related measures (De and Lee, 1990). Moreover, by appropriately choosing search parameters, the user can decide to get a reasonably good (but not necessarily optimal) solution quickly, or spend more computational overhead to generate solutions that are likely closer to the optimal.

Although our KBS is currently implemented in a MAC SE/30 using Common-LISP, it is by no means the only choice of hardware and programming language for system implementation. The use of a PC environment here allows room to improve the performance of our KBS by simply using a more powerful piece of hardware such as a SUN workstation or a main frame. The use of Common-LISP enables us to develop our KBS in a structured and modular fashion with the potential of taking advantage of the built-in object-oriented programming support. The current user interface in our KBS is in the form of a menu, which makes the system user-friendly and can be easily tailored to suit different needs and requirements of individual users. The fact that our KBS separates the knowledge base from its problem solving components allows, for example, the updating of the knowledge base without affecting the functioning of the problem solving components. The user can access and modify the knowledge of the KBS easily, even during the solution process. The knowledge base can be refined and extended as the user learns more about the scheduling process, or new knowledge about scheduling theory becomes available (e.g. new dispatching heuristics or better evaluation functions).

Although the system focuses on solving short-term production scheduling problems at the operational level, the knowledge base can serve as the beginning of recomputation in real time rescheduling stage. This system can be extended not only to solve other problems such as FMS system set-up planning but that such a set of inter-related production planning decisions (i.e. system set-up planning, static scheduling and real time rescheduling) can be made in an integrated manner through a common knowledge base. The implication of a particular decision for other decisions can be examined; the interaction between subproblems can be studied and exploited. It is the common knowledge base which will make an integrated FMS production planning system more than the sum of its parts. We hope to demonstrate that in our future research.

REFERENCES

Bobrow, D.G. and Winograd, T. (1977) An overview of KRL, a knowledge representation language. *Cognitive Science*, **1**(1), 3–46.

Brachman, R.J. (1978) A Structure Paradigm for Representing Knowledge. *Report 365*, Bolt Beranek and Newman, Inc., Cambridge, MA.

De, S. and Lee, L. (1990) FMS scheduling using filtered beam search. *Journal of Intelligent Manufacturing*, **1**(3), 165–83.

De, S. and Lee L. (1991) Flexible assembly scheduling using a knowledge-based approach, *Expert Systems with Applications: An International Journal*, **6**(3), 309–26.

Goldstein, I.P. and Roberts, R.B. (1977) The FRL Primer. *Report AIM-408*, Artificial Intelligence Laboratory, MIT, Cambridge, MA.

Hu, T.C. (1961) Parallel sequencing and assembly line problems. *Operations Research*, **9**, 841–8.

Kusiak, A. (1989) Aggregate scheduling of a flexible machining and assembly system. *IEEE Transactions on Robotics and Automation*, **5**(4), 451–9.

Maxwell, W.L., and Mehra, M. (1968) Multiple-factor rules for sequencing with assembly constraints. *Naval Research Logistics Quarterly*, **June**, 241–54.

Minsky, M. (1975) A framework for representing knowledge, in *The Psychology of Computer Vision*, (ed. P. Winston), McGraw-Hill, New York, pp. 211–77.

CHAPTER 12

'Optimal' rule switching for flow shops with random workloads

James C. Chen and Arne Thesen

12.1 INTRODUCTION

Production systems where parts of various types pass through the same sequence of machines are often referred to as flow shops (Fig. 12.1). Assembly lines and transfer lines are typical examples of such systems. A problem that frequently arises when operating a flow line is the problem of determining the sequence in which parts should be processed at each station such that some overall performance measures (say, parts produced per hour) is maximized. Unfortunately, it has long since been established that this is an exceptionally difficult problem that, for most practical situations, cannot be solved within the time constraints available.

Fig. 12.1. A three-machine flow shop with four, five and three buffer spaces. Parts of three different types are being processed. Machine 3 is idle.

In order to overcome this difficulty, most practitioners resort to the use of simple sequencing heuristics when dealing with flow shop scheduling problems. For example, the part with the shortest processing time (SPT) may be processed first. This approach leads to good results in many situations and a number of researchers have reported on the relative benefits if different sequencing heuristics for different situations. However, since different heuristics perform differently under different situations and since situations change over time, it is reasonable to expect that a system that uses between different scheduling heuristics for different situations [e.g. SPT when the queue is long and longest processing time (LPT) when the queue is short] may perform better than one that uses a single heuristic for all situations. In this chapter we show that this is indeed the case and we present the development of a knowledge base that facilitates the selection of recommended scheduling rules for different situations.

Scheduling algorithms that use different scheduling rules under different conditions are referred to as state-dependent scheduling (SDS) algorithms. The idea of SDS is not new. For example, it was discussed in Conway, Maxwell and Miller (1967). However, recent development in real-time shop floor control systems has renewed interest in the idea, as it now is possible to implement such a scheduling system on the shop floor. With this in mind, researchers have been applying SDS to such diverse areas as flexible manufacturing cells e.g. Seidmann, 1987), automated guided vehicle systems (e.g. Hodgson *et al.*, 1987), Kanban systems (e.g. Hodgson and Wang, 1991), electroplating lines (e.g., Thesen and Lei, 1990) and steel production systems (e.g. Buzacott and Callahan, 1973). We are not aware of any state-dependent rule base for flow shops with random workloads.

An SDS algorithm is implemented in two phases. In the *rule base generation* phase, a taxonomy of all possible system state is developed and the recommended decision for each individual state is determined. These decisions are then placed in a knowledge base. In the *schedule implementation phase*, we then simply go to the knowledge base and retrieve the correct procedure for the present situation. Needless to say, the resulting productivity of the flow shop depends upon the quality of the knowledge base being used.

A number of different approaches to the problem of developing a knowledge base for the flow shop scheduling problem are available. Among these are:

1. Analytic optimization (e.g. Seidmann and Schweitzer, 1984).
2. Exhaustive simulations (e.g. Thesen and Lei, 1990).
3. Use of experts (e.g. Yih and Thesen, 1991).
4. Automated learning (e.g. Pierreval and Ralambondrainy, 1990).

Each approach has advantages and disadvantages. For example, while analytic models can only be applied to small and well-defined problems, the resulting knowledge base may be proven to lead to optimal schedules. Likewise, expert opinions are applicable to larger problems, but the resulting schedules cannot be guaranteed to be optimal.

A hybrid approach is used in this chapter. An optimal knowledge base is first analytically developed for a small two-machine flow shop. This knowledge base is then heuristically extended to deal with larger problems. Extensive simulations are then used to demonstrate that the resulting knowledge base results in a more productive flow shop than the one observed using five other well-known scheduling rules.

12.2 A TWO-MACHINE FLOW SHOP

We will first develop a knowedge base for the efficient operation of a two-machine flow shop as shown in Fig. 12.2. Parts of two separate types are processed first at machine one, then at machine two. Inter-arrival intervals and processing times are exponentially distributed, with means of A_0, A_1, A_2 and B_0, B_1, B_2. Both machines have an input buffer with room for only two parts. Parts arriving when the buffer for machine one is full are lost. Parts wishing to enter machine two when its buffer is full remain at the first station (i.e. blocked) until room is available in the buffer.

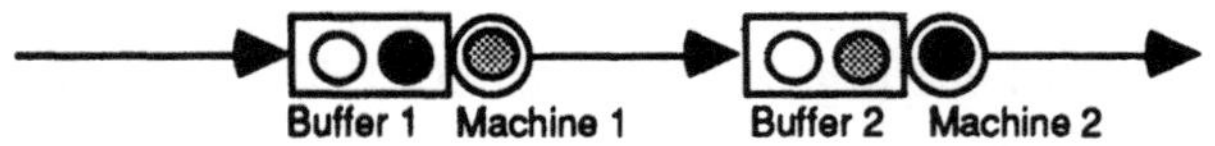

Fig. 12.2. A two-machine flow shop with two buffer spaces in front of each machine. Parts of two different types are being processed.

If there is only one part waiting in the buffer when it is time to start processing a new part, then the machine simply starts to process this part. However, if there are two parts waiting in the buffer, then the machine must decide which part to process first. This problem is typically 'solved' by first rank ordering the parts according to some attribute (e.g. due date and processing time) of each part and then by selecting the parts with the highest (or lowest) value. Panwalkar and Iskander (1977) presented an extensive survey of such scheduling rules. For the two-machine flow shop shown here, one of the rules shown in Table 12.1 is typically used.

Table 12.1 Five scheduling rules commonly used for the two-machine flow shop.

Rule	*Description*
FCFS	first come, first served
SPT	shortest processing time first
LPT	longest processing time first
LWR	least work remaining
MWR	most work remaining

It has been known for a long time that the desirability of such rules depends to some extent on the performance measure being used and on the attributes of the current job stream. For example, SPT may lead to shorter queues and FCFS may lead to lower variability of sojourn times. However, no rule has been shown to lead to performance that dominates the performance obtained using other rules.

One reason for the failure of a single rule to be the best is the fact that the 'best' choice at a given instant often is a function of the overall objective and of the state of the system at that time. For example, SPT may be the best if the next machine is empty and LPT may be the best if the next buffer is full. This suggests that the performance of the flow shop may be improved if the machine uses different decision rules for different system states. Using the techniques described in Section 12.3 it can in fact be proven that simple switching rules shown in Table 12.2 improves the throughput of the two-machine flow shop up to 8% over the throughputs observed when single rules are used. Similar improvements are also observed for larger systems. However, the analytic models of these systems are so complex that the optimality of the solutions cannot be proven at this time

Table 12.2. A simple rule-switching policy for the two-machine flow shop shown in Fig. 12.2

Machine	*Condition*	*Rule*
1	buffer 2 is not full	SPT
1	buffer 2 is full	LPT
2	all	SPT

12.3 DEVELOPING THE KNOWLEDGE BASE

Here we will use the two-machine flow shop discussed above to illustrate the process of building a knowledge base for an expert scheduling system. This is a four-step process:

1. Develop a state taxonomy.
2. Determine transition parameters for each state in the taxonomy.
3. Establish the 'optimal' decision rule(s) for each state.
4. Use this information to develop the rule selection knowledge base.

These steps are discussed below.

12.3.1 The taxonomy

The purpose of the state taxonomy is to provide a structure for selecting recommended scheduling rules for different situations. The taxonomy must therefore contain sufficient information to distinguish between states where different rules are appropriate. However, the taxonomy should not be *too* rich, as this will lead to an excessively large state space. (Each point in the state space can be thought of as a 'scheduling problem' that must be solved in advance.) As shown in Table 12.3, the taxonomy used in our example is quite simple. Four separate *locations* (the two buffers are the two machines) are identified. The states of these locations are independently described. In

Table 12.3. States used in the state taxonomy for the two-machine flow shop

Location	*States*
Buffer	empty holding one part of type *a* holding one part of type *b* holding one part of type *a* and one part of type *b* holding two parts of type *a* holding two parts of type *b*
Machine 1	empty working on part *b* working on part *a* blocking, holding part *a* blocked, holding part *b*
Buffer 2	empty holding one part of type *a* holding one part of type *b* holding one part of type *a* and one part of type *b* holding two parts of type *a* holding two parts of type *b*
Machine 2	empty working on part *b* working on part *a*

A given state in the taxonomy is the union of the states of the four locations listed here. The overall taxonomy contains 241 feasible states. Note that information about remaining service times at the machines and the sequence of parts in the buffers is not included.

order to reduce the number of states in the taxonomy, a large amount of possibly useful information is discarded. For example, the sequence of parts in a buffer is ignored (i.e. we do not know if part *a* follows part *b* or vice versa). We see that each buffer has six states, machine one has five states and machine two has three states. This results in a taxonomy with a total of 540 $(=6*5*6*3)$ states. However, a number of these states are meaningless. For example, machine one cannot be blocked when buffer two is not full. Another example is that a machine should not stay idle when there exist parts waiting in its buffer. When all such states are eliminated, the resulting taxonomy includes 241 states.

12.3.2 State transition parameters

After the system enters a given state, it remains in this state until one of the following four events takes place (the time that the system stays in a state before next transition is called the *waiting time*):

1. Machine one completes the processing of a part.
2. Machine two completes the processing of a part.
3. A part of type *a* enters buffer one from the outside.
4. A part of type *b* enters buffer one from the outside.

In most cases we do not know in advance which of the above events are going to occur. For example, if both machines are busy and buffer one is not full, then any one of the four events could take place. On the other hand, if buffer one is full and machine one is blocked, then the only feasible event is the end-of-service event for machine two.

Our goal is to identify the states following the present state and to determine the parameters (i.e. waiting times and transition probabilities) of state transitions. Knowledge of the current state and the state changing events is often sufficient. However, there are a number of states (referred to as *decision states*) where the next state is a function of a decision made by the machine. For example, if there is one part of type *a* and one part of type *b* in the second buffer at the time that machine two finishes processing one part, the next state depends on whether the machine chooses to start processing part *a* or part *b* (Fig. 12.3). For our example, there are 26 decision states for machine one and 50 decision states for machine two.

It is always possible to identify, for each state in the taxonomy, the probability that any one of the four events listed above takes place. Here, based on the memoryless property of exponential distributions, we give an example of the derivation of waiting time and state transition probabilities. Let *-ba;bab* indicate the state (shown in the left side of Fig. 12.6) that buffer one holds one part *b*, machine one is working on one part *a*, buffer two holds one part *b* and one part *a*, and machine two is working on one part *b*. The mean waiting time (let us call it *W*) of this state is the mean time until the first event (of the four possible events) happening.

$$W = \frac{1}{\frac{1}{A_0} + \frac{1}{B_0} + \frac{1}{A_1} + \frac{1}{B_2}}.$$

The next state of state *-ba;bab* depends on which event happens first. Table 12.4 shows the next states and their related transition probabilities. For example, if an end-of-service event at machine one occurs first, the next

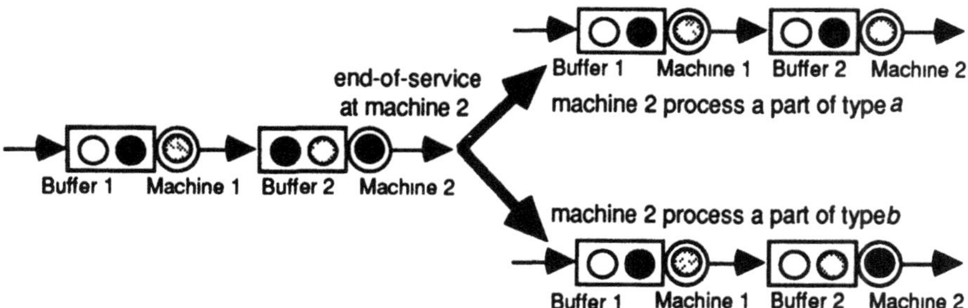

Fig. 12.3. State -ba; bab is one decision state. Its next state depends on the decision (made when an end-of-service event at machine two occurred) of picking up one part of type a or b.

Table 12.4. State transition probabilities of state *-ba;bab* and their relative events and next states

Event type	*Transition probability*	*Next state (decision)*	
End-of-service at machine one	$\frac{1}{A_1} * W$	*-bA;bab*	
End-of-service at machine two	$\frac{1}{B_2} * W$	*-ba;-ba*	(machine two picks one part of type *a*)
		-ba;-ab	(machine two picks one part of type *b*)
Arrival of one part *a* to buffer one	$\frac{1}{A_0} * W$	*aba;-ba*	
Arrival of one part *b* to buffer one	$\frac{1}{B_0} * W$	*bba:-ba*	

state is *-bA;bab* where one part *a* is blocked at machine one, and the transition probabilities to this state is $1/A_1*W$. Here, no decision needs to be made. On the other hand, when an end-of-service event at machine two occurs first, machine two needs to choose between parts of type *a* and *b* to process next. If machine two selects one part *a*, then the transition probability to state *-ba;-ba* is $1/B_2*W$. Otherwise, when one part *b* is chosen, the next state will be *-ba;-ab* with the same transition probability of $1/B_2*W$.

12.3.3 The 'optimal' decision rules

The quality of the resulting schedule depends on the decisions made at the decision point identified in the state taxonomy. In the flow shop, the decision point is the instant that, after an end-of-service event has just occurred, a machine has to choose one part to process next. For small problems, it is possible to analytically derive all state transition parameters and then use semi-Markov decision processes (SMDPs) to find the optimal decision rule for each state (e.g. machine two picks one part of type *a* at the decision point of state *-ba;bab*). For large problems, however, the exploding state space prohibits the direct generation of such rules.

An SMDP (for details, see Jewell, 1963a,b; Howard, 1971) is an outgrowth of semi-Markov process and dynamic programming. Howard (1971a,b) and Schweitzer (1971) have presented iteration algorithms for deriving the optimal policy based on state transition parameters such as transition probabilities, waiting times and transition rewards. The transition reward indicates the profit that a process gains when moving from one state to another. For example, if we want to maximize the system throughput, we may set the transition reward to 1 whenever the flow shop produces a good product.

After the implementation of SMDP, the optimal decision rules for 26 decision states of machine one and 50 decision states for machine two are generated (see Appendices A and B). Decision points and next states (resulted from the implementation of the optimal decisions) are also listed.

12.3.4 The knowledge base

Here, on the basis of the insight gained from the relation between state taxonomy and their optimal decision rules, we generalize the results from SMDP to a knowledge base.

Among the 26 decision states for machine one, six states belong to a group where both types of jobs are in buffer one and buffer two is not full. For this group of decision states, the optimal rules suggest that machine one picks one part of type *a*. Since A_1 (the processing time of part *a* at machine one) is less than B_1, this indicates that SPT should be used by machine one when its immediate downstream buffer is not full. The other group of decision states for machine one has 20 states where both types of jobs are in buffer one and buffer two is full. The optimal decision rules suggest that machine one picks one part of type *b*. That is, machine one switches to using LPT when its immediate downstream buffer is full.

For machine two, there are 50 decision states. All of these states have the characteristics that both types of jobs are in buffer two and machine two is the last machine in the system. The optimal decision rules advise that machine two picks one part of type *a*. This suggests that machine two uses SPT to pull jobs out of the system as soon as possible.

From the above analysis, the generalized rule-switching policy listed in Table 12.5 is developed. This policy is simple and intuitively appealing, as it suggests a machine to use SPT to avoid starvation of the downstream machine (when the immediate downstream buffer is not full) and to use LPT to avoid blocking at the current machine (when the immediate downstream buffer is full).

Table 12.5. An extended rule-switching policy for flow shops

Rule	*Condition*
SPT	a machine has neither downstream buffer nor downstream machine
SPT	a machine's immediate downstream buffer is neither full nor 'almost full'
LPT	a machine's immediate downstream buffer is full or 'almost full'

'Almost full' is interpreted as 100% full for low CV (e.g. less than 0.33), 85% full for medium CV (e.g. between 0.33 and 0.67), and 71% full for high CV (e.g. larger than 0.67). CV is the coefficient of variation of a part's processing time distribution at a machine.

12.3.5 Extension to larger problems

Admittedly, the two-machine flow shop problem studied here is quite simple and 'optimal' knowledge bases for optimal scheduling of this problem have very limited practical value. However, this problem has the advantage that it contains most of the elements of larger problems and the 'optimality' of the resulting schedule can be guaranteed. In the next section we will postulate extensions to the knowledge base for somewhat larger problems and we will give simulation results that suggests that the resulting knowledge base still performs very well indeed.

In order to extend the rule base introduced in Fig. 12.4 to larger problems, we must accommodate larger buffer sizes and more machines. We also need to take into account the effects of distributions of machine processing times on the performance of the knowledge base. Figure 12.8 shows the extended knowledge base after the implementation of sensitivity analysis.

12.4 EVALUATION

We will use simulation to evaluate the performance of the state-dependent knowledge base (shown in Fig. 12.8) as the complexity of the problem increases. Our performance measure will be throughput, measured in parts produced per unit of time. For larger systems it is very difficult to determine the optimal production rate of flow shops with random service times. We therefore compare the performance of SDS with the ones observed when the following conventional sequencing rules are used:

1. FCFS
2. SPT
3. LPT
4. MWR
5. LWR

12.4.1 The experiment

The evaluation is set up as a factorial design. The response variable is system throughput. The factors are (see Table 12.6):

1. Number of machines.
2. Buffer sizes.
3. Number of job types.
4. Ratio of arrival rates to service rates.
5. Variability of processing times.
6. Scheduling rules.

Table 12.6. Factors used in evaluating the extended knowledge base

Variable	*Factor description*	*Level*			
		−1	*0*		*+1*
M	number of machines	2			7
B	buffer size	2			7
J	number of job types	2			7
R	ratio of arrival rates to service rates	0.71	1		1.43
C	coefficient of variation of processing time distribution	0	0.5		1
S	scheduling rules	SDS FCFS	SPT	LPT	MWR LWR

For simplicity, we assume that all buffers are of the same size and that all parts arrive at the same rate. Also, some scenario specifies arrival rates larger than the sevice rates. This is not a realistic real life scenario. However, this scenario is included to give insight into the behavior of highly saturated systems.

Simulations of 432 scenarios of flow shops are undertaken (for detailed results, see Chen and Thesen, 1991). For each scenario, we make five replications with run length 500 000. The common random numbers technique is applied to simulation replications of different scenarios.

12.4.2 Results

Simulation results show that SDS is robust. SDS is, in terms of system throughput, always better than FCFS, LPT, MWR and LWR by up to 4, 8, 6 and 3%, respectively. Moreover, SDS performs at least as well as SPT. The former outperforms the latter by up to 2% against system throughput. In fact, the improvement is not very substantial. However, we highlight the following two points. First, since there are so many constraints in flow shops (e.g. neither merging, nor branching, nor alternate routing is allowed), much improvement should not be expected. SDS achieves more significant improvements (up to 50%) in robotic manufacturing systems (see, e.g. Chen and Thesen, 1992). Second, this improvement can be obtained without much investment. For example, we can use a computerized part coding scheme or a vision system (which are also needed in the implementation of rules such as SPT and MWR) to identify job types and use a sensor or a switch to trace the current buffer contents. We can then use a mirocomputer or programmable controller to receive signals from these peripheral devices (through serial or parallel communication) and implement SDS.

By the consideration of only the main effect of each factor, we find that the superiority of SDS becomes more significant when

1. The average buffer size increases (e.g. from 2 to 7; see Fig. 12.4).
2. The number of job types increases (e.g. from 2 to 7; see Fig. 12.5).

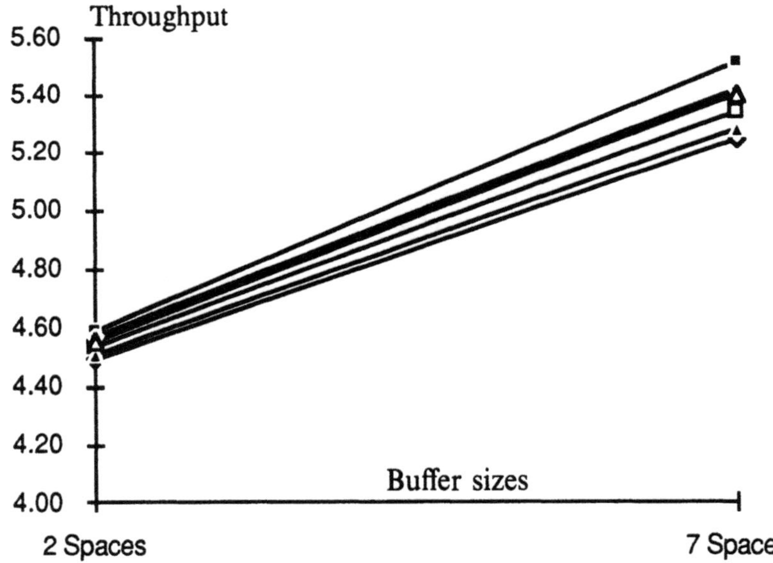

Fig. 12.4. The effects of scheduling rules and the buffer size on the system throughput of flow shops. ■ SDS; □ FCFS; ◆SPT; ◇LPT; ▲ MWR; △ LWR.

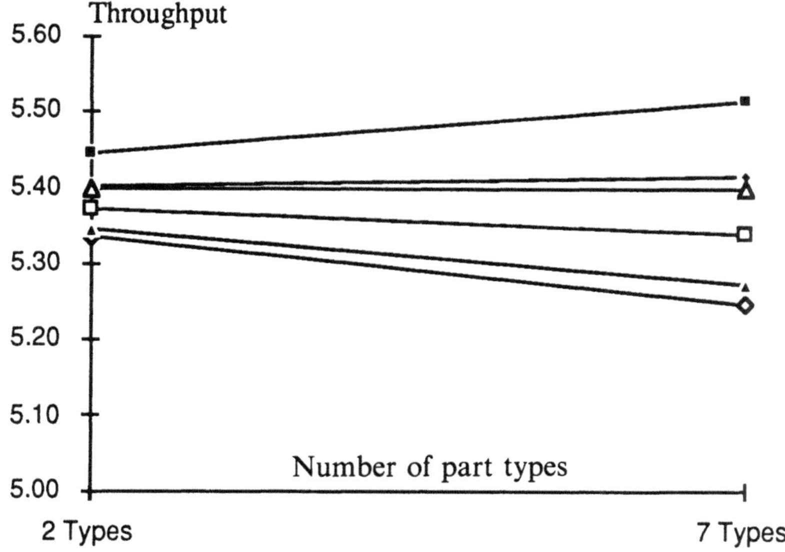

Fig. 12.5. The effects of scheduling rules and the number of job types on the system throughput of flow shops. ■ SDS; □ FCFS; ◇ LPT; ▲ MWR; △LWR.

3. The variability of processing time distribution increases (e.g. from constant to Erlang distribution and to exponential distribution; see Fig. 12.6).
4. The system utilization (reflected by the ratio of mean service time to mean interarrival time) increases (e.g. from 71 to 100%).

The increase in the number of machines does not result in increased superiority of SDS. This indicates that SDS is effective in both small system (e.g. with two machines) and large system (e.g. with seven machines).

It is worth noting that LPT has significant contribution to SDS, even though it is generally treated as a 'hopeless' rule. This is because LPT is complementary to SPT, and it needs to be used selectively with SPT to avoid starvation and blocking in flow shops.

12.4.3 Additional experiments

More studies have been conducted for 'larger' flow shops scheduling problems to explore if the optimal decision rules directly derived from the study of such problems are similar to the extended rule-switching policy shown in Fig. 12.8. For example, we examined a three-machine flow shop with 1179 states and a four-machine flow shop with 22 023 states. We observe that the decision rules of most of the decision states are in consistence with our rule-switching policy.

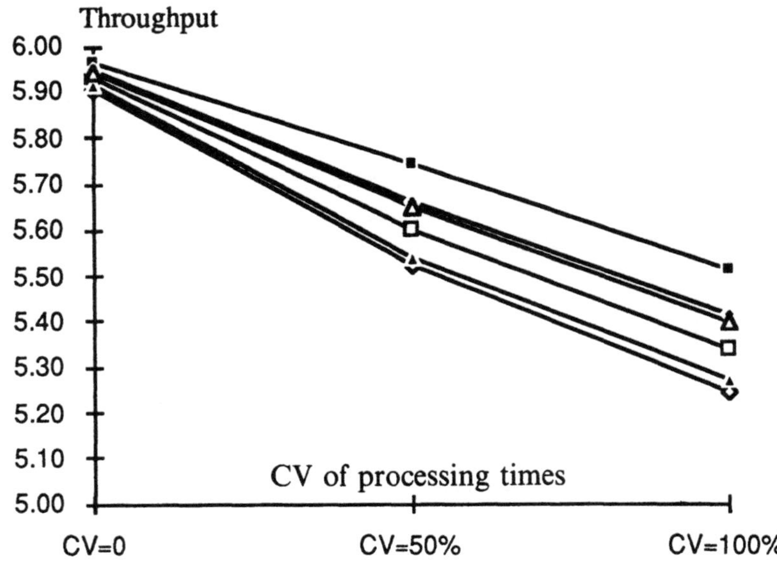

Fig. 12.6. The effects of scheduling rules and variability of processing times on the system throughput of flow shops. ■ SDS; □ FCFS; ◇ LPT; ▲ MWR; △ LWR.

12.5 CONCLUSIONS

A state-dependent knowledge base is built to determine the 'optimal' rule-switching for two-machine flow shops with random workloads. It suggests that different machines use different scheduling rules under different system states. A systematic approach is presented to facilitate the development of the knowledge base. This approach has four steps: develop a state taxonomy, determine state transition parameters, establish the optimal decision rules and develop the knowledge base.

The knowledge base is further generalized and extended to a rule-switching policy applicable to large flow shops. This policy suggests that a machine needs to switch rules from SPT to LPT when its immediate downstream buffer is full or 'almost full'. As a rule-of-thumb, 'almost full' is interpreted as 100% full for low coefficient of variation (CV; e.g. less than 0.33), 85% full for medium CV (e.g. between 0.33 and 0.67), and 71% full for high CV (e.g. larger than 0.67).

Our study shows the feasibility and effectiveness of SDS in flow shop scheduling problems The knowledge base developed in SDS can be consulted in real time on the shop floor. We observe that the rule-switching policy outperforms individual scheduling rules by up to 8% (in terms of system throughput) in our case studies. A reduction in throughput was never observed. The superiority of SDS may become more significant as the average buffer size, the number of job types, the system utilization or the variability of processing time distribution increase.

ACKNOWLEDGMENT

J.C.C. was supported in part by an IBM Manufacturing Reseach Fellowship (1991–1992).

REFERENCES

Buzacott, J.A. and Callahan, J.R. (1973) The pit charging problem in steel production. *Management Science*, **20**(4), 665–74.

Chen, J.C. and Thesen, A. (1991) Real-time state-dependent scheduling in dynamic and stochastic flow shops. *Tech. Rep. 91–10*, Department of Industrial Engineering, University of Wisconsin-Madison.

Chen, J.C. and Thesen, A. (1992) *State-Dependent Robot Scheduling Using Simulated Semi-Markov Decision Processes.* ORSA/TIMS Joint National Meeting, Orlando, Florida, April, 1992.

Conway, R.W., Maxwell, W.L. and Miller, L.W. (1967) *Theory of Scheduling*, Addison-Wesley, Reading, MA.

Hodgson, T.J. and Wang, D. (1991) Optimal hybrid push/pull control strategies for a parallel multistage system: part I. *International Journal of Production Research*, **29**(6), 1279–87.

Hodgson, T.J., King, R.E., Monteith, S.K. and Schultz S.R. (1987) Developing control rules for an AGVS using Markov decision processes. *Material Flow*, **4**, 85–96.

Hodgson, T.J. and Wang, D. (1991) Optimal hybrid push/pull control strategies for a parallel multistage system: part II. *International Journal of Production Research*, **29**(7), 1453–60.

Howard, R.A. (1971a) *Dynamic Probabilistic Systems. Volume I: Markov Models*, John Wiley & Sons, New York.

Howard, R.A. (1971b) *Dynamic Probabilistic Systems. Volume II: Semi-Markov and Decision Processes*, John Wiley & Sons, New York.

Jewell, W.J. (1963a) Markov-renewal programming. I: formulation, finite return models. *Operations Research*, **11**, 938–48.

Jewell, W.J. (1963b) Markov-renewal programming. II: Infinite return models, example. *Operations Reseach*, **11**, 949–71.

Panwalker, S.S. and Iskander, W. (1977) A survey of scheduling rules. *Operations Research*, **25**(1), 45–61.

Pierreval, H. and Ralambondrainy, H. (1990) A simulation and learning technique for generating knowledge about manufacturing systems behavior. *Journal of Operational Research Society*, **41**(6), 461–74.

Schweitzer, P.J. (1971) Iterative solution of the functional equations of undiscounted Markov renewal programming. *Journal of Mathematical Analysis of Applications*, **34**, 494–501.

Seidmann, A. (1987) On-line scheduling of a robotic manufacturing cell with stochastic sequence-dependent processing rates. *International Journal of Production Research*, **25**(6), 907–24.

Seidmann, A. and Schweitzer, P.J. (1984) Part selection policy for a flexible manufacturing cell feeding several production lines. *IIE Transactions*, **16**, 355–62.

Thesen, A. and Lei, L. (1990) An expert scheduling system for material handling hoists. *Journal of Manufacturing Systems*, **9**(3), 247–52.

Yih, Y. and Thesen, A. (1991) Semi-Markov decision models for real-time scheduling. *International Journal of Production Research*, **29**(11), 2331–46.

CHAPTER 13

An expert system approach to surface mount pick-and-place machine selection

Chung-Yu Liu, C.R. Emerson and K. Srihari

13.1 INTRODUCTION

The machine selection problem in printed circuit board (PCB) assembly using surface mount technology (SMT) involves complex decision making. The amount of time spent and the possibility of a sub-optimal decision could both adversely affect the manufacturing economics of production operations. The machine selection problem in the PCB assembly domain requires the use of multi-alternative and multi-attribute decision-making principles (Shieh, 1991).

A significant portion of the information used in the machine selection problem in the PCB assembly domain is not exact. It is fuzzy rather than precise (Shieh, 1991). The machine selection problem can be satisfactorily handled by an expert. However, personnel with adequate levels of expertise are not very common. Besides, the presence of uncertainties and trade-offs in the decision-making process have caused an increasing demand for tools to improve the decision maker's capability (Weber and Coskunoglu, 1990). The almost continual evolution of equipment coupled with the dynamic status of the PCB assembly domain mandates the need for a constant review of domain knowledge to ensure that the machine selection needs of a specific manufacturing scenario are best satisfied. The decision-making process during machine selection for the PCB assembly domain is often subjective. As a result, the decision makers have a critical need for interactive, informative and reliable multi-attribute decision support systems.

This paper describes a multi-attribute decision support expert system which utilizes fuzzy reasoning in selecting a pick-and-place machine for a surface mount PCB assembly line. The need for uncertainty management in this application is discussed. The problem statement and the research objective are delineated. The multi-attribute decision support expert system's design and specifications are described with special emphasis on the uncertainties considered and the uncertainty management method used. Also described are the system's architecture, the construction and content of the knowledge base, the design of the inference mechanism, and the details of the user interface. The inputs and outputs of the system are described. The limitations of this research and ideas for future research are presented.

13.2 CONCEPTS USED IN THIS RESEARCH

This research used concepts relating to expert systems and their application in manufacturing, fuzzy set theory and the management of uncertainty, surface mount manufacture of PCBs, placement techniques in SMT, and factors used in choosing a placement machine (or process) for a specific placement scenario.

13.2.1 Knowledge-based expert systems

The manufacturing domain has witnessed the widespread use of artificial intelligence (AI)-based expert systems. An AI-based expert system uses problem specific domain knowledge to suggest expert quality solutions to problems. Expert systems have been used in a variety of applications within the manufacturing domain (Abdou and Dutta, 1990; Kumara, 1986). These include interpretation, prediction, diagnosis, design, planning, monitoring, repair and control. The knowledge used by an expert system to perform these tasks, in general, has been extracted from the human experts in the domain and encoded in a formal language (Luger and Stubblefield, 1989). As with highly skilled humans, expert systems tend to be specialists focusing on a narrow set of problems. Also like humans, their knowledge tends to be both theoretical and practical having been perfected through experience. Unlike humans, however, current expert systems are not capable of learning from their own experience.

Knowledge-based expert systems can be used to assist the user in every facet of the PCB assembly using the SMT domain (Srihari and Emerson, 1992). The heuristic nature of this domain combined with the lack of widespread in-depth knowledge makes this an ideal domain for expert system development and use. This is further mandated by the continual need to enhance productivity and by increasingly competitive markets. The global objective of expert system use in PCB assembly is to improve manufacturing efficiency, enhance product yield, and positively impact quality and productivity. As expert system applications continue to contribute to corporate profitability, their role will expand.

Many tasks in surface mount PCB assembly, e.g. setting up and debugging the reflow process, solder paste selection, identifying solder paste printing parameters, etc., require the use of expert knowledge because of the large number of factors involved. There are few experts with the required knowledge. An expert system solution to these problems will formalize the complicated set of informal procedures understood by the experts alone and make this expertise available to a wider range of people at many different locations on the shopfloor (Srihari and Emerson, 1992). Consequently, the expert will be able to concentrate on other areas and not remain tied up with mundane on-line parameter setting. In addition, expert system solutions will result in production consistency, i.e. there will be a reduction in shift to shift variation. Moreover, valuable expertise will not be lost through employee turnover.

A review of the literature indicates that AI has not been widely used to assist in multi-attribute decision making (Levine, Pomerol and Saneh, 1990). Of the numerous expert systems developed in the manufacturing realm, only a few have been developed purely for management support (Shaw, 1988). There is a tremendous potential for the use of AI techniques in the solution of multi-attribute decision-making problems. Expert systems can be used as decision support systems. They can help users rate and weight the attributes in multi-attribute decision-making problems (Negoita, 1985). Through the use of rules and facts, expert systems can improve the quality of decisions. A great deal of uncertainty is involved in a semi-structured decision-making atmosphere (Canada and Sullivan, 1989).

13.2.2 Fuzzy sets

Fuzzy set theory can be applied to the multi-attribute decision problem. In these problems the source of imprecision is the absence of sharply defined attribute for class membership rather than the presence of random variables (Maiers and Sherif, 1985). Various methods and models have been suggested to solve the multi-attribute decision-making problem using fuzzy models. They differ by their assumptions concerning the input data and by the measures used for aggregation and ranking. Also, they either concentrate on the aggregation of rating, the ranking or both.

Fuzzy sets are extensions of ordinary sets (Zadeh, 1987). Fundamental to this basic set theory is the notion that an item is either a member or it is not a member of a set. For example, x belongs to set A or x does not belong to A. However, in the real world membership in a set is not always well defined. Fuzzy set theory permits precise formal manipulation of sets where membership is in terms of degrees rather than just belonging or not belonging (Maiers and Sherif, 1985).

A fuzzy set is a class of objects with a continuum of grades of membership. More precisely, a fuzzy set is characterized by a membership function, defined as a real number in the interval [0, 1]. It is clear that fuzzy sets can be reduced to ordinary sets by constraining the membership to the extremes of the range 0 or 1. For example, a membership measure $\mu_A(x) = 0.8$ suggests that x is a member of set A to a degree 0.8 on a scale where 0 is no membership at all and 1 is complete membership. To illustrate, consider a universe of discourse X with subset A. If the transition between full membership ($\mu=1$) and non-membership ($\mu=0$) of the set A is gradual rather than sharply defined, then A is a fuzzy set. $\mu_A(x)$ thus represents the grade of membership of the element x in the set A. Let $X=\{x_i|i=1,\ldots,n\}$ be a set of objects, then a fuzzy set A in X is a set of ordered pairs such that:

$$A=\{(x,\mu_A(x)\},\quad x\in X,\quad 0\leqslant\mu_A(x)\leqslant 1$$

The assignment of the membership grade is subjective in nature and reflects the context in which the problem is viewed. Although the membership function of a fuzzy set has some resemblance to a probability function when X (space of objects) is a countable set (or a probability density function when X is a continuum), there are essential differences between these concepts. In fact the notion of a fuzzy set is completely non-statistical in nature (Zadeh, 1987). Therefore, fuzzy sets are used to describe information which is non-statistical in nature and consequently cannot be described by the standard techniques of probability theory.

13.2.3 The fuzzy set theory approach to uncertainty management

Fuzzy set theory uses the theory of possibility (Zadeh, 1987). Unlike probability, the concept of possibility does not involve the notion of repeated experimentation. Possibility is not a substitute for probability but rather another kind of uncertainty (Zimmerman, 1985). Zadeh (1987) proposed possibility theory as a measure of vagueness, just as probability theory measures randomness. Consequently, the concept of possibility is non-statistical in character. It is a natural concept to use when the imprecision or uncertainty in the phenomena under study are not susceptible to statistical analysis or characterization (Zimmerman, 1985).

The main contention of fuzzy set theory is that, although probability theory is appropriate for measuring randomness of information, it is inappropriate for measuring the meaning of information (Canada and Sullivan, 1989). Fuzzy set theory expresses the lack of precision in a quantitative fashion by introducing a membership function that can take on real values. It is not concerned with how these distributions are created, but rather with the rules for computing the combined possibilities. Thus, it includes rules for combining possibility measures for expressions containing fuzzy variables. A few AI-based expert systems employing fuzzy set theory in manufacturing have been reported in the literature (Negoita, 1985).

13.2.4 Pick-and-place machines in PCB assembly

The pick-and-place machine combines advanced technologies including electronics, computers, photo-electronics and mechanics to build a surface mount component (SMC) placement machine. The SMC placement machine, which is commonly referred to as a 'pick-and-place machine', is the most important piece of manufacturing equipment in a surface mount PCB assembly line. This essential equipment is expected to place components reliably and accurately to meet throughput and quality requirements.

There are numerous surface mount placement machines in the market today (Shieh, 1991). Various models differ from each other in machine capacity and capability. Some manufacturers emphasize higher output speed of their products while the others emphasize the flexibility of their products. Since no pick-and-place machine is best for all applications, the effort required in selecting an appropriate SMC placement machine should constitute a major part of the effort spent in selecting surface mount capital equipment (Prasad, 1989). As a result, the first task of the pick-and-place machine selection problem is to realize how many potential alternatives there are in the market. Some of the criteria for pick-and-place machine selection have been previously identified (Lea, 1988; Prasad, 1989). Shieh (1991) considered four different pick-and-place models in his research.

13.2.5 Types of pick-and-place machines

In general, an SMC placement machine is composed of a body base, a board handling system, component feeders, placement heads, placement tools, and a vision system. Based on machine characteristics, surface mount placement equipment falls into four categories. These are defined below.

In-line placement equipment involves a moving board with a fixed head. Each head places one component with a cycle time of 1.8–4.5 s per board. The number of placements is equal to the number of components on the board. In-line placement equipment typically uses a short cycle time and works on high volume placement for a single product. Each head can only place one component for each setup.

Simultaneous placement equipment uses a fixed table and a fixed head where all components are placed simultaneously. The cycle time per placement is 7.0–10.0 seconds. Each head can be fitted with multiple tools. Change over times between different board types is excessive due to the rigidity of the method. Simultaneous placement machines are difficult to reprogram. They are typically used for high throughout production of a single board.

In sequential placement equipment, the table can move in the X–Y direction. The placement head can also move in the X, Y and Z direction. Components are placed in succession individually with a cycle time of 0.3–1.8 s per component. These placement systems can be equipped with one or two heads. These machines are flexible and easy to program. Sequential placement equipment has a software-controlled X–Y moving table or head system, variable speed and the capacity for placing fine pitch components.

Sequential/simultaneous placement equipment allows the table to move in the X–Y direction. It has a fixed head and allows sequential or simultaneous placement. The component placement cycle time is usually about 0.2 s cycle time per component. Sequential/simultaneous equipment can be equipped with multiple heads and a software-controlled X–Y moving table. It is not highly flexible.

13.2.6 Factors to be considered in selecting pick-and-place machines

In surface mount PCB assembly, the pick-and-place machine(s) constitutes a significant investment. Also, a poor SMC placement machine can negatively impact process yields (Prasad, 1989). A buyer of SMC placement equipment must define current and future needs before embarking on the evaluation and selection of a placement machine. Some of the factors to be used in machine selection have been introduced by Shieh (1991). Machine selection factors can be classified into three categories – accuracy, speed and flexibility.

Accuracy in an SMC placement machine includes X—Y accuracy, rotational accuracy, resolution, repeatability along the X–Y axis and rotational (theta) repeatability. Speed is often evaluated by the placement rate. Flexibility includes the ability to automatically load and unload substrates, an ample feeder capacity, the ability to reassign feeders, and the availability of different feeder types. Factors that define the flexibility of an SMC placement machine include the software for the optimization of feeder placement, programmable Z axis pressure during placement, adhesive dispensing capacity, a communications interface, the component (missing/incorrectly oriented/wrong size) verification mechanism, vision assisted placement, and automatic tool and/or nozzle changing. The flexible placement machine should offer off-line and teach mode programming. It should be compatible with computer aided design systems. Flexible SMC placement machines would allow bi-directional PCB movement. Substrate registration could be based on the edge of the board, the tooling hole, or the vision system used. Other factors that affect flexibility are the types of components that can be placed including rectangular and cylindrical chips, capacitors, small outline devices and all other active devices. The placement machine must be able to deal with a large range of substrate dimensions and component sizes.

13.3 PROBLEM STATEMENT AND RESEARCH OBJECTIVE

The increasing use of SMT within the PCB assembly domain has been well documented. SMT helped PCB designers and manufacturers cope with the need for increasing component placement density, decreasing component lead pitch and enhanced electrical requirements. The leads of components used in SMT are placed on the PCB instead of being inserted through the substrate as in conventional insertion mount PCBs.

Increased use of SMT has enhanced the importance of selecting an acceptable (near optimal) placement machine or a specific PCB assembly line. The decision process is often subjective, and requires an in-depth knowledge of the domain. Also multiple objectives often need to be satisfied. Very few personnel working in industry today are experts in SMT placement machine(s). A poor or less than adequate decision for this selection could result in a significant negative economic impact.

The selection of a pick-and-place machine in the SMT domain is a multi-attribute decision-making process which is driven by attributes (criteria) and alternatives. Over 50 different attributes need to be considered in the SMT placement machine selection process (Shieh, 1991). These attributes coupled with the large number of equipment choices available today make it almost impossible to evaluate and select a placement machine using the traditional approaches. Therefore, there is a pressing need for an efficient decision support system which uses multi-attribute decision-making techniques.

The objectives of this research were:

1. To develop an uncertainty management methodology for machine selection in the SMT-PCB assembly domain incorporating fuzzy set theory in the decision-making algorithm.
2. To develop an inference mechanism which models the inexact reasoning of a human expert in pick-and-place machine selection.
3. To develop a knowledge base for pick-and-place machine selection.
4. To develop a machine selection decision support expert system.

The decision support expert system described uses fuzzy set theory to manage uncertainty. It allows a user to update the knowledge base or to add new knowledge bases. The system developed is user friendly with help screens and an explanation mechanism.

13.4 SYSTEM DESIGN AND SPECIFICATIONS

The decision support expert system, named EXPS or Expert system for Pick-and-place machine Selection, was designed and developed to deal with placement machine selection. EXPS assists the decision maker in selecting an appropriate pick-and-place machine among the models currently available. It requires all machine related attributes to be evaluated (rated and weighted) by the user during the decision process. Using a comprehensive knowledge base, EXPS uses an inexact reasoning mechanism based on fuzzy set theory to manipulate the knowledge in the knowledge base to suggest a possible solution(s) while considering the inter-relationships between multiple factors that affect the final choice(s). The system selects an appropriate model of the pick-and-place machine under specific combinations of work part restrictions, machine specifications and machine capabilities.

Because of the applied nature of this study, every attempt has been made to investigate the problem within a realistic framework by taking into consideration various technological considerations.

13.4.1 System input and output

The user is required to answer questions in order for the system to acquire sufficient information about the user's needs. The system is menu driven requiring the user to make choices from structured question and answer menus. It is assumed that the user's input is an accurate reflection of the user's needs. The input provided by the user is assimilated and utilized by the inference mechanism. The user, through the input process, must define current and future needs before embarking on the evaluation and selection of a pick-and-placement machine. User inputs to the system are either machine related, PCB substrate related or component related pieces of information.

Inputs to EXPS include:

1. Machine related information including required production capabilities, capacities, functions and specifications.
2. Substrate restrictions including maximum and minimum substrate dimensions.
3. Component restrictions including the component and feeder types, and the maximum and minimum component dimensions.

The outputs are the conclusions drawn by the system using the information in the working memory and the criteria in the knowledge base. The conclusions drawn have a numerical value attached. The value is a fuzzy number representing the grade of membership of the specific machine in the perspective (machines) alternatives set. The outputs provided by the system includes three groups of placement machines selected from the knowledge base. The first group is a list of placement machines which are qualified considering the user's requirements with respect to machine specifications and work part restrictions. The second group is a list of placement machines which are qualified with respect to the user's requirement for a machine's capabilities. The third group is a list of placement machines with comparable basic capabilities.

13.4.2 Uncertainty considered

Uncertainties result from the use of abduction inference as well as from attempts to reason with missing or unreliable data. Uncertainties are caused by the factors existing in the SMT placement equipment domain when machine selection criteria are introduced. These factors are the attributes which affect the PCB assembly process, which restrict the work part processing capacty and which relate to machine specifications. There are four categories of factors of concern. They are comparable basic capabilities, competitive capabilities, machine specifications and work part restrictions.

Both quantitative and qualitative factors are involved in the selection criteria. Machine specification and work part constraints fall in the category of quantitative factors while machine capability (basic and competitive) is a qualitative factor. These factors are the main source of uncertainty considered by the system.

Two kinds of uncertainties arise in the machine selection process. The first kind is attribute uncertainty, which comes from the vagueness (fuzziness) of both the experts' and users' knowledge. The second is the alternative uncertainty, which is evident in situations where there are suitable alternatives available other than the optimal alternative.

Comparable basic capabilities are found in all models' specifications. They are all qualitative factors. Comparable basic machine capabilities include the programmable *Z* axis pressure during placement, automatic PCB load and unload capacity, missing component verification using a vacuum nozzle, vision assisted board location, automatic tool and nozzle replacement, and the availability of tape and reel feeders. Other comparable machine features include off-line programming, ability to various active and passive component types, and the least component lead pitch which can be accurately placed.

Machine parameters that are termed competitive functions are all qualitative factors. They include adhesive dispensing capacity, communications interface, the component geometry and electrical function verification, centering method used, vision aided placement, placement optimization methodology, multiple placement heads, waffle pack and tube component feeder mechanisms, on-line teach programming, and vision-aided fine pitch component placement.

Machine specifications are quantitative factors. They include linear and rotational accuracy, resolution, linear and rotational repeatability, placement rate, and feeder capacity. Work part (or assembly PCB) restrictions are also quantitative factors. They include maximum and minimum substrate length, width, and thickness, component dimensions (maximum and minimum) that can be handled and the least component lead pitch that can be placed.

13.4.3 Uncertainty management methodology

Uncertainty management techniques have been applied in this system on the basis of adaptability. Uncertainty management has not been applied to all areas of the decision-making process of EXPS. The inference mechanism used employs fuzzy set theory to deal with both attribute uncertainty and alternative uncertainty. The possible fact generator implemented in this system generates all available possible facts to manage alternative uncertainty. The mathematical model of inexact reasoning utilizing fuzzy set theory and the possible fact generator are discussed below.

(a) *Mathematical model*

A basic assumption in the use of EXPS is that it interacts intelligently with the user. While the domain knowledge resident in the system is obtained from domain experts, the necessary facts (fuzzy data) are supplied by the system user. The system combines the expert's rules with the user's knowledge of the current environment to deduce suggested decisions.

The rules are initially formulated in natural language. Experts then weight each factor to reflect whatever level of imprecision or vagueness is appropriate to the actual state of the expert's knowledge. The attribute weight can be treated as certainty factors (CFs). A set of fuzzy numbers is extracted from these CFs. For each alternative, the expert's CF is:

$$\mathrm{CF} = \mathrm{MIN}(\mathrm{CF}_1, \mathrm{CF}_2, \mathrm{CF}_3, \ldots, \mathrm{CF}_N).$$

The fuzzy data mirror the degree to which the user's knowledge of the state of the world varies from vague to precise. Users rate each attribute according to their needs. The score an attribute secures is a measure of confidence (MC). These MCs are another fuzzy set. The user's MC for each alternative is:

$$\mathrm{MC} = \mathrm{MIN}(\mathrm{MC}_1, \mathrm{MC}_2, \mathrm{MC}_3, \ldots, \mathrm{MC}_N).$$

Fuzzy numbers can be multiplied and a suitability fuzzy number (SFN) can be calculated for each alternative. Since the CFs and MC are from different sets, they are independent of each other. Hence, for each alternative the SFN is:

$$\mathrm{SFN} = \mathrm{CF} * \mathrm{MC}$$

Grouping the SFN of each alternative, a suitability fuzzy subset can develop where:

$$\text{Suitability} = \{(\text{Alternative\_1}, \mathrm{SNF}_1), (\text{Alternative\_2}, \mathrm{SNF}_2), (\text{Alternative\_3}, \mathrm{SNF}_3), \ldots, (\text{Alternative\_}N, \mathrm{SNF}_N).$$

According to Zadeh's min–max rule (Zadeh, 1987), the most appropriate candidate should be the greatest value of membership in the suitability fuzzy subset.

$$\text{Optimum Selection} = \mathrm{MAX}(\mathrm{SNF}_1, \mathrm{SNF}_2, \mathrm{SNF}_3, \ldots, \mathrm{SNF}_N).$$

Nevertheless, a decision support system supports but does not replace managerial judgment. Consequently, the system lists all available alternatives followed by their SNF. The final decision is still essentially made by the user.

(b) *Possible fact generator*

A possible fact generator is used by EXPS to manage uncertainty among alternatives. This uncertainty is related to those available facts other than those directly related to the user's needs. The possible fact generator is designed for quantitative facts. These facts are pieces of evidence or information that are necessary to arrive at conclusions when solving quantitative problems. Currently in EXPS, the possible fact generator works only for machine specifications and work part restrictions. To illustrate, let us consider two simple rules:

Rule 1: SELECT G1 IF C1 is greater than M1 AND
C2 is greater than M2 AND
C3 is less than M3 AND
C4 is equal to M4.

Rule 2: SELECT G2 IF C1 is greater than N1 AND
C2 is greater than N2 AND
C3 is less than N3.

Where G1 and G2 are the alternatives, and
C1, C2, C3 and C4 are the attributes (or 'facts' in terms of the knowledge base).

If $M1 = 0.2$, $M2 = 14$, $M3 = 7$ and $M4 = 120$,
$N1 = 0.1$, $N2 = 12$ and $N3 = 10$, then

Then for these example rules, $M1 > N1$, $M2 > N2$ and $M3 < N3$. Therefore when the user's demand meets the criteria in Rule 1, alternative G1 and G2 should be selected concurrently since Rule 2 also satisfies the demand mathematically.

One of the major functions of decision support systems is to suggest all qualified and available alternatives to the user. The possible fact generator needs to do this. When the user's inputs fulfill the criteria of Rule 1 which means facts M1, M2, M3 and M4 will be added to the working memory of EXPS, then the possible fact generator will introduce facts N1, N2 and N3 into the working memory. Thus, both rules (Rule 1 and Rule 2) will fire. In EXPS, specific facts generated by the possible fact generator are quantitative, machine specification related data on SMC placement machines available in the knowledge base.

13.4.4 System framework, knowledge base, inference mechanism and user interface

EXPS is a rule-based expert system in which knowledge is represented in terms of predicate calculus. The knowledge represented has been extracted from domain experts working in industry, documents published by SMT placement equipment manufacturers and from domain-related literature. The system's operation is controlled by selecting the rules to use, accessing and executing those rules with exact reasoning, and determining when an acceptable solution has been found. If an exact solution is not available, some alternatives might be offered by the system with inexact reasoning. In order to facilitate system operation, knowledge is represented using fuzzy production rules in the knowledge base.

The overall system framework of EXPS components is depicted in Fig. 13.1. Components of EXPS include the user interface, a knowledge update mechanism, a knowledge base, a possible fact generator, an inference engine

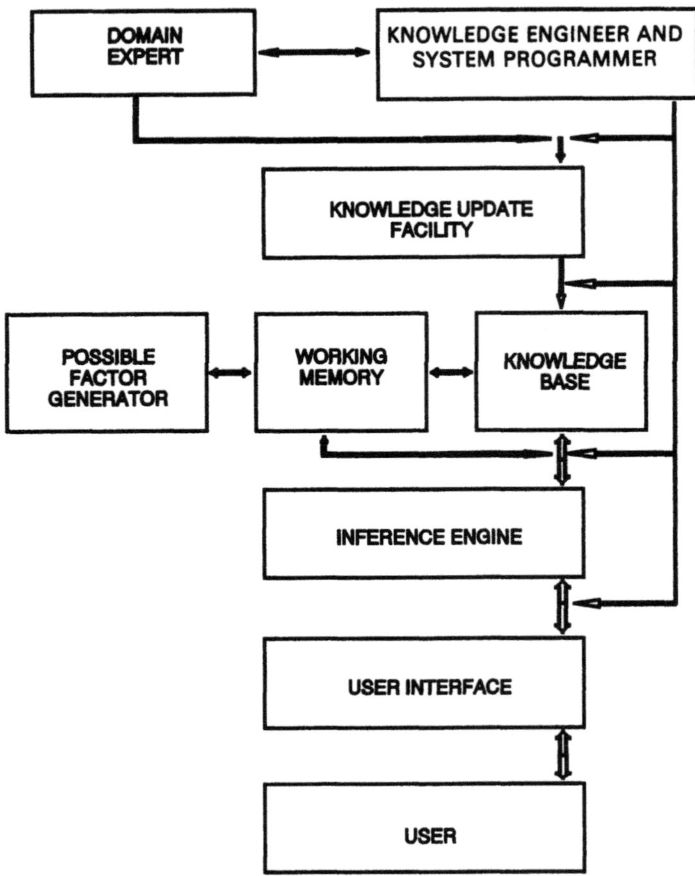

Fig. 13.1. The system framework of EXPS.

and the working memory. The inter-relationships and the information flows between components are represented by arcs. The knowledge is derived from the domain expert and coded by the knowledge engineer (system developer) into the knowledge base.

The knowledge base development process requires knowledge extraction or acquisition, formulation and subsequent representation. The domain-related selection criteria mentioned above are represented in the knowledge base. EXPS employs a rule-based knowledge base. The knowledge is represented solely in terms of predicate calculus in Prolog. The pick-and-place machine selection knowledge is derived from domain experts and documents of SMC placement equipments. The system operation is controlled by selecting the rules to use, accessing and executing those rules with exact reasoning, and determining when an acceptable solution has been found. If an exact solution is not available, some alternatives might be offered by the system with inexact reasoning.

EXPS also uses fuzzy production rules to represent knowledge. Facts and rules apply fuzzy quantifiers to deal with uncertainties. In some cases, a frame data structure is used within a rule to formulate the representation of rules and facts. The frame data structure is comprised of slots, and can result in the construction of a moderately complex rule. For a qualitative factor, a frame would be:

```
Frame Name:          select
Alternative:         (alternative or solution)
Fuzzy Quantifier:    (CF)
Relation 1:          exclude
Attribute 1:         (qualitative attribute 1)
Relation 2:          exclude
Attribute 2:         (qualitative attribute 2)
   ⋮
Relation N:          exclude
Attribute N:         (qualitative attribute N)
```

A frame for quantitative factors would be:

```
Frame Name:          choose
Fuzzy Quantifier:    (CF)
Alternative:         (alternative or solution)
Relation 1:          greater_than (or less_than)
Attribute 1:         (quantitative attribute 1)
Relation 2:          greater_than (or less_than)
Attribute 2:         (quantitative attribute 2)
   ⋮
Relation N:          greater_than (or less_than)
Attribute N:         (quantitative attribute N)
```

For rules containing qualitative attributes, the 'exclude' is a necessary term to construct the rules. The representation of fuzzy production rules in a frame data structure in EXPS is:

```
Rule = select (alternative or solution) IF
        exclude (qualitative attribute 1) AND
        exclude (qualitative attribute 2) AND
        exclude (qualitative attribute 3) AND
           ⋮
        exclude (qualitative attribute N). (CF = μ)
```

For rules in which quantity attributes are involved, either 'greater_than' or 'less_than' is a necessary term to construct the rules. The representation of fuzzy production rule in a frame data structure in EXPS is:

Rule = choose (alternative or solution) IF
greater_than (quantitative attribute 1, corresponding value 1)
AND
greater_than (quantitative attribute 2, corresponding value 2)
AND
less_than (quantitative attribute 3, corresponding value 3)
AND
⋮
less_than (quantitative attribute N, corresponding value N).
(CF = μ)

Two kinds of frames are implemented. The first type are the default data frames containing information on SMC placement machines on the market. The second type are the input frames which store the information supplied by the user.

In EXPS, knowledge update can be performed by the domain expert. There is no need for software programming knowledge. The person performing the knowledge update (e.g. the domain expert) has to be familiar with the procedure required. This feature present in EXPS allows the system to keep up with the rapidly evolving surface mount domain knowledge.

The inference mechanism used in EXPS employs the backward chaining approach in a depth-first searching scheme. It utilizes the uncertainty management methodology in its reasoning requiring the use of fuzzy quantifiers in terms of the measures of confidence and certainty factors. The fuzzy quantifiers allow the system to deal with uncertainty in a consistent manner, and come up with an ordered list of likely alternatives (i.e. advice).

The backward chaining algorithm used in EXPS works as follows:

1. Find the first related rule (goal) which contains a certain alternative A (object) in its proposition.
2. If there are no such rules, stop the search.
3. Otherwise, for each such rule (R), use the following logic to determine/evaluate the proposition of the rule:
   For each clause (C) in the proposition of R:
      Find the CF (Certainty Factor) of R.
      Look in the working memory for the data in the clause.
      If there is no fact in the working memory related to that data,
         Start from the top with that fact as the current fact.
         Collect data through knowledge acquisition (from the user).
         If the data is quantitative,
            The possible fact generator is activated.
            Scan the knowledge base to find other data which are also qualified.
            If other qualified data does exist,

The possible fact generator generates a list of the qualified data as a unique fact; then add this fact into the working memory.
Collect the MCs from the user; then choose the smallest one.
Evaluate the suitability of rule for the alternative using the information (MC and CF) in the working memory.
If the evaluation of the rule is a non-zero numerical value make the conclusion (given in the rule) shown.
Discard the foregoing rule.
4. Discard related facts.

The user interface in EXPS is menu driven and uses English-like inquires. A user will interact with EXPS in an interactive manner usually through choices in a menu. Inputs to EXPS could be qualitative or quantitative. An explanation facility and a context specific help facility are also provided for EXPS.

13.4.6 Software architecture

The core program in EXPS is a system controller that manages other application programs with different functions. There are five application modules in EXPS. EXPS uses a hierarchical structure to fulfill the objectives of the knowledge base, the inference engine, the possible factor generator, the user interface and the system utilities.

The overall system architecture of the EXPS (Fig. 13.2) is modular. These modules are classified into management level (level I), system level (level II) and operation level (level III). The main controller module in level I controls all operations in the system. The software modules in level II implement the system's support functions and the modules in level III sustain functions run by the user.

The main controller module fulfills two functions, i.e. (computer) memory management and operation management. The user operates the system by providing certain instructions to the system. By using the functions performed by the modules in level II, the main controller module in level I activates the required working modules in level III according to the user's instructions. The information flow in the system is from level I via level II to level III.

EXPS uses three knowledge bases. The first, SPEC.KBA, contains rules (i.e. capability criteria) related to quantitative factors and attributes. Each rule is designed to select a specific pick-and-place machine based on the user's machine specification requirements as well as the work part restriction requirements. The certainty factor for each rule in this knowledge base is unity since the specifications and restrictions must meet a user's (buyer's) requirements without any variation. The second knowledge base, FUNC.KBA, contains rules (i.e. specification criteria) related to qualitative factors/attributes. Rules in this knowledge base are developed to manage

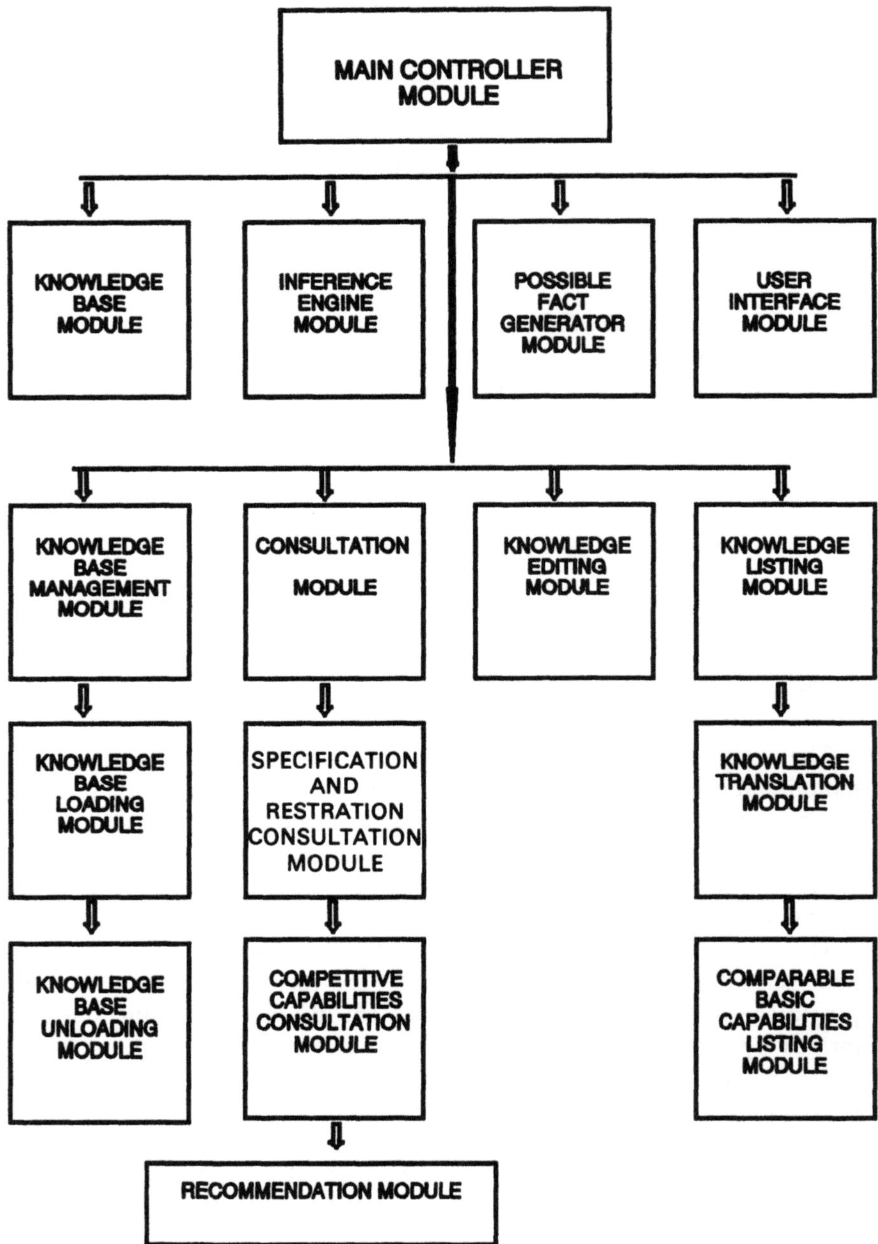

Fig. 13.2. The overall system architecture of EXPS.

qualitative criteria. Each rule is capable of selecting an appropriate pick-and-place machine according to the machine capabilities required by the user. The certainty factor of each rule is identified based on the mathematical model introduced previously. The third knowledge base,

FACT.KBA, contains facts (i.e. existing data) concerned with machine specifications and work-part restrictions. The FACT.KBA stores facts (i.e. information) on available SMC placement machines.

13.5 ACCOMPLISHMENTS AND LIMITATIONS

This research researched, designed and developed an expert system-based decision support system to assist in a multi-attribute decision-making scenario involving the selection of an SMC placement machine. This problem involves semi-structured decision making. Traditional computerized decision support systems approach the problem either with mathematical methods or with relation charts. Computerized data base systems assist in the decision support function, but are not an efficient and effective decision support mechanism. The decision support system should be easy to learn and to operate, and must reach a conclusion efficiently. The system is also expected to offer an optimal selection among the alternatives. The decision support expert system (EXPS) developed in this research meets all the criteria mentioned above.

Fuzzy set theory shifts the traditional probability-based methodology to the possibility theory-based methodology in problem solving. Therefore, the fuzziness as well as the uncertainty in a specific problem can be measured without the support of statistical data collected previously. EXPS uses the fuzzy set theory for problem solving in a simple and explicit manner.

The capabilities of EXPS can be enhanced. Some uncertainty factors which might affect the flexibility of an SMC pick-and-place machine have not been considered. These factors could be classified as process factors, machine design factors, SMT technology factors, technical support (from the machine supplier) factors and economic factors. The memory management scheme used can be enhanced. The knowledge update mechanism used can be improved. The search strategy used can be made more efficient. A valuable additon to EXPS would be a module to execute sensitivity analysis during the machine selection process.

13.6 CONCLUSION

The research presented in this paper is based on two techniques, i.e. fuzzy set theory and expert systems. The development of expert systems has resulted in the need to reason with imprecise data. Fuzzy set theory provides an alternative method of uncertainty management to those of classical logic and probability. This research coordinates the use of fuzzy sets and expert systems through the use of fuzzy set theory in an application on selecting an SMT pick-and-place machine.

REFERENCES

Abdou, G. and Dutta, S.P. (1990) An integrated approach to facilities layout using expert systems. *International Journal of Production Research*, **28**(4), 685–708.

Canada, J.R. and Sullivan, W.G. (1989) *Economic and Multiattribute Evaluation of Advanced Manufacturing Systems*, Prentice-Hall, Englewood Cliffs, NJ.

Kumara, S.R.T., Joshi, S., Kashyap, R.L. *et al.* (1986) Expert systems in industrial engineering. *International Journal of Production Research*, **24**(5), 1107–25.

Lea, C. (1988) *A Scientific Guide to Surface Mount Technology*, Electrochemical Publications, Ayr, UK.

Levine, P., Pomerol, M.J. and Saneh, R. (1990) Rules integrate data in a multicriteria decision support system. *IEEE Transaction On Systems, Man. and Cybernetics*, **30**(3), 678–85.

Luger, G.F. and Stubblefield, W.A. (1989) *Artificial Intelligence and the Design of Expert Systems*, Benjamin/Cummings, Menlo Park, CA.

Maiers J. and Sherif, Y.A. (1985) Application of fuzzy set theory, *IEEE Transaction On Systems, Man. and Cybernetics*, **15**(1), 175–89.

Negoita, C.V. (1985) *Expert Systems and Fuzzy Systems*, Benjamin/Cummings, Menlo Park, CA.

Prasad, R.P. (1989) *Surface Mount Technology Principles and Practice*, Van Nostrand Reinhold, New York.

Sarin, S.C. and Salgame, R.R. (1990) Development of a knowledge-based system for dynamic scheduling. *International Journal of Production Research*, **28**(8), 1499–512.

Shaw, M.J. (1988) Knowledge-based scheduling in flexible manufacturing systems: an integration of pattern-directed inference and heuristic search. *International Journal of Production Research*, **26**(5), 821–44.

Shieh, H. (1991) Developing a selection model for pick-and-place equipment in SMT. *Masters Thesis*, SUNY-Binghamton, New York.

Srihari, K. and Emerson, C.R. (1992) Knowledge based management of technology for PCB assembly, in *Management of Technology III*, Industrial Engineering And Management Press, Norcross, GA, pp. 838–47.

Weber, E.U. and Coskunoglu, O. (1990) Descriptive and prescriptive models of decision making: implications for the development of decision aids. *IEEE Transaction On Systems, Man. and Cybernetics*, **20**(2), 310–16.

Zadeh, L.A. (1987) *Fuzzy Sets And Applications: Selected Papers*, John Wiley, New York.

Zimmermann, H.J. (1985) *Fuzzy Set Theory And Its Applications*, Kluwer-Nijhoff Publications, Hingham, MA.

CHAPTER 14

FIXPERT: A rule-based system for workholding device selection of rotational parts

B. Bibanda, P.H. Cohen and C. Tunasar

14.1 INTRODUCTION AND SYSTEM OVERVIEW

Process planning is the determination of machining and assembly operations needed to produce a part. Computer-aided process planning (CAPP) systems are software tools to help automate this discipline. Of late, there has been considerable research effort directed towards the development of generative CAPP systems.

Workholding device selection, which is also an important issue, but has not received much emphasis in the past, directly influences workholding time, which usually makes up a large portion of unproductive time in a machining operation. The workholding time for part is defined as 'the time to unload and load a part and the pro-rated time to mount the workholder on the machine tool'. The high investment costs for tooling of numerically controlled machine tools emphasizes the importance of minimizing the unproductive time such as work holding time. The work holding time and the type of workholder to be used during an operation is not taken into account by most CAPP systems (Bidanda and Cohen, 1987).

The process of fixture or workholding device selection for rotational parts has traditionally been an iterative, trial and error procedure. Since a variety of fixtures (or workholding devices) for concentric, rotational parts for turning operations already exists, the primary issue here is the selection of the best, or 'optimal' fixture for a part given the following situation-specific information:

1. Part geometry.
2. Dimensional & geometric tolerances.
3. Machining operation on a given surface(s) of the part.

4. Cutting parameters for the operation.
5. List of fixtures available.
6. Specifications for each of the fixtures.

This paper aims at describing the development of FIXPERT, a computer aided fixture selection system for rotational parts that can serve as a 'post processor' for existing CAPP systems. The focus is on the development of the structure of the rule-based system, its rules and associated data bases.

FIXPERT is a rule-based system coded in Prolog, an expert system development tool. The overall structure of FIXPERT and its interface with the other modules is shown in Fig. 14.1. As can be seen, FIXPERT interfaces with an existing computer-aided design (CAD) system (AUTOCAD™ by AUTODESK Inc.). The interface to the CAD systems allows not only the insertion of tolerance information into the CAD drawing but also allows this information to be transferred to the knowledge base, thus taking into account the tolerance of the part when making a workholding decision.

As part of the computer-aided fixture selection package, the following software modules were developed:

1. A knowledge-based system to make fixturing decisions; these decisions are based on qualitative rules, quantitative models and heuristics.
2. A program to compute the minimum force needed to prevent a part from slipping in a three jaw chuck.
3. A program which uses a heuristic algorithm to choose a set of workholding devices to minimize non-productive time from a file containing a list of feasible workholding devices and associated holding surfaces for each operation in the process plan.

The file structure and formats needed to develop each of the software modules is discussed in Section 14.3. We now detail background issues in automated fixture design.

14.2 BACKGROUND

This section is reproduced from Bidanda and Muralikrishnan (1992). The process of fixture design is a combination of rule-based logic and appropriate expert input. The major work packages in fixture design include the identification of clamping and locating points, and then the selection of specific standard elements from a set of available base plates, locators and clamps. Traditionally, expert input is provided by the tool designer. An artificial intelligence (AI) based approach is relevant not only in the context of computerization to improve efficiency but also as an integrating link between CAD and computer-aided manufacture (CAM). One of the reasons for the relatively slow progress in automated fixturing is that CAD data bases containing product information are ill suited to fixturing needs. The

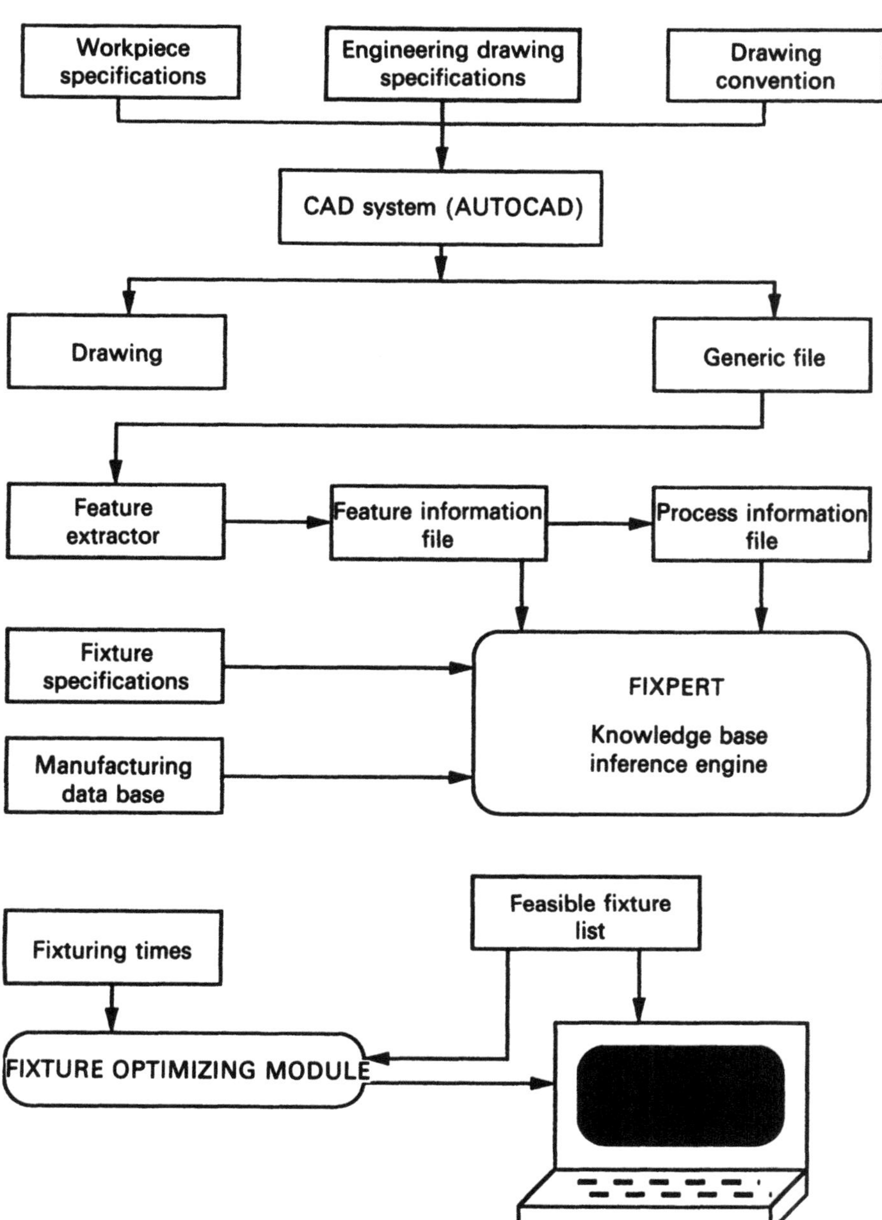

Fig. 14.1. System structure of FIXPERT.

kind of information needed for fixturing decisions (features, tolerances, surfaces needed to be machined, etc.) are seldom in a form that aids manufacturing analysis/planning. Also, while most drafting systems represent two- or 2.5-dimensional models, a fixture design involves three-dimensional interpretation of objects. For rotational parts, a three-dimensional interpretation from a two-dimensional model is not very complicated; the process is more complex when considering prismatic components.

Early efforts to computerize fixture design focused on providing a graphical interface and a library of standard components, from which a designer could choose elements to build a fixture. These decisions are made by the fixture designer and designs often vary with the style of the individual designer.

The focus of research in this area has shifted to the application of AI-based techniques to automate a larger portion of the design process. Typical inputs to an automated fixture design system are workpiece description including geometry, features, tolerances, machining allowances, machining operations and machine information. Outputs to be expected are fixture configurations for each machining operation. The processing involved includes interpretation of part features from the drawing, generation of workpiece orientation, calculation of fixturing and clamping coordinates, generation of alternative fixture configurations, and analysis of fixtures for sufficient location and stability. These problems are non-trivial; thus individual research efforts typically tend to address a subset of this area. The issues involved in automated design of fixtures can be classified as (1) schemes for representing the shape and structure information of a workpiece, in a form that will aid manufacturing analysis, and (2) establishment of rules for workholding, development of computer-based inference systems for fixturing decisions and implementation in the form of automated assembly.

14.2.1 Establishment of rules for workholding

We have classified the knowledge base required for fixturing decisions (for rotational components) into three distinct types of rule categories (Bidanda and Cohen, 1987):

1. Qualitative rules.
2. Quantitative rules.
3. Heuristics or rules of thumb.

Examples of qualitative rules used include:

1. If the holding surface is hexagonal, then a four jaw chuck cannot be used.
2. A part can be held between centers only if it has no through hole.

Quantitative rules relate machining and workpiece parameters. For example, if the diametral error due to tool and workpiece deflection (a function of cutting force, length of overhand, modulus of elasticity, etc.) is greater than the tolerance, then the workholding scheme is not feasible.

Examples of heuristic rules used include:

1. If a process plan specifies a speed greater than 400 r.p.m., then evaluate feasibility of using a counter centrifugal chuck (to prevent slippage).

2. When turning between centers, if the length/diameter ratio of the workpiece is greater than 6 and less than 10, add one steady rest to minimize sag and deflection.

Once a set of rules have been developed and stored in a knowledge base, they are applied depending on workpiece characteristics. Rule-based systems can be developed for such problems. Software programs based on recursive languages such as LISP and Prolog are especially well suited for these tasks.

In the case of rotational parts, the problem is predominantly one of fixture selection rather than of design. An algorithm for optimal selection of workholding devices for rotational parts was developed by Bidanda and Rajgopal (1990). This algorithm is based on dynamic programming and chooses a fixture for each operation of the manufacture of a part (based on the process plan). Optimal fixtures are chosen from a set of feasible fixtures for each operation, such that set-up time is minimized.

The problem of fixturing prismatic parts poses a greater challenge because the parts are more varied in geometry and the expert rules need to have more of a design and analytical basis. Chou, Chandru and Barash (1989) developed a mathematical theory for the analysis and synthesis of fixtures for prismatic parts based on screw theory. Forces applied by locators and clamps and those generated by machining operators are modeled as wrenches (a load vector and a couple along the same axis). Fixturing solutions are then tested for four functional requirements: locational stability, deterministic workpiece location, clamping stability and complete restraint.

A similar approach is also suggested by Gandhi and Thompson (1986). Here machining, clamping, frictional forces and torques are expressed by a resultant wrench. The system of forces is solved for a static equilibrium to determine reactions at the locators. The choice of the support points is dictated in part by these solutions. Fixture design is then evaluated to calculate the deflections of the part and the result of this analysis may necessitate redesign of the fixture if necessary. Modular fixtures provide an opportunity to configure and automatically assemble fixtures for prismatic parts from a combination of standard elements. Much of the AI work on prismatic fixture design is centered around the use of modular fixtures, although the practical realization of automatic assembly to a reasonable degree of accuracy is not yet economically feasible.

Nnaji and Lyu (1990) developed rules for expert fixturing of regular polygonal prisms for face milling operations. The primary input to the program is the workpiece description (which is constrained due to the assumption of a polygonal prism); system output includes a data file containing the coordinates of fixturing points and a list of the modular fixture elements. The 3-2-1 method of locating and clamping is adopted, and equations to calculate the six fixturing locations and three clamping points are derived. The first locating plane (three locators) is constructed such that the area of the triangle formed by the three points is large enough to maximize stability. For prisms with an even number of faces, the plane that is opposite and parallel to the surface to be milled is chosen. However, the prisms with an odd number of faces have an edge that is opposite to the surface to be machined. In this case, the first plane is constructed based on the two adjacent faces of the opposite edge. This ensures that the surface to be milled remains perpendicular or parallel to the milling cutter axis. The second plane, having two locators, is chosen in a manner that maximizes the distance between the locators. Also, these locators have the same X axis and coordinate as two of the locators in the first set. Their Y and Z coordinates are selected so that they do not interfere with the tool path. The third plane is related to the second in that the third locator should be at the same height as in the second plane. Finally, clamping is effected against the locators in the second and third planes.

Markus *et al.* (1984) developed an expert system to assist in planning the initial configuration of fixture modules for simple prismatic parts. Workpiece description and clamping/support points are input to the system and result in the generation of fixture configurations The fixture is composed of seats (rest buttons) and towers (a functional unit of the fixture holding one or more seats) and clamps. The base of the fixture is a pallet with slots machined in the form of an X–Y grid. Towers and clamps are positioned along these slots in a manner similar to commercial modular fixtures. Tower types are differentiated based on whether they locate in the horizontal direction or the vertical direction or are used for clamping. A tower is characterized by its type, orientation and direction of the slot (X or Y) where it is fixed. In the first step of the design process, a collection of towers is identified based on the clamping and support coordinates entered. Subsequent design is completed using these towers only. A designer directing the search process cannot set priorities and discard alternative solutions. Once designed, the fixture is assembled manually.

Fixture configuration involves realization of the relative positions of the modules and the workpiece. Gandhi and Thompson (1986) demonstrate that spatial relationships between two objects, such as 'fit', 'against' and 'coplanar', can be modeled mathematically. Geometric features such as face, edge, vertex, shaft and hole are associated with the contact locations between the workpiece and the fixture, and the spatial relationship between them is defined analytically (e.g. an edge is against a face if it lies completely in the plane of the face). Once the locator positions are determined using the forced analysis described earlier, this knowledge of spatial relationships is used to select appropriate fixture modules. For example, locating a cylinder on a V-block can be considered to be satisfying the relations 'face (of the V) against shaft'. Fixture modules are selected such that they satisfy the spatial relationships determined.

Feasibility of fixture layout depends on the absence of element interference. While this is true even of manually assembled designs, it is especially significant in an automated assembly approach, where, in addition to the interference between fixture elements, there may be interference between the set-up hardware (typically the robot) and the fixture elements. Youcef-Toumi, Bausch and Blacker suggest a 'holistic' approach to determine layout feasibility for modular fixtures. Workpiece layout requirements are transformed to element constraints. Their analysis centers around the physical characteristics of the modular elements of the fixturing system with particular attention on the vertical support modules used for location. For a given fixturing system, parameters such as 'minimum distance between support points that can be supported by non-interfering modules' and 'furthest possible distance between two support points where their candidate modules could interfere' are defined. Candidate modules for any support point are analyzed for interference with candidates for other support points. Modules are classified as interacting if there is a possibility of their interfering with other modules based on comparisons with the parameters defined. For example, if two support points are closer than the furthest possible distance described above, then they would be termed interacting. Actual interference analysis of interacting modules is accomplished by looking for intersections at boundaries of the respective CAD models. All such interactions are considered and the layout is deemed infeasible if there is at least one support point for which there is no single non-interfering module. Although this approach has been used in the analysis of sheet metal fixture, it is general enough to be applied to other modular fixturing applications.

Our focus was to develop a workable expert system for fixture decisions of concentric, rotational parts. We now discuss the structure and format of the main modules of the FIXPERT system.

14.3 THE FIXTURING DECISION PACKAGE

In order to make a fixturing decision, the following data bases and files were developed:

1. A feature information file for each component part.
2. The process information file of the component part.
3. A manufacturing data base containing information on different materials.
4. A fixture information file.

Each of the above modules are stored as ASCII text files. The first two files (feature information and process information) are part dependent. The third file (manufacturing data base) is a generic file, that is both part and shop-floor independent. The last file (fixture information) is a shop-floor dependent file. The fixtures available will be specific to a particular situation (or shop floor). The rule-based system was developed in Prolog and thus the system structures of the entire package are described in readable pidgin Prolog. The interface modules of FIXPERT listed above are now described.

14.3.1 Feature information file

After an engineering drawing of the part is drawn on a CAD system (AUTOCAD™ was chosen here), a feature information file (FIF) of the component part is built up using feature extraction algorithms (Bidanda and Cohen, 1987). The Part FIF contains information on each surface of the workpiece stored as a clause (or fact). This information includes dimensional and geometric tolerances based on ANSI Y14.5 (American National Standards Institute, 1987). Each of these clauses contains the predicate name followed by information on different objects in this clause (or fact) within parentheses. Feature (or surface) information is stored in a predicate name 'feature' and the format of this clause is shown below:

feature (feature number,
  surface type,
  surface location,
  starting point *x* coordinate,
  starting point *y* coordinate,
  end point *x* coordinate,
  end point *y* coordinate,
  length of the surface,
  diameter of the surface,
  height of the surface,
  information on whether machining is required,
  the tolerance associated with the surface,
  the cross section of the surface).

A typical feature clause containing the information described above is shown below:

feature(2,"hor", "ext", 5.5, 4.5, 6.5, 4.5, 1.0, 1.5, 1.5, y, "[ve50]", "cir").

This clause can be interpreted as follows: this is the feature representation of surface #2 of the sample component part. The surface is horizontal; it is an external surface, and this surface starts at the coordinates 5.5,4.5, and ends at the coordinates 6.5,4.5. In addition, the surface has a length of 1.0 units, a diameter and height of 1.5 units (the diameter equals height since the surface is horizontal), the surface needs to be machined; the tolerance associated with the surface is a vertical external tolerance of 0.005 units, and the cross-section of the surface is circular.

Since each clause describes one feature (or surface), a part with say 11 features will need a Part FIF containing 11 clauses. It may be noted however, that since the class of parts considered here is limited to concentric, rotational parts, only surfaces above the principal axis of the part are deemed necessary to completely describe part geometry due to the inherent symmetry of this class of parts.

A component part can contain a variety of 'features'. In the current implementation of FIXPERT, the features of the component part are stored in a file with a name specified by 'Part Name' and a suffix '.FTR'.

14.3.2 Process information file

This file contains the previously developed process plan of the component part. The process information file (PIF) may be manually generated or generated as an output of a CAPP package (Chang and Wysk, 1985; Wang and Wysk, 1987). It is ideally suited as an interface for a microcomputer-based process planning package such as Turbo-CAPP (Wang and Wysk, 1987). In its current implementation, FIXPERT requires the PIF to consist of a list of clauses, with each clause containing details of an individual operation. The clause is named 'operation' and its structure is as shown below:

operation (operation number,
  operation type,
  operation description,
  list of surfaces to be machined during the operation,
  speed,
  feed,
  depth of cut).

An example operation clause is listed below:

operation(60,'turn','rough',[5,6],220,.0154,.0100).

This clause is interpreted as follows: the clause represents an operation numbered 60 in a process plan. The operation is a rough turning operation; the surfaces to be machined here are surfaces 5 and 6; the speed of the machine is to be set at 220 feet per minute, and a feed rate of 0.0154 in per revolution, and a depth of cut of 0.100 in are the cutting parameters for the operation.

14.3.3 Manufacturing data base

Information about different types of raw materials are stored as clauses in the manufacturing data base. The clause is named 'material' and its format is as shown below:

material (name of the material,
  specification of the material,
  the specific horsepower of the material, and
  modulus of elasticity of the material).

A typical 'material' clause will read as follows:

material(steel,1010,0.580,30).

This clause can be interpreted as follows: this represents a steel with a specification of 1010, a specific horsepower of 0.580 h.p./in^3/min, and a modulus of elasticity of 30×10^6 p.s.i. While many more material details can be useful and normally need to be added to a manufacturing data base, the manufacturing data base used here contains only information relevant to making fixturing decisions.

14.3.4 Fixture information file

This file contains information on all the fixtures available in a specific situation (e.g. a particular shop floor). Each type of workholding device is described as a separate predicate. Sample predicates used in the knowledge base and associated pieces of information contained in each predicate are ennumerated below.

1. Three jaw chuck (brand of the chuck,
   maximum outside diameter that can be held by the chuck,
   smallest outside diameter that can be held by the chuck,
   maximum inside diameter that can be held by the chuck,
   smallest inside diameter that can be held by the chuck,
   the length of each chuck jaw,
   the rated speed of the chuck).
2. Collet (brand of the collet,
   maximum outside diameter that can be held by the collet,
   minimum outside diameter that can be held by the collet).
3. Mandrel (dummy variable since the brand is unimportant to decision
   minimum diameter of hole that can be used as holding surface
   maximum diameter of hole that can be used as a holding surface).
4. Toolholder (brand of toolholder,
   length of tool overhang,
   material of the toolholder,
   material specification of toolholder, material name,
   cross-section of the toolholder,
   diameter or width of the toolholder).
5. Machine tool (brand of machine tool,
   machine tool error specified by manufacturer).

14.4 SYSTEM STRUCTURE

The opening menu of FIXPERT and options available to the user are shown in Fig. 14.2. As can be seen, the fixture decision package itself is divided into five different levels and other assorted utilities. The user has the

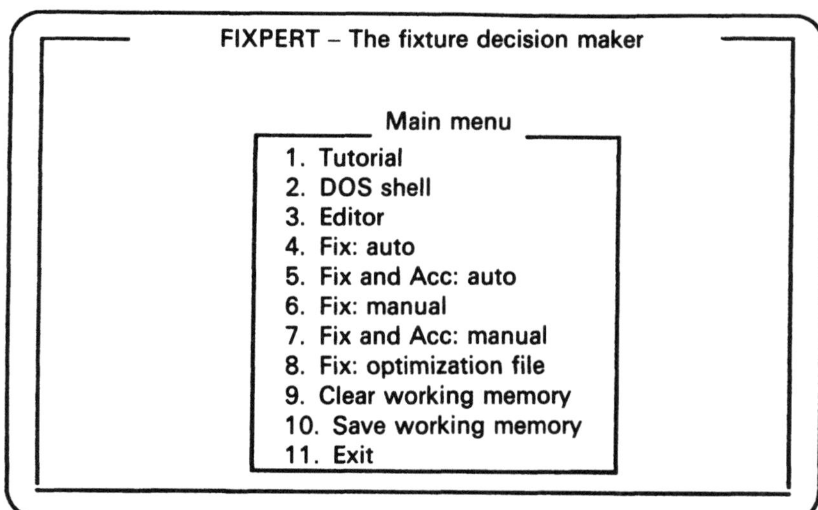

Fig. 14.2. Opening menu.

option of choosing the level of detail needed in making the fixture selection decision. The different levels range from obtaining a comprehensive list of feasible fixtures for all operations in the process plan (as contained in the PIF) and all feasible holding surfaces (from information contained in the FIF) to a detailed list of fixturing accessories and recommendations for a specific operation and holding surface. Each of these options (Fig. 14.2) of FIXPERT is described below:

1. Tutorial. A description of FIXPERT and instructions required to successfully execute the rule based system is contained here. The tutorial is an ASCII based file and may easily be modified or updated using a text editor or word processor.
2. DOS Shell. This option uses the SHELL command to enable the user to perform tasks (within the DOS environment) outside of FIXPERT. However FIXPERT remains in the RAM and the user can return to FIXPERT by simply typing 'EXIT' at the DOS command prompt.
3. Editor. This feature allows the user to modify the rule base during execution, if required.
4. Fix: Auto. Here the program, after prompting the user for the part name, takes each operation in the process plan of the part, displays the surfaces to be machined and prompts the user for a list of preferred holding surfaces. The program then searches its rule base, determines the workholding devices that can be used, and then displays a list of all feasible fixtures and fixture accessories that can be used (taken from the list of available fixture and accessories). This cycle is repeated for all operations in the process plan. This level is primarily used when the user

is looking to evaluate holding surfaces for all operations in the process plan.

5. Fix & Acc: Auto. The main difference between this and the previous level is that, instead of only listing the feasible fixtures for preferred holding surfaces, the program also supplies a list of recommended fixture accessories and also checks to see if the workholding device is feasible from the point of view of the tolerance specified in the part drawing for the machining surfaces under consideration. Thus, this level provides a more detailed evaluation of workholding possibilities.
6. Fix: Manual. This level is similar to option 4, the main difference being that it is more interactive in that, instead of the program cycling through the entire process plan, operation by operation, the user has the choice of specifying the operation and holding surface. Thus, this level is primarily useful if workholding possibilities for a single operation are to be ennumerated and evaluated as a preliminary search.
7. Fix & Acc: Manual. This is similar to option 5; however, it gives the user the choice of choosing the operation number and preferred holding surfaces. It then provides a list of feasible fixtures, fixturing accessories and checks the feasibility of the workholding device. The tolerance feasibility checks in this package were limited to three jaw chucks.
8. Fix: Optimization File. At this level, the program execution is fully automatic and the user only needs to input the part number. Then, for each operation in the process plan, all possible workholding devices and all possible holding surfaces are written to a disk file. This level is used when the fixture optimization module, which is described later, is to be utilized.
9. Clear Working Memory. The computer RAM is initialized by clearing it.
10. Save Working Memory. The computer RAM is written to an ASCII file.
11. Exit. The user can exit FIXPERT to return to DOS.

FIXPERT uses three different windows to communicate with the user. These are (1) a question and answer processor, (2) a data window, and (3) a window when FIXPERT's decisions are communicated. These are shown in Fig. 14.3.

14.4.1 System details

The rule-based system is focused around a predicate named 'fixtures'. This predicate consists of the following objects:

fixtures (operation number, fixture name, brand, holding surface number).

Since the operation number and holding surface(s) are either user supplied or program supplied, these two parameters are already instantiated. Thus, if a set of rules is defined for a particular type of workholding device, say a

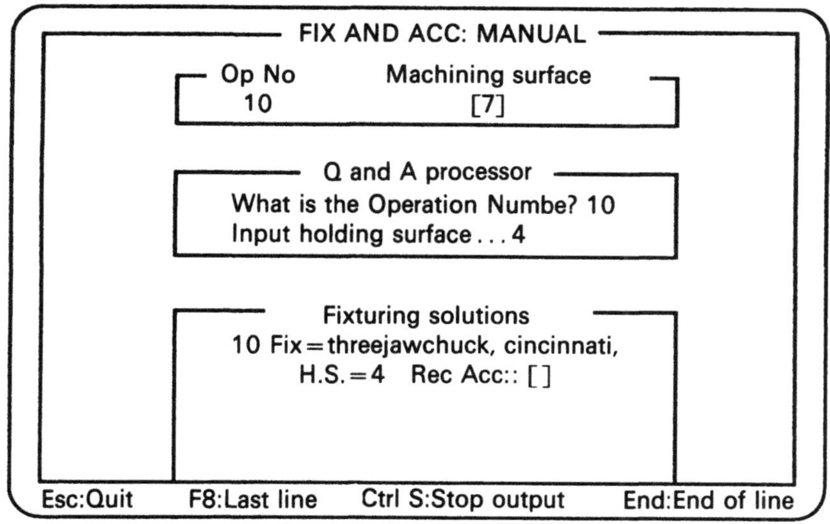

Fig. 14.3. Working menu – option 7.

fixture type named 'three jaw chuck' which has a brand name 'cincinnati', the structure of the rules will be as follows:

fixtures(operation #, threejawchuck, cincinnati, holding surface #):-
  Rule 1 .... and
  Rule 2 .... and
  .... and
  Rule $(n-1)$ .... and
  Rule n ....

Now, for a given operation number and holding surface, this predicate will succeed if Rule 1, Rule 2,..., Rule n, are true. Moreover, the program backtracks to determine all possible (fixturing) solutions. Hence, for a given operation number and holding surface number, the program deduces a list of all possible fixtures since there is generally more than one feasible solution.

Thus, the rule-based system can be given a goal with the operation number and holding surface already instantiated, and the fixture name and brand uninstantiated. For example, if the rule-based system is given the goal of finding all possible fixtures for operation number 10 and a holding surface 2, then the goal can be formulated as follows:

fixtures(10,X,Y,2).

Due to Prolog's unification function, the systems searches for all possible solutions to this goal, and thus X and Y are instantiated, in turn, to all feasible fixtures.

The program levels 1, 2 and 3 (described earlier) formulate the goal 'fixtures'. However, for levels 4 and 5, additional predicates to recommend a list of fixture accessories are formulated. These are considered to be devices such as steady and follower rests which are used to minimize workpiece deflection and hence machined part error. A predicate for fixture accessories 'fix_acc' is defined as follows:

fix_acc (operation number, fixture name,
holding surface, name of fixture accessory).

If this goal succeeds, then this fixture accessory is added to a list of fixture accessories associated with the operation number, holding surface and fixture.

A third class of predicates for assessing the feasibility of three jaw chucks from a tolerancing perspective named 'fixcheck' is defined. This predicate computes the estimated deflection of the workpiece and the tool, and then calculates the estimated diametral and runout errors. Hoever, this predicate is different from the other predicates described earlier. It always succeeds when a feasible fixture is a three jaw chuck and computes the estimated error. Thus, the output of this predicate (the computation of error) is a side effect, i.e. secondary to success of the predicate.

14.4.2 Coding details

The code was developed in the Borland International Inc. version of Prolog – Turbo Prolog, version 1.1. The entire system runs on a PC-DOS compatible microcomputer with at least 384 K of memory. Since all of the associated files (process plan file, fixture information file, feature file and manufacturing data base file) are ASCII text files, they can be edited using any standard text editor.

14.4.3 Modularity of the package

One of the primary objectives in structuring the fixturing decision module was to make it as modular as possible. It is possible to add a new fixture to the data base. For example, say a new three jaw chuck is to be included. A clause describing this workholding device can easily be added to the existing data base. The clause would have the following format:

```
ThreeJawChuck (threejawchuck,New_Tool,
  /** other relevant information **/).
```

Moreover, if a new kind of fixture, say a six jaw chuck, is introduced in the future, this can be added to the rule-based system in the manner described below.

First, a clause for a six jaw chuck will need to be defined. This clause will be defined such that it will contain all of the information needed to make fixturing decisions. This definition will be part of the knowledge base itself. Then, a set of rules governing the use of this type of chuck will need to be formulated and this can be included in the knowledge base of the system using the following format:

fixtures(Operation Number, six jaw chuck, brand, Holding surface):-
{rules governing the use of this work
holding device will be placed here}

Lastly, information on each individual six jaw chuck available will also be included in the file containing fixture information. The format of this information will follow the format of the clause defined in the rule-based system. Now, the inclusion of new type of six jaw chuck is complete and will be taken into account when making fixturing decisions.

14.4.4 Chucking force module

The chucking module is a microcomputer-based package and is meant to provide supplementary information to the user. The program requires the following user input:

1. Part name.
2. Raw material of the part.
3. Material specification.
4. Operation number.
5. Coefficient of friction between the part and chuck jaw.

The program estimates the cutting force by consulting the manufacturing data base and determining the specific horsepower of the workpiece material, and from the cutting parameters (found in the process plan). It then estimates the weight of the part based on the average diameter (from the file containing feature information) and density of the raw material.

These values are then substituted into equations which evaluate the holding force needed by each of the three jaws for all possible positions of each chuck jaw. This value is the minimum force with which each jaw will need to clamp the part.

The program then estimates the loss in force due to centrifugal force (from the speed of the operation, extracted from the process plan) and thus a final minimum force with which each jaw will need to clamp the part is computed.

14.4.5 Fixture optimization module

As mentioned earlier, the 'Fix: optimization file' option in FIXPERT ennumerates a list of all feasible fixtures with all feasible holding surfaces. This is written to a disk file. A fixture optimization module which utilizes a heuristic algorithm to minimize the total load/unload time and set-up time for the entire process was formulated, coded and implemented. The rationale behind the concept of fixture optimization is described below.

A typical list of feasible fixtures is shown in Table 14.1 for two different operations. Each fixture has a load/unload time associated with it. This includes time required to load and/or unload the component part between operations. In addition, each fixture will have an associated set-up time, i.e. time to remove the fixture and to load the fixture on a lathe. Each operation in the process plan will thus have a load/unload time, and a setup time.

Table 14.1. Sample list of feasible fixtures

Operation	*Fixture name*	*Brand*	*Holding surface*
10	threejawchuck	cincinnati	2
10	threejawchuck	cincinnati	4
10	threejawchuck	cincinnati	6
20	threejawchuck	cincinnati	2
20	threejawchuck	cincinnati	4
20	turn_between_centers	xx	1

(*a*) *Problem formulation and algorithm*

The objective here is to minimize $\Sigma_{i=1}^{n}[(L/UL)_i+S_i]$. Thus, we seek to minimize the total workholding time for a batch of parts during their sequence of manufacture as specified in the process plan. Assuming there are '*n*' operations in the process plan, the following notation is employed.

F_i = fixture for the *i*th operation.
S_i = set-up time for the *i*th operation.
$(L/UL)_i$ = load/unload time for the *i*th operation.
$(HSurf)_i$ = Holding surface for the *i*th operation.
B_i = fixture brand for the *i*th operation,
M = a binary viariable which denotes the machining direction with reference to a holding surface and the surfaces to be machined during the operation. It can take the value 'left' or 'right'.
PM = Previous machining direction. This is also a binary variable, similar to '*M*'. However, it denotes machining direction in the previous operation, which is governed by (1) the holding surface of the fixture of choice and (2) the surfaces to be machined, for the previous operation.

The rules that govern the set-up and load/unload time for the *i*th operation are as follows:

Rule 1: The setup time for operation $i=0$, if the fixture used for the operation is the same as the one used in operation $(i-1)$.

$$S_i = 0 \text{ if } \quad F_{i-1}=F_i, \text{ and } B_{i-1}=B_i.$$

Rule 2: The set-up AND the load/unload time for operation $i=0$ if the fixture AND holding surface used for the operation is the same as the ones used for operation $(i-1)$.

$$(L/UL)_i = 0 \text{ if } F_{i-1} = F_i, \text{ and}$$

$$HSurf_{i-1} = \text{HSurf}_i, \text{ and}$$

$$B_{i-1} = B_i, \text{ and}$$

$$M = PM.$$

A heuristic algorithm that will provide a solution to this problem was developed. This algorithm optimizes the workholding time for each step (or operation) in the process plan taking into consideration the feasible fixtures in the current operation and the fixture selected in the previous operation. The different steps of the proposed heuristic are described below:

1. For the first operation in the process plan, evaluate the list of feasible fixtures and choose the fixture which, for a given batch size will have the smallest set-up plus load/unload time.
2. Evaluate each subsequent operation. For each subsequent operation, select the fixture which has the minimum setup plus load/unload time, taking into consideration the fixture selected in the previous operation.

Thus, if the fixture and holding surface combinition selected for operation $i-1$, with a holding surface H_{i-1}, is also a feasible fixture and holding surface combinition for operation i, then this fixture will also be the fixture of choice for operation i, as long as the machining direction is the same as the previous machining direction, since now, the marginal set-up and load/unload times are 0.

However, if the fixture selected for operation $i-1$ is also a feasible fixture for operation i, with a different holding surface, then this fixture may or may not be the next best choice for operation i. This is because even though marginal set-up time $=0$ if the original fixture is retained, there might be another feasible fixture whose load/unload time is low enough to offset its set-up time.

Step 2 is repeated for all subsequent operations in the process plan. In the next section, the fixtures of choice for four sample parts using the algorithm described above will be shown. This solution assumes the batch of parts are completed each operation at a time. In other words, if a batch of 50 parts had to go through operations 10 and 20, then all the 50 parts complete operation 10 before operation 20 is begun.

This algorithm appears to provide the optimal fixture sequence and workholding times for most realistic cases.

14.5 SAMPLE SESSION

This section presents a sample session demonstrating the use of FIXPERT optimization module. A sample part named 'S4' is considered (see Fig. 14.4). This part has eight features and 11 operations to be performed as can be seen from Tables 14.2 and 14.3, respectively.

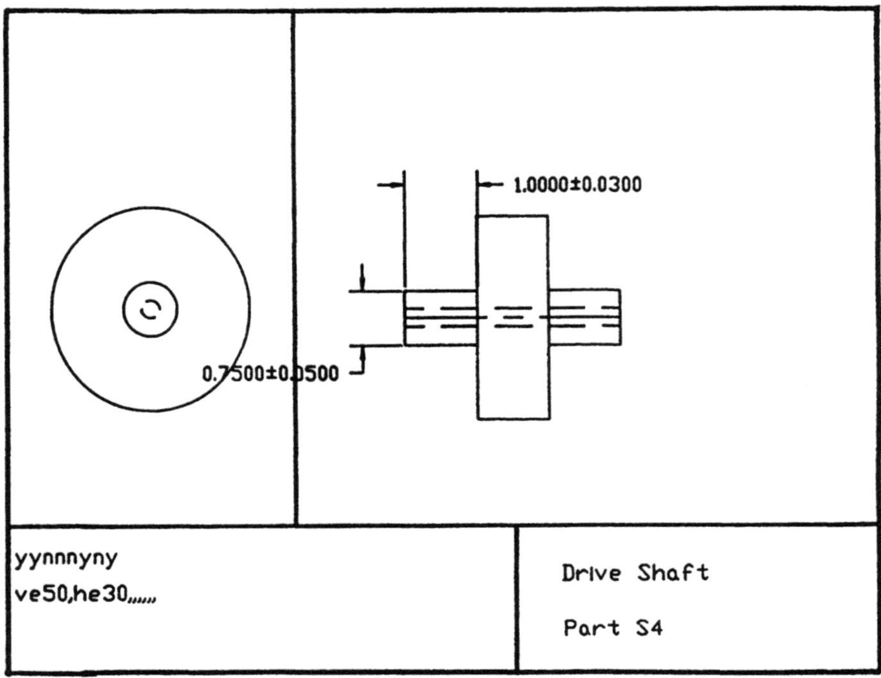

Fig. 14.4. Drawing of part S4.

Table 14.2. Feature information file for part S4

```
feature(1,"fce","lft",5.5000,4.8750,5.5000,5.2500,0.00,0.00,0.38,"y",["ve50"],"cir")
feature(2,"hor","ext",5.5000,5.2500,6.5000,5.2500,1.00,0.75,0.00,"y",["he30"],"cir")
feature(3,"ver","ext",6.5000,5.2500,6.5000,6.2500,0.00,0.00, 1.00, "n",[""],"cir")
feature(4,"hor","ext",6.5000,6.2500,7.5000,6.2500,1.00,2.75,0.00,"n",[""],"cir")
feature(5,"ver","ext",7.5000,6.2500,7.5000,5.2500,0.00,0.00, 1.00, "n",[""],"cir")
feature(6,"hor","ext",7.5000,5.2500,8.5000,5.2500,1.00,0.75,0.00,"y",[""],"cir")
feature(7,"fce","rgt",8.5000,5.2500,8.5000,4.8750,0.00,0.00, 0.38, "n",[""],"cir")
feature(8,"hor","int",5.5000,5.0000,8.5000,5.0000,3.00,0.25,0.00,"y",[""],"cir")
```

Program execution begins either by typing 'FIXPERT' which invokes the executable file or through Prolog by compiling and running. The opening menu having 11 choices is displayed (recall Fig. 14.2). From this main menu, option 8 is selected which will execute the fixture optimization

Table 14.3. Process information file for part S4

```
operation(10,"face","blank",[7],100,0.015,0.050)
operation(20,"turn","rough",[6],300,0.0236,0.100)
operation(30,"turn","finish",[6],500,0.0015,0.020)
operation(40,"turn","finish",[5],400,0.0015,0.030)
operation(50,"turn","finish",[4],400,0.0015,0.002)
operation(60,"turn","rough",[2],300,0.0236,0.100)
operation(70,"turn","finish",[2],400,0.0015,0.020)
operation(80,"turn","finish",[3],400,0.0015,0.020)
operation(90,"drill","blank",[8],140,0.0236,0)
operation(100,"ream","blank",[8],200,0.0236,0)
operation(110,"part","blank",[1],175,0.0188,0)
```

routine that creates a listing of all possible workholding devices. A working menu is displayed next (recall Fig. 14.3). This level is fully automatic, the user only needs to input the part number inquired by the question and answer window. Typing in the part number of 'S4' starts execution. During execution of this level, all possible workholding devices and all possible holding surfaces for every operation are written to a disk file named 'S4.OPT'. The user can see the decisions made through data and solution windows. After the completion of this process, menu option 11 exist from FIXPERT. Table 14.4 contains the output of this optimization module containing all possible fixture–holding surface combinations for each operation of part 'S4'.

Table 14.4 Feasible fixtures for part S4

Operator	*Fixture name*	*Brand*	*Holding surface*
10	threejawchuck	cincinnati	2
10	threejawchuck	cincinnati	4
10	threejawchuck	cincinnati	6
20	threejawchuck	cincinnati	2
20	threejawchuck	cincinnati	4
20	turn_between_centers	xx	1
30	threejawchuck	cincinnati	2
30	threejawchuck	cincinnati	4
30	turn_between_centers	xx	1
40	threejawchuck	cincinnati	2
40	threejawchuck	cincinnati	4
40	threejawchuck	cincinnati	6
40	collet	milwaukee	6
40	collet	acetools	6
40	turn_between_centers	xx	1
50	threejawchuck	cincinnati	2
50	threejawchuck	cincinnati	6
50	collet	milwaukee	6
50	collet	acetools	6
50	turn_between_centers	xx	1
60	threejawchuck	cincinnati	4

Table 14.4.—*Contd.*

Operator	*Fixture name*	*Brand*	*Holding surface*
60	threejawchuck	cincinnati	6
60	collet	milwaukee	4
60	collet	acetools	4
60	collet	milwaukee	6
60	collet	acetools	6
60	turn_between_centers	xx	1
70	threejawchuck	cincinnati	4
70	threejawchuck	cincinnati	6
70	collet	milwaukee	4
70	collet	acetools	4
70	collet	milwaukee	6
70	collet	acetools	6
70	turn_between_centers	xx	1
80	threejawchuck	cincinnati	2
80	threejawchuck	cincinnati	4
80	threejawchuck	cincinnati	6
80	collet	milwaukee	2
80	collet	acetools	2
80	collet	milwaukee	2
80	collet	acetools	2
80	collet	milwaukee	4
80	collet	acetools	4
80	collet	milwaukee	6
80	collet	acetools	6
80	turn_between_centers	xx	1
90	threejawchuck	cincinnati	2
90	threejawchuck	cincinnati	4
90	threejawchuck	cincinnati	6
90	collet	milwaukee	2
90	collet	acetools	2
90	collet	milwaukee	2
90	collet	acetools	2
90	collet	milwaukee	4
90	collet	acetools	4
100	threejawchuck	cincinnati	2
100	threejawchuck	cincinnati	4
100	threejawchuck	cincinnati	6
100	collet	milwaukee	2
100	collet	acetools	2
100	collet	milwaukee	2
100	collet	acetools	2
100	collet	milwaukee	4
100	collet	acetools	4
110	threejawchuck	cincinnati	2
110	threejawchuck	cincinnati	4
110	threejawchuck	cincinnati	6
110	collet	milwaukee	4
110	collet	acetools	4
110	collet	milwaukee	6
110	collet	acetools	6

Having obtained all possible fixture-holding surface combinations, the next logical step is to select the best workholder for each operation in terms of minimizing overall set-up time. This is achieved by a separate algorithm described in section 14.3.5 and coded in Turbo-Pascal on a PC-DOS machine. The program reads in the optimization file that is shown in Table 14.4 (created by FIXPERT) and a data file containing set-up and load/unload times for each fixture type. It then finds the best fixtures for each operation as listed in Table 14.5.

Table 14.5. Fixture optimization for part S4

******THE FIXTURE OPTIMIZATION MODULE******

Pt Name\Dir: s4		*Batch size:*	*5.0E+01*	*Set-up load*
Operation	*Fixture type*	*brand*	*H.S*	
10	threejawchuck	cincinnati	2	2.80 37.50
20	threejawchuck	cincinnati	2	0.00 0.00
30	threejawchuck	cincinnati	2	0.00 37.50
40	threejawchuck	cincinnati	2	0.00 0.00
50	threejawchuck	cincinnati	2	0.00 0.00
60	collet	milwaukee	4	3.20 12.50
70	collet	milwaukee	4	0.00 0.00
80	collet	milwaukee	2	0.00 12.50
90	collet	milwaukee	2	0.00 12.50
100	collet	milwaukee	2	0.00 0.00
110	collet	milwaukee	4	0.00 12.50

14.6 SUMMARY AND CONCLUSIONS

FIXPERT is a user friendly rule-based system that selects workholding devices for concentric, rotational parts, using inputs from a CAD package, a CAPP package and a manufacturing data base. The system is modular and can easily be modified or updated to include new workholding devices, new materials and new inferential rules. It also interfaces with a fixture optimizer which uses a heuristic algorithm to choose the 'best' fixture for a given operation, taking the entire process plan of the component part into consideration. This is an important step towards the integration of CAD and CAM.

REFERENCES

American National Standards Institute (1973) *ANSI Standards Y14.5*, American National Standard Institute.

Bidanda, B. and Cohen, P.H. (1987) *An Integrated CAD–CAM Approach for the Selection of Workholding Devices for Concentric, Rotational Components.* Proceedings of the 14th Conference of the National Science Foundation (NSF) Production Research and Technology Program. Society of Manufacturing Engineers, Dearborn, MI.

Bidanda, B. and Muralikrishnan, C.K. (1992) Flexible fixtures for intelligent manufacturing, in *Intelligent Design and Manufacturing* (ed. A. Kusiak), John Wiley & Sons, New York.

Bidanda, B. and Rajgopal, J. (1990) Optimal selection of workholding devices for rotational parts. *IIE Transactions*, **22**(1), 65–72.

Borland International Inc. (1986) *Turbo Prolog Owners Handbook*, Borland International Inc., Scotts Valley, CA.

Chang, T.C. and Wysk, R.A. (1985) *An Introduction to Automated Process Planning Systems*, Prentice-Hall, Englewood Cliffs, NJ.

Chou, Y.C., Chandru, V. and Barash, M.M. (1989) A mathematical approach to automated configuration of machining fixtures: analysis and synthesis. *Journal of Engineering for Industry*, **111**, 299–306.

Gandhi, M.V. and Thompson, B.S. (1986) Automated design of modular fixtures for flexible manufacturing systems. *Journal of Manufacturing Systems*, **5**(4), 243–52.

Nnaji, B.O. and Lyu, P. (1990) Rules for an expert fixturing system on a CAD screen using flexible fixtures. *Journal of Intelligent Manufacturing*, **1**(1), 31–48.

Markus, A., Markuz, Z., Farkas, J. and Filemon, J. (1984) Fixture design using PROLOG: an expert system. *Robotics and Computer Integrated Manufacturing*, **1**, 167–72.

Wang (Ben), H.P. and Wysk, R.A. (1987) Intelligent reasoning for process planning. *Computers in Industry*, **8**, 293–309.

Youcef-Toumi, K., Bausch, J.J. and Blacker, S.J. (1989) Automated setup and reconfiguration for modular fixturing. *Robotics and Computer Integrated Manufacturing*, **5**(4), 357–70.

CHAPTER 15

Learning in robotic task planning

Nina M. Berry and Soundar R.T. Kumara

15.1 INTRODUCTION

Planning has been an intricate part of robotic development beginning with the fixed programs of the zero generation manipulator robots and the first generation programmable robots. The second generation adaptive robots began the conventional planning by incorporating sensor input and appropriate planning technics. However, with the advancements in sensors and microcomputer capabilities, the third generation robot ushered in an era of improved planning technics for robotics (Vukobratovic, *et al.*, 1988).

As a result of continued years of research the generation of commands needed to complete a given task has begun incorporating aspects of artificial intelligence (AI) into the planning architecture. The advent of AI began in the 1960s with the general problem solver (GPS), which led to the planning technics of robotic problem solving systems like STRIPS, ABSTRIPS and PLanex. These early planning systems used the state-space search technics of AI; the progressive area of machine learning (ML) holds the next evolution of robotic planning systems.

This chapter's major objective is to explore how ML is being combined with robotic planning systems to improve the decision and reactive processes. First we will explore the motivation behind robotic task planning systems (RTPS). Then the chapter will discuss some of the problems of task planning that are being addressed through current research. This research invokes the combining of reactive and search-based architectures with learning algorithms resulting in new approaches to robotic planning schemes.

The remainder of the chapter will be a discussion of the robot control architecture TheoAgent, presently being created at Carnegie Mellon University by Tom Mitchell (Mitchell, 1990). This section will explore the motivations and the implementation technics behind TheoAgent, as well as how this architecture resolves some problems within task planning. The chapter will conclude with a discussion of what issues still remain to be addressed in the area of RTPS through the analysis of TheoAgent and other approaches to reactive architecture.

15.2 ROBOTIC TASK PLANNING

In this section we will briefly discuss how the traditional RTPS is implemented in a three-phase approach.

15.2.1 Motivation

The basic motivation behind a robotic planning system is to produce a series of instructions to solve a predefined task (goal or task generation). In addition to the creation process the system monitors the execution of the commands by making any adjustments based on current world configurations. Thus a robot's actions change it from one state or configuration of the known world into another. An RTPS requires extensive knowledge about the surrounding environment to produce a concise plan of motion.

This knowledge is represented in Table 15.1 as five separate items: description of object(s) to be manipulated, robots structure, initial state, final (goal) state and task environment (Fu, Gonzalez and Lee, 1987). Table 15.1 also gives a brief description of each item and the planning phase that uses the information. Together, these items collectively yield the beliefs of the working world for the robot.

Table 15.1. Description of task environmental information

Information	*Purpose*	*Phases*
Description of objects to be manipulated	Objects to be moved in the environment	Modeling task specifications
Robot structure	Description of robots physical and sensor capabilities	Modeling task specifications
Initial state	Design/layout of starting world. Based on task environment and objects to be manipulated	Modeling task specifications
Final state	How the world should appear upon completion of the task. Might contain all of the items in the initial state.	Task specifications Program synthesis
Task environment	Description of all objects that cannot be moved (e.g. walls, tables, doors)	Modeling task specifications Program synthesis

The fundamental aspects of a task planning system can be viewed as three phases for creating a plan (program) to achieve the perspective task. The three phases are modeling, task specifications and program synthesis (Fu, Gonzalez and Lee, 1987). In the sections to follow we will briefly elaborate on each of the three phases listed above.

15.2.2 Modeling

Although the domain of modeling is an important aspect of the planning phase, this section will only present a brief preview of this topic. The depth, size, location and sensing capability are the factors in the creation of the model of a robotic world. These factors aid the planning system in determining its speeds of motion, grasping strength and basic mobility while accomplishing its task. When presented with a task the system must construct a world model based on the task environmental information given in the previous section.

This information provides the planner with the ability to develop a world model containing all of the objects' geometrical and physical descriptions, kinematic descriptions, robot sensor characteristics and robotic structure. This descriptive process is an important aspect of the overall world model that is traditionally obtained through computer-aided design (CAD), computer vision and advanced sensing technics. Another method of obtaining a world model is with a primitive task-level language. This language permits the user to aid in the description process through routine commands and applied parameter values.

The descriptions obtained from the word model is stored in a data base for retrieval in the task specification phase. Although the structure of the database varies, 'physical statements' describe the geometrical/physical design of the objects within the world.

15.2.3 Task specifications

The basic idea behind task specifications is to begin the generation of a high-level language that can be further refined into a program in the next phase. A task is 'actually defined by a sequence of states of the original world model', which can be generated by three methods (Fu, Gonzalez and Lee, 1987). The first method requires the usage of a CAD system to display the objects in the correct locations from state to state. Another method utilizes advanced robotic sensor technics to specify the relationship between the robot and the objects of that states.

The third method uses 'symbolic spatial relationships' describing the sequences of states, generated from the decomposition of tasks (from a subtask system). This representation is defined between individual objects and permits the user to specify the different configurations of the world. The configurations must conform to the constraints placed on them by the symbolic relations.

Regardless of the implementation method, the task is sufficiently defined as a sequences of states, with the results being filtered into the program synthesis phase of the RTPS. The data base for the planner would now contain the original world model physical statements and sets of factual statements representing the different states needed to achieve the goal state.

15.2.4 Program synthesis

The generation of the program to control the robot's motions is the last and most difficult phase of a RTPS. As the robot planner is converting the task specification into the finalized program the following three issues of programming must be considered.

1. Motion planning.
2. Grasping planning.
3. Error checking.

These three effect how the robot can pick up objects, move around the environment, avoid collision with objects within the environment and be free of erroneous information during the execution of the program. For example, the process of motion planning is accomplished by applying the following four items during its creation.

1. A guarded departure from the current configuration.
2. A free motion to the desired configuration without collision.
3. A guarded approach to the destination.
4. A compliant motion to achieve the goal configuration (Fu, Gonzalez and Lee, 1987).

The finalized robot program to accomplish the given task will be expressed as a sequence of grasping commands, several different motion specifications and testing commands to insure the progressive success of the program (Fu, Gonzalez and Lee, 1987).

15.3 IMPLEMENTATIONS OF PLANNING IDEAS

The RTPS is a simple problem solver that has fascinated the AI community for years. There have been numerous variations of the RTPS, all based on different technics of implementation. Among these methods are systems designed on the structured environments of explanation based learning (EBL) and Reactive Architecture. This section will emphasis the motivation and implementaton of these two technics.

15.3.1 EBL motivation

Although AI has progressed planning system technology, the restrictive controlled environment of robots presented a complex domain not easily represented in an encoded format (Segre, 1988). The importance of ML to the robotic community is due to the continuing research to improve knowledge representation to express a domain 'explicitly and with recognizable encoding' (Segre, 1988). Furthermore, these domains must encompass current and past knowledge of the environmental situations and plans.

The introduction of learning schemes to planning systems began with the fundamental trial-and-error learning of early search-based technics. The popularity of the STRIPS system which incorporated GPS to reduce the state search environment marked the successful use of AI in robotic planning. Another early learing method used in RTPS was similar-based learning (SBL), which is defined by the skill of obtaining new knowledge that did not exist in the knowledge implicitly before. This inductive learning concept incorporates a variation of the learning by example, known as classification. SBL retains knowledge by processing numerous good and bad examples supplied to the system. This form of kowledge retention is data driven since it depends on incoming data for analysis. The ideas of SBL later emerged in the EBL algorithms that extracted their knowledge from single examples. The next section will discuss the EBL approach in detail.

(*a*) *EBL environment*

The EBL system is classified, by the AI community, as a knowledge-based system that uses domain theory to drive the analysis process of the given example. EBL is a learning scheme developed from deductive learning concepts, which are detailed as the improvement of existing knowledge into a form easier or more efficient to use. The EBL algorithm would generally be viewed as an addition to the task specification. Instead of the RTPS having to be programmed from one step to the next, the EBL algorithm creates the steps or subgoals as it is evaluating the task.

The basic motivation behind the EBL system is an algorithm that increases the existing domain knowledge by utilizing domain theory, operation criterion, a training example and a task goal definition.

1. Goal definition. Describes the conditions of when the goal state being constructed is acceptable.
2. Goal example. This is an example of a desired task definition that is similar to what the system is trying to achieve. This is equivalent to the goal states discussed earlier in RTPS.
3. Domain theory. Information that presently exists in the knowldge base, used to explain the correctness of the created goal state. This concept consists of known facts or newly acquired input to justify the goal state.
4. Operational criteria. This concept determines the form that the goal state will appear in once it has been determined (Mitchell *et al.*, 1986).

The output of an EBL system consists of a desired goal state and an explanation of why this goal is acceptable. The operational criteria permits the output to directly connect the task specification and programming synthesis phases of an RTPS. The format for the EBL system consists of receiving the above input and then analyzing how it can obtain the desired output.

As stated before the analysis is based on existing data within the knowledge base. During the analysis the desired goal may require the creation of subgoals before the goal state can be realized. Lastly, the system alters the goal state to meet the criteria of the operational phase. The completed task generation is now returned to the planning system for execution.

(*b*) *An EBL example*

The input for the EBL algorithm consists of the goal definition and goal example, shown below:

⟨⟨GOAL DEFINITION⟩⟩	"Place *x* ON-TOP of *w*"
ON-TOP (*x*, *w*) ⇔	[(ISA-CLEAR (*x*) AND ISA-CLEAR (*w*))
	AND (STACK-HEIGHT (*x* < 4)
	AND STACK-HEIGHT (*w* < 3))]

⟨⟨GOAL EXAMPLE⟩⟩	
CLEAR (A)	ARM (0)
CLEAR (C)	LOCATION (A, ⟨1,0,0⟩)
SURFACE (A, FLOOR)	LOCATION (C, ⟨5,0,0⟩)
SURFACE (C, FLOOR)	LOCATION (AGENT, ⟨3,0,0⟩)

The goal definition is a predicate calculus statement which is justifiable by the goal example and internal domain theory. The goal example is very similar to the entries used in the traditional RTPS, during the task specification phase.

The task to 'place box A on box C', is similar to the structures in RTPS; however, the EBL system must create subgoals to achieve this task. For simplicity we shall assume that the same subtask created in the traditional RTPS will be produced by the EBL system during the evaluation of the original goal (task) definition. When each subtask is created, a justification of the subgoal accompanies the results.

Some of the domain theory used in this example is illustrated in the predicate calculus statements shown below. This domain theory pertains specifically to the justification of the goal definition only. The theory needed to evaluate the subgoals created by the system is not shown here.

⟨⟨DOMAIN THEORY⟩⟩
SURFACE(vv,FLOOR)→STACK-HEIGHT (vv < 2)
STACK-HEIGHT (*vv* < 3)→(LOCATION (vv, x-axis,1,*z*-axis)
OR LOCATION (*vv*, *x*-axis,2,*x*-axis))
AND CLEAR(*vv*)

ISA-CLEAR (vv)→CLEAR (vv)
Note: *vv* – some object name.
x-axis and *y*-axis – legal values for the axes in question.

The process of proving (justifying) the goal definition is listed in Table 15.2 as a four-step process of evaluating the known domain theory and goal examples. Table 15.2 shows that unknown domain theory utilizes the existing theory and goal examples to create explanations for the new theory. Table 15.2 also shows that domain theory is evaluated until it reaches the level of known domain examples.

Table 15.2. Evaluation steps of domain theory and goal examples

1. Evaluate goal definition statement(s).
2. Apply domain theory. If necessary statement does not exist goto step 4.
3. Can the implication be justified with the available goal knowledge?
   (yes) – Replace variables with legal value from goal example.
   (no) – Goto step 2.
4. Apply known domain theory and goal example knowledge to create a new rule, then goto step 2.

The evaluation of the domain theory can be further illustrated using a LISP pseudo-code algorithm shown in Fig. 15.1. The algorithm Explanation_code(*x*,*w*) is an example of the proving process for the EBL algorithm. The known domain theory is represented as functions that evaluate the known goal examples. These functions return true if the clause is correct; otherwise, the domain theory needs to be further evaluated before the goal example knowledge can be accessed. This derivation causes the creation of subgoals which are illustrated as the false clause of each 'IF' test. The subgoals simply recall the Explanation_code algorithm to be evaluated.

The axioms are referenced when domain theory is not available for the statement in question. The axiom in this example requires the goal statement, the operator and the variable used in the statement. Not shown in this is the decide_process function, which evaluates the domain theory and goal examples to create the explanation clause. However, the explanation for the STACK_HEIGHT ($vv<4$) is shown in Fig. 15.1 as a result of the decide_process function.

Figure 15.2 illustrates the EBL systems' proving process as a tree structure. The subgoals are represented as unproven ovals on the tree. The root of the tree is represented by the predicate ON-TOP (*x*,*w*) and each level further justifies why the goal of 'placing A on C' is correct. The objects ⟨*x*,*w*⟩ are replaced by ⟨A,C⟩, respectively, due to the contents of the goal example. The first level details the evaluation of the height location of the boxes via the STACK-HEIGHT clauses. The EBL system successfully learns the definition of STACK-HEIGHT ($x<4$) by combining the domain theory

```
Explanation code ( Place x on w):

 (Explanations_code (x w)
    ( Preconditions
       (and      (STACK-HEIGHT (x < 4)) (STACK-HEIGHT ( w < 3))
       (and      (ISA-CLEAR (x))  (ISA-CLEAR (w)))
       (if  ('condition unknown')  *nothing*  Unknown_Domain_theory(goal op var)) ))

Explanations code functions:
(ISA-CLEAR (vv)
   (if (member (CLEAR(vv)  goal_list))  true  nil))
(STACK-HEIGHT (vv)
   ( if (and (ISA_CLEAR(vv))
                   (or     (member (LOCATION (vv, x-axis,1,z-axis)   goal_list)
                           (member (LOCATION (vv, x-axis,2,z-axis)   goal_list))   true
nil))

Axiom 1:  Apply known domain theory
(Unknown_Domain_theory (goal op var)
   ( if (and (SURFACE (var floor))
          ( if (ISA_CLEAR (var)) *nothing* subgoal)
          ( if (member (LOCATION(var, 1, y-axis, z-axis) goal_list))   *nothing*
subgoal))))
```

Fig. 15.1. Pseudo-code for domain theory and goal examples.

and goal example. This is illustrated in the tree in the leaves by SURFACE (A,FLOOR) and CLEAR (A). The domain theory started by testing SURFACE (vv, FLOOR)→STACK-HEIGHT ($vv<2$), since this evaluated to the STACK-HEIGHT <2. The system applied ISA-CLEAR(vv)→CLEAR(vv), thus the system knows A is clear and on the floor. By utilizing the clause for STACK-HEIGHT ($vv<3$) as a model it created a new clause to cover ($vv<4$). One possible domain theory is STACK-HEIGHT ($vv<4$)→LOCATION (vv, x-axis,1,z-axis) AND CLEAR (vv). The second level of this tree shows the evaluation of ISA-CLEAR. Once again the EBL system falls back on the domain theory to evaluate these clauses.

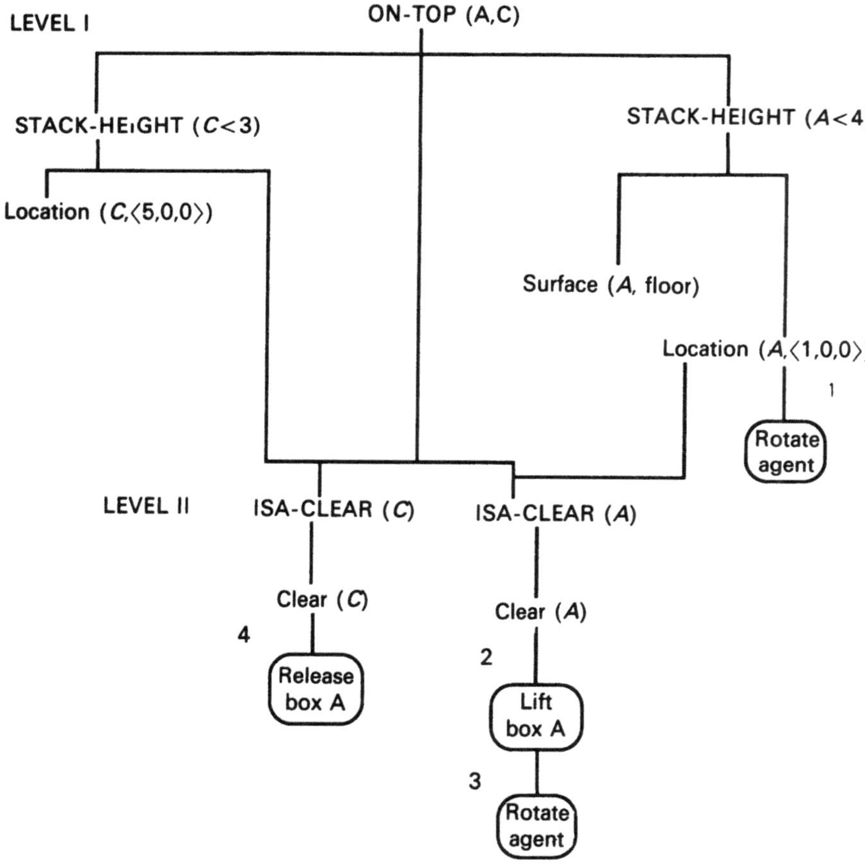

Fig. 15.2. Proving tree for EBL psudeo-code illustrated in Fig. 15.1.

Once the goal descriptions have been proven to be correct, the resulting explanations and task generations are run through the operational criteria. This transforms the task into a usable form for the system. In this example the form is equivalent to the executable code created by the program synthesis phase. The newly acquired knowledge remains part of the EBL domain theory system.

(c) *Advantages and disadvantages of EBL*

Regardless of the implementation technic used, an RTPS using the EBL algorithm provides some advantages that greatly improve the original operation. Some of these advantages are listed below:

1. EBL requires only a single example.
2. EBL can learn from a less optimal solution (final state), just an adequate problem solution is required.

3. EBL need not be invasive. This implies that the EBL algorithm does not have to alter the goal example to create the desired results.
4. EBL generations are correct if the domain theory is complete and correct (Segre, 1988).

The above advantages and current reseach using the EBL algorithms has led to several new approaches to planning system architectures. The EBL algorithm improved the robot's reaction time and developed independence from constant human input. With the aid of EBL the overall flexibility of the robotic planning systems is now being explored.

Even with these advances the EBL system still remains an open-loop architecture, thereby limiting the flexibility of the sytem. This lack of flexibility makes the EBL system unable to react to changes in the environment while planning for a specific task. Therefore, the resulting plan may not be successful when the environment is altered greatly. Replanning for this type of change reduces the speed of the EBL system, when a completely new plan must be created. A more adventurous architecture is needed to permit instant updating the reaction to new situations. This method is known as 'reactive architecture' and will be discussed in greater detail in the next section.

15.3.2 Reactive architecture motivation

Both the RTPS and the EBL systems represent a form of open-loop robotic architecture. The open-loop occurs during the non-interactive gap between the robot and the task planning system. This lag in interaction causes the robot to 'blindly follow a program or plan without verifying that its operations are having their intended effects' (Kaelbling, 1986).

The legitimacy of task generation becomes a critical factor in unpredictable environments, where changes in the robot's world easily effects the success of the task. In the early 1980s research began on a different type of closed-loop system to permit interaction between the planner and sensor input from the robot world. This new architecture exemplifies reactive behavior in a form of a dynamic environment.

Reactive architecture is the process of creating appropriate actions from environmental situations without making predictions about future events. This architecture was introduced as an alternative approach to the traditional planning that was implemented in robotic controllers. The planning scheme differed in the approach for solving the goal state of a given task. While traditional planning depends on extended predictions in the future, one possible reactive planner might simply 'repeat a cycle of evaluation the environment and determining an appropriate action' (Gervasio, 1990). Another implementation of a reactive system is similar to an operating system using 'message passing with a round robin scheduler', permitting the planner to receive data from the sensors as it plans (Kaelbling, 1986). The next section will further explore the environmental aspects of reactive architecture systems as presented by Kaelbling (1986).

(a) Reactive architecture environment

Regardless of the technics used to implement reactive architecture, all of the methods have three major purposes to fulfill. These three criteria are listed below (Kaelbling, 1986):

1. Modularity. The system should be built incrementally from small components that are easy to implement and understand.
2. Awareness. At no time should the system be unaware of what is happening; it should always be able to react to unexpected sensory data.
3. Robustness. The system should continue to behave plausibly in novel situations and when some of its sensors are inoperative or impaired.

These criteria will be discussed in detail in the sections to follow. A general reactive system has an embedded planner to construct plans based on the initial state, world model and the goal (task) to be accomplished. This planning system is sustained in a dynamic environment with a timed cycle. This form of implementation ensures that the plan being generated is based on the most recent environmental conditions. Upon completion of the task the planner submits the plan of execution and a flag signaling to the system that the planner has accomplished its task.

What happens if the environmental conditions change the initial state conditions? The planner stores the important initial values away, while doing constant checking to ensure that these conditions hold throughout the planning stage. However, if the conditions change, the planner replaces the old values and begins planning again.

A crucial aspect of the reactive system is the perception component. This component determines the system's ability to react to situations in a reflex-like action. The ability for a robot to maneuver around a room without requesting a plan is accomplished by the perceptual data that reside in the system. A good example of this is a robot avoiding a wall it is about to hit. The following peceptions and actions might appear to the system as possible commands the robot would implement without planning (Koelbling, 1986):

1. Hit (object)→stop
   ⟨⟨robot believes it will collide with the object⟩⟩
2 Hit (not (object))→go
   ⟨⟨robot belief's it will not collide with the object⟩⟩

   Not (1) .AND. Not (2)→stop .And. determine location of wall
   ⟨⟨robot does not know if it will collide with the wall or not⟩⟩

The perception component may be constructed of several sublevels of hierarchical control interconnected by mediating devices. The mediator acts as a referencing agent combining the abilities of several levels of behavioral control into a complex cohesive behavior component. The behavior refers to the 'mapping of sets of input to the perception component and outputting the desired commands to the controls of the system' (Kaelbling, 1986).

The levels of hierarchical control effect the levels of behavior presented in the system. The block world example for the reactive system will further explore a simplistic illustration of the hierarchical control used in a reactive system. As stated earlier all reactive architectures function with three major criteria to achieve. The next two sections will further elaborate on these ideas.

(*b*) *Modularity*

The modularity of the system increases the reliability and understanding of the system's implementation. There are two forms of modularity, horizontal and vertical decomposition. The vertical format requires the system to be subdivided into components handling individual tasks such as modeling, execution, planning and perception. This type of structure handles the simplistic behavior of the system by evaluating virtually each component every time.

The horizontal decomposition is a true structure where the action and the perception component are connected in a manner similar to an implication axiom. Although each structure may exemplify a different behavior a given structure might depend on other behavioral aspects of the system. The popularity of the horizontal decomposition is a result of the system to establish simple behavior and construct complex behavior on top of the existing system.

While the horizontal system is preferable, this system tends to be more 'dependent on conditions in the world, rather than on the particular properties of sensor readings' (Kaelbling, 1986). This modified system would divide the traditional perception and action fields of the horizontal method into two vertical sections. However, the components would still harbor horizontal decomposition with each section.

Illustrations of the traditional horizontal and vertical/horizontal systems are shown in Fig. 15.3 (a and b). The advantage of a vertical/horizontal system is the complete control of sensor ability by the perception component. Since the sensor commands are also redirected back into the perception component any changes in the environment directly effect the legitimacy of the sensor commands, which is the output of the actions component. The modularity aspect of reactive architecture effects the system's speed and overall performance; however, there are other aspects that must be considered. In the next section the awareness and robustness of the reactive system will be discussed in detail.

(a)

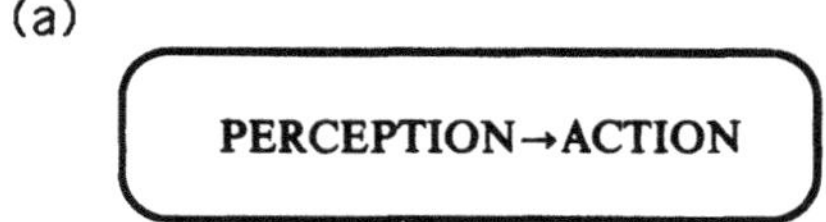

(b)

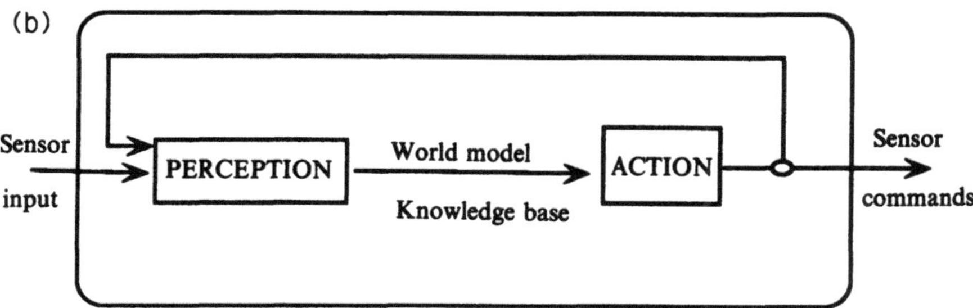

Fig. 15.3. (a) Horizontal, and (b) vertical decomposition of the system control.

(c) *Awareness and robustness*

It has already been stated that the awareness of an RTPS is crucial if the robot is to function in an unpredictable environment. The awareness of a reactive system is usually based on the bounding of time for a given process. Therefore, a system might declare a cycle for which any process must complete execution or signal to continue into the next time cycle.

Closely linked to the awareness of the reactive architecture is the robustness of the system. A robust system can react to sensor input like human reflexes in a known situation. Of course, quick actions such as these are based on the amount of available knowledge within the system. This form of robustness is known as behavioral and is usually accompanied by 'high-level path-planning modules that generate actions based on a strong model of the environment' (Kaelbling, 1986).

The next form of robustness is known as perceptual. Like human peception, this apect of the reactive architecture is an accumulation of all sensory information and known knowledge about the robot's world. The failure of a sensor does inhibit the system's ability to plan for a task if the knowledge cannot be supplied by another sensor. Together these two forms of robustness and awareness complete the criteria needed for a system being supplemented with reactive architecture.

(d) *Reactive example*

The reactive system for the robot consists of the three behavior levels shown in Fig. 15.4. Each level corresponds to a different abstraction of behavioral input governed by the sensory data. The goal of this limited system is to

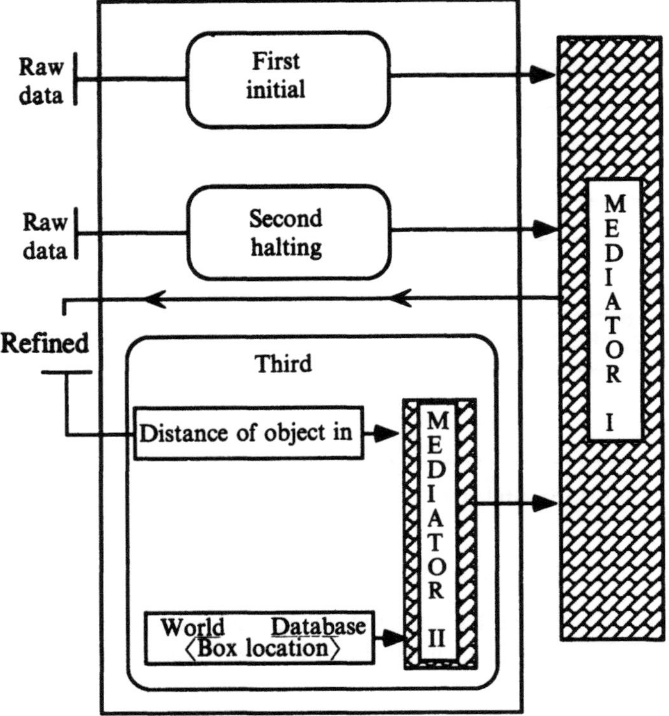

Fig. 15.4. Three levels of decomposition behavior for a reactive system (Kaelbling, 1986).

'place box A on box C'. First the robot must maneuver toward the boxes before contact can be established. Remember the robot is equipped with sensors telling it the location of the arm and the closeness of the robot to objects and walls.

The first behavioral level receives raw sensor data which are stored in the initial variable. If these values become increasingly small, meaning the robot is very close to an object, the agent will halt. The second behavioral level reads accumulated raw sensor data and evaluates control in case the robot must halt its motion. In this situation the second level may be turning the robot until the halting conditions change enough for the robot to move forward. However if a successful motion cannot be made this level of abstraction returns the command *nocmd* instead. Furthermore, the halting of motion in both levels may occur if the agent does not refresh its sensor input within one time cycle.

The third behavioral level represents a higher level of abstraction that evaluates the closeness of the agent to the desired box in question. For this level of abstraction to work the system must know the location of the box from the world model data base. This knowledge is combined with the robot's known location to see if the robot is within lifting position. The two horizontal components in level three interact through mediator_II and all of the levels are combined via mediator_I to produce a desired command.

The basic design of a mediator consists of sensor commands and motor commands. Figure 15.5 illustrates the structure design of the behavioral system shown in Fig. 15.4. The second level of horizontal decomposition is listed under halting solution. The third level is broken down into the two sections of world model and distance-to-object. The mediator_II references the sensor and motor command output from the third behavior level. Mediator_I connects the behavioral outputs of all of the levels.

The robot is located in unit ⟨3,0,3⟩ of the environment. The location of box A is in unit ⟨1,0,0⟩ and the system receives sensor data signaling that the agent may move straight forward without collision. Level two outputs the servo_1_unit command moving the robot ahead, while level three evaluates the closeness of this action to the desired object. After checking the world data base, level three continues the movement of the robot until the location statement of the robot equals ⟨1,0,0⟩.

The sensors indicate the robot's success in locating the box A and knowledge interaction module permits the system to complete the task by issuing the commands: lift box A, rotate box A to +90°, release box A and rotate back to 0°. Of course, the system's ability to successfully accomplish this example is far more detailed in an actual system. However, this simplified example serves to illustrate the advantage of reflexive rules to a RTPS.

(*e*) *Simple simulation of a reactive architecture system*

We have developed a limited simulation of a reactive architecture system called reactive architecture for mobile implementation (ReAMI). This system is capable of planning a collision-free path within a dynamically changing laboratory room occupied by at least one chair and a box. The planning system is faced with the problem of creating a path from a starting point to a destination without colliding with any of the objects or boundaries of the room.

The ReAMI system combines the technics of sensory input and reasoning methodology to create a unique system capable of reacting to changing environmental situations without modifications to the reasoning actions of the system. The planning system would have to react to changes within the environment even during the planning phase. Therefore, the system would need the ability to modify any existing plan currently under construction. The ReAMI system uses traditional RTPS architecture enhanced with reactive architecture to present a closed-looped structure, capable of interrupting the planning phase for a given task (or subtask) and introduce a new belief about the environment. The ReAMI code, written entirely in LISP, is presented in the Appendix.

```
Halting solution

sensor-cmd =
     if (left_sensor_weak) then left_sensor
     if (right_sensor_weak) then right_sensor
     if (front_sensor_weak) thn front_sensor

motor-cmd =

     if (THE-loc-front = *clear* and
        THE-loc-left = *clear* and
        The-loc-right = *clear*) then sevo_1_unit
     else
     if (loc-travel <> *clear*) then *obstacle-trouble*

World Model

input-cmd_data base =
     box-loc = *access data base*
     table-loc = *access data base*

sensor-cmd =

     if (right-sensor = pick-up (box-loc)) then
     *at correct location*

motor-cmd =

     if (world-model-pick-up (box-loc) = true) then
        Rotate-arm (90)
        Extend-arm
        Grip-arm
        Lift-box (1 2 units)
        *noop*

Distance to object

sensor-cmd =

     if (agent-loc = THE-loc-front) and (THE-loc-front = *clear*) then
        *loc_robot_set*
     else *nocmd*

motor-cmd =

     if (THE-loc-robot = *loc_robot_set*) then linear-speed
     else *noop*

Mediator II (between distance to obj  and world model)

sensor-cmd =

     if (distance-to-obj-sensor-cmd = *loc_robot_set*) then
        pick-up-action-sensor-cmd
     else world_model-sensor-cmds

motor-cmd =

     if (distance-to-obj-motor-cmd = *noop* and
        pick-up-action-motor-cmd = *noop*) then *nocmd*
     else re-establish (distance-to-obj and world-model)

Mediator I (between level 2 and level 3)

sensor-cmd =

     if (halting-solution-sensor-cmd = *noop*) then halting-solution
     else if (mediator_II-sensor-cmd = *nocmd*) then *nocmd*

motor-cmd =

     if (mediator_II-motor-cmd = *nocmd*) then *await-cmd*
     else if (halting-solution-sensor-cmd = *obstacle_trouble*) then
        *sound-alarm-motion-inhibited*
```

Fig. 15.5. Pseudo-code representation of a three layered decomposition behavior system (Kaelbling, 1986).

(*f*) *Advantages and disadvantages of reactive architecture*

The advantage of the reactive architecture is the increased flexibility of the RTPS. This improvement in flexibility is a direct result of the system's ability to react to unanticipated situations. The speed of the reactive architecture is also an advancement in RTPS, due to the direct action response to certain perceptions.

Although the reactive system has many advantages there are some problems incurred by using this architecture. The abiity to react to a sensor failure slows down the robustness of the system. The dependency on sensory input keeps the system from utilizing prior knowledge, and forces the relearning of old rules and actions. The repetitiveness of reactive planning may produce a wrong reactive rule causing delays determining a result or aborting the process (due to time-out) without a solution. This flawed approach makes reactive planning a good candidate for machine learning strategies like EBL. In the next section we will discuss briefly what advantages a system that combines EBL and reactive architecture can offer.

15.3.3 Combining EBL and reactivity

This paper began with a discussion on the current status of RTPS, followed by a brief overview of the two expanded task planning systems (EBL and reactive architecture). Although both expanded systems added some significant attributes to the original ideas, the system still lacked important functions. If the two aspects were combined a new architecture would be produced to form a new complete autonomous planning system. What can a reactive architecture enhanced with EBL provide for an RTPS?

The reactivity would permit the system to adjust its inputs to changing situations instantly. While the EBL system could modify the planning abilities of the system in a learning fashion without having to be retrained. The EBL system could further assist the reactive architecture in the usage of existing knowledge, since it is domain driven. This aspect alone would permit the system to utilize rules and actions more efficiently.

The combining of EBL and reactive architecture yields the best of two worlds – the versatility of adjusting for the unknown, while learning from the experience at the same time. The idea of combining these ideas has yielded some interesting systems. In the second half of this paper we will explore in detail one of these system, known as TheoAgent (Mitchell, 1989).

15.4 THEOAGENT

The philosophy, motivation and implementation technics of a reactive/EBL robotic task planning system are explored in this section. The project is centered around a frame-based, problem-solving architecture, known as

Theo. This architecture combined with several learning technics and an external agent (robot) comprise the complete TheoAgent system, presently being implemented at Carnegie Mellon University by Tom Mitchell (Mitchell, 1989).

15.4.1 Theo's objective

The TheoAgent project is an attempt at designing a general-purpose learning robot architecture by a cohesively combined knowledge representation, problem solving, reactive architecture and search-based learning. The knowledge representation used is based on a frame-slot structure similar to the RLL system (Bourbakis, 1991). The system implements problem solving with a leveled inferencing structure that accesses the appropriate slots. The search-based architecture provides the problem solver interfacing, traditionally used in RTPS. Although search-based planning is slower, the enhanced EBL algorithm being used does provide the ability to deal with 'diverse sets of unanticipated goals and situations' (Mitchell, 1989). The reactive architecture adds the advantage of quick decision making based on current enviromental situations. Unlike search-based planning, reactive systems operate faster even without an advance model of the environment.

The major objective of TheoAgent is to produce a new learning robot architecture that permits the planning system to react to known situations immediately and plan for the unknown quickly. The designers of TheoAgent plan on implementing the following three objectives into the architecture (Mitchell, 1989):

1. It must become increasingly reactive, by reducing the time required for it to make rational choices; that is, the time required to choose actions consistent with the above predictions and its goals.
2. It must become increasingly correct at predicting the effects of its actions in the world.
3. It must become increasingly perceptive at distinguishing those features of its world that impact its success.

These objectives are based on incorporating the work of three different approaches to learning from unknown situations. The second objective, becoming increasingly correct, is based on a robot arm learning from non-deterministic actions. The robot learns from a limited domain and a vision system about how to tilt objects in a tray without dropping them (Christiansen *et al.*, 1990). The ability to become increasingly perceptive is based on a mobile robot that identifies a set of training objects with the aid of sensing routines (Tan and Schlimmer, 1990). The robot maneuvers around the room observing and identifying each object to produce a 'decision tree structure for sensing and classifying subsequent objects' (Mitchell, 1990). The last two objectives have not yet been completely accomplished in the TheoAgent system.

The present architecture of TheoAgent has successfully achieved the first objective of 'increased reactivity' via the creation of stimulus–response rules. The stimulus–response rules 'map directly from observed world features to the recommended action' (Mitchell, 1990). However, if the rules do not exist the demand planning sequence invokes the EBL planner to develop new rules. The newly created rules are incorporated into the rule base component to enhance the system's ability to react faster. Together these robot control structures use 'stimulus–response' rules to react to situations in subseconds.

15.4.2 The fundamentals of Theo

This section will represent an overview of the functional (knowledge) and inferencing technics used within the software structure of TheoAgent. The software architecture behind the TheoAgent system is known as Theo.

15.4.3 Knowledge representation in Theo

'Theo is a frame-oriented architecture that supports a particular approach to knowledge representation, provides a collection of inference methods, and includes mechanisms for generalizing from examples (Mitchell *et al.*, 1990). The structure of this software is a slot referencing mechanism based on prior stored values in a cache. If the value is not available, the system infers the need to invoke the planning structure to obtain the value instead. This design is based on the interaction between learning, knowledge representation, and problem solving.

An important organizing aspect of Theo is the *beliefs clauses* used to represent the knowledge within the system. The beliefs exhibit what Theo knows about a given assertion. The notation actually used in the software resembles the format of (⟨entity⟩ ⟨slot⟩ ⟨value⟩) or (⟨slot⟩ ⟨value⟩ ⟨entity⟩, where the slot represents an entity. The value part of the belief clause is inferred or returned as the clause is being evaluated. This format allows the beliefs in Theo to reside in a 'frame-based representation, in which both frames and their slots correspond to entities' (Mitchell *et al.*, 1990).

The process of evaluating a belief is called a *problem instance* and is of the form (⟨entity⟩ ⟨slot⟩). This formula represents a question (instance of the relation) within the belief clause, whose answer can be found in the desired relation (slot). When the problem is evaluated, the results yield an 'element of the relation's range' that corresponds to the one-to-one mapping between the belief and the problem (Mitchell *et al.*, 1990).

A collection of problem instances are called *problem classes* and appear in either the form (⟨entity⟩ ⟨slot⟩) or ⟨slot⟩. In the (⟨entity⟩ ⟨slot⟩) format *entity* represents a *class of elements* that reside within the domain of the relational name ⟨slot⟩. The notation for this format is usually (?*e* ⟨slot⟩)

where *e*? = ⟨entity⟩. The ⟨slot⟩ format signifies the class of problem instances of the form (*x*? ⟨slot⟩), where *x*? is a variable representing a member of the domain of ⟨slot⟩. The prior format of (⟨entity⟩ ⟨slot⟩) is a 'subset of the ⟨slot⟩ relation restricted to the subdomain specified by ⟨entity⟩' (Mitchell *et al.*, 1990).

Figure 15.6 represents the nesting of the basic structure of beliefs and illustrates how the other two main structures build upon this idea. The usage of the three structures permits Theo to represent all of the data and stored knowledge in a form easily manipulated by the software.

Together, the three structures can be defined to represent the general knowledge base format 'description ::= (name value description*), where name can be any Lisp atom, value can be any Lisp expression and * denotes 0 or more iterations' (Bourbakis, 1991). This notation permits the system to represent name as either a top leveled entity or as a subslot that resides within the problem instance.

15.4.4 Inference

The inference structure within Theo controls the decisions used to obtain a desired value for the slot in question. For the slot, S, values are obtained from the 'beliefs stored in the subslot' of S (Mitchell *et al.*, 1990). This multilayered inference mechanism is shown in Fig. 15.7. The first layer is entered by utilizing the subslot TOGET, which has the following format: (entity slot TOGET). The TOGET subslot then references a LISP function (the entity) to evaluate the contents of the slot. (The TOGET function permits user interaction by also permitting user supply LISP functions.) Upon completion of the evaluation, Theo returns a value for the subslot from the cache. Although all of the inferencing levels return a value for the problem instance given, only layers 2 and 3 might involve the use of EBL. Therefore, an explanation of the methods used to create the value will accompany these results.

The second layer is invoked through the TOGET command when no LISP function is provided. This layer evaluates each address within the Available.Methods slot until an answer is obtained. The user may also interact with the second layer by using the inference methods INHERITS and DEFAULT.VALUE when predefining slots.

The inference mechanism enters the third layer by, evaluating the Available.Methods statement DEFINE. This layer's primary purpose is to 'attempt to utilize the DEFINITIONS and SPECIALIZED.DEFINITIONS subslots to infer a value. The DEFINITIONS slot is a list of expressions which are evaluated in sequence in an attempt to infer a value for the slot' (Mitchell, *et al.*, 1990). The user can also interact with this layer of inference by, creating their own DEFINITIONS for slots. If SPECIALIZED.DEFINITIONS are used the returned value is 'inferred via the EBL by TMAC' (Bourbakis, 1991).

Beliefs

Form

(<entity> <slot>) = <value>

Example

(blk_A floor) = <value> * blk_A is a member of relation floor? answer if t or f

Entities

Problem Class

Form

(<entity> <slot>)

Example

(blk_A floor) = t * is the entity blk_A a member of the floor relational domain.

Problem Instance

Form

<slot> or (<entity> <slot>)

Example

(?x floor) * ?x represents any member of the floor relational domain.

(?unit_1 floor) * ?unit_1 represents an entity set in the relational domain. The entity returned by ?unit_1 is a member of the unit_1 entity set.

Application

(blk_A floor difficult?) = t *(blk_A floor) = entity/problem instance
problem instance -> blk_A = entity
floor = slot
(difficult?) -> slot

This example states that it is difficult to determine if blk_A is a member of the relational domain of floor. Note that the the entity above is a member of the relation difficult.

Fig. 15.6. Belief structure of TheoAgent (Mitchell, 1989).

15.4.5 Learning scheme I – caching

The caching of values is the heart of the reactive architecture scheme and the simplest learning mechanism. All inferred values and an explanation are presently cached by the system when the WHENTOCACHE slot is

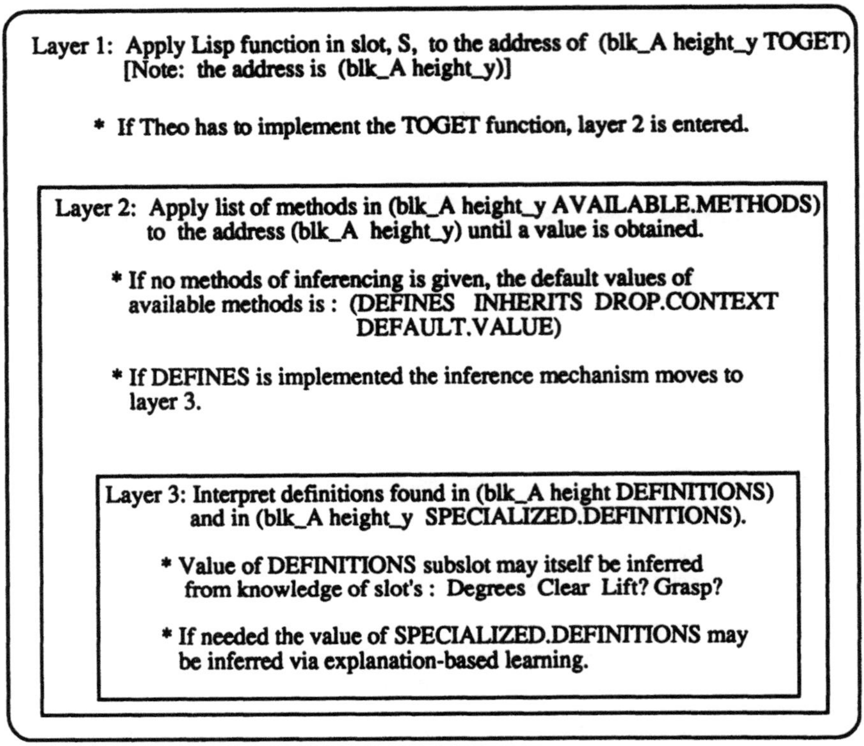

Layer 1: Apply Lisp function in slot, S, to the address of (blk_A height_y TOGET)
[Note: the address is (blk_A height_y)]

* If Theo has to implement the TOGET function, layer 2 is entered.

Layer 2: Apply list of methods in (blk_A height_y AVAILABLE.METHODS) to the address (blk_A height_y) until a value is obtained.

* If no methods of inferencing is given, the default values of available methods is : (DEFINES INHERITS DROP.CONTEXT DEFAULT.VALUE)

* If DEFINES is implemented the inference mechanism moves to layer 3.

Layer 3: Interpret definitions found in (blk_A height DEFINITIONS) and in (blk_A height_y SPECIALIZED.DEFINITIONS).

* Value of DEFINITIONS subslot may itself be inferred from knowledge of slot's : Degrees Clear Lift? Grasp?

* If needed the value of SPECIALIZED.DEFINITIONS may be inferred via explanation-based learning.

Fig. 15.7. Layered inferencing mechanism (Mitchell, 1989).

evaluated. The evaluation is caused by the DEFAULT.VALUE setting currently implemented in Theo as the value of the WHENTOCACHE slot. Accessing these prestored values resembles the horizontal decomposition, by producing immediate actions for the robot to execute. The terminology of the stimulus-response rules are a direct result of this process. If Theo access the slot within the cache and a value is prestored it responds immediately with an action, otherwise the system must infer the value with further inference technics and perhaps the EBL algorithm.

It was stated earlier that each value is accompanied by an explanation of the inference process. This expanation is an important aspect of the Theo system. The 'explanations are used for a simple form of truth maintenance. If any slot value in the system is changed, then the values of all dependent slots are uncached' (Mitchell *et al.*, 1990). The result of this action causes the system to reset these values to *NOVALUE* and will have to be recomputed if referenced.

15.4.6 Learning scheme II – SE

The learning architecture of Theo is also responsible for reordering the Available.Methods slots of the frame structure. Traditionally the methods listed in this slot are tested according to a first-come-first-used basis. However, the traditional transverse of the list does not always produce the best approach to the inference of the slot in question. To improve the ordering of the methods Theo is equipped with an 'inductive learning system which can order the AVAILABLE.METHODS for any slot in Theo', this system is known as SE (Mitchell, 1990).

The SE program is based on the statistical sampling work of Abramson and Korf (1987). The ordering is accomplished by evaluating random samples of computations (applying of a method) based on a given problem class. The function used by SE to sort the available methods is: cost of success + (probabilities of failure/probabilities of success) × cost of failure (Etzioni, 1988). The computation clusters are ordered with respect to computational cost. 'That is, the cost of computations in the cluster is more or less the same' (Mitchell, 1990). This evaluation of the cluster holds a good prediction of the ordered method and replaces the present ordering of the Available.Methods. Reordering the AVAILABLE.METHODS also occurs when specialized definitions are created, to reflect a better plan of inference.

15.4.7 Learning scheme III – TMAC

Theo traditioally makes every attempt to first use the available stimulus–response rules before resorting to the inferring of values and accessing the EBL aspects of the system. The TMAC component is called explanation-based generalization (EBG) and is entered via level 3. The program used to 'examine explanations of previously successful slot inferences', is called TMAC (Mitchell, 1990). The macro-methods are derived from the explanations returned with the inferred values. The EBL algorithm uses these explanations as examples of correctly inferred instances and creates the desired macro. These specialized macros are then stored in the 'generalization/inheritance hierarchy' at a position determind by the 'macro's domain of applicability' (Mitchell, 1990). The macro resembles the slot structure and is stored under the SPECIALIZED.DEFINITIONS subslot of the slot ^S. Placing the macro in this area permits slots that are specializations of ^S to inherit the new macro-method also.

15.4.8 Learning scheme IV – CHUNCKER

The last learning mechanism can 'compile Theo's problem solving experience into general rules which can help it infer related slot values in the future' (Toshikazu and Mitchell, 1991). The CHUNKER (based on the original CHUNKER system in SOAR) facility within Theo resembles an EBL chunking system that 'supports development of self-modifying problem solving systems' (Bourbakis, 1991). However, unlike TMAC, CHUNKER is applicable to any slot within all three levels of inferencing. CHUNCKER receives a belief and an explanation tree, which is searched using backtracking, for operational beliefs. These beliefs are collected, evaluated and formed into rules that are applicable to the TheoAgent rule structure. Each rule is accompanied by a set of beliefs which support the rule and maintain the truth of the evaluation.

15.4.9 The Theo structure

Theo is a problem solver system like the original RTPS; however, the new stimulus–response architecture provides the system with a self-improving quality. The structure of Theo consist of three major parts: the inductive inference, reactive component and search-based planning component. Figure 15.8 represents a typical control cycle within the stimulus–response architecture from the initial sensing stage through the agent implementing the resulting action/plan. The cycle is a four step process utilizing the three parts of Theo as described below:

1. Sensory data (either lazy or eager) being returned by the agent.
2. Decision process begins the problem solver with inference-based rules.
3. Reactive component for a quick response action to a situation.
4. Search-based planner for unanticipated situations that have no predefined rule(s).

The system begins the process by deciding which type of sensing it will implement. The eager sensing represents data that will effect the current

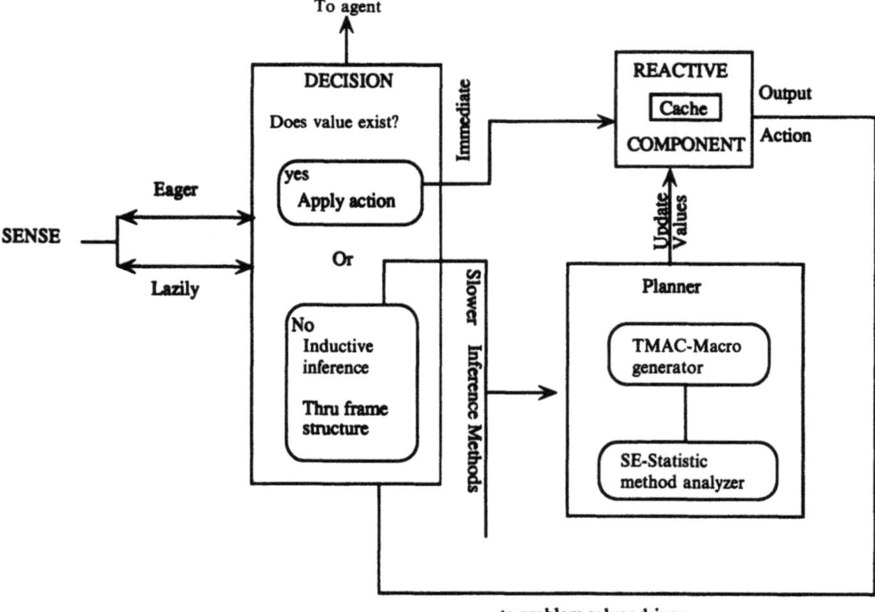

Fig. 15.8. Control cycle for TheoAgent.

world model and the concluding action for the agent. The 'lazily recomputing' sensing refers to data that are deleted during each cycle and recomputed only if requested by the problem solver (inference structure).

Once the data are received the decision component must decide if the response to the task will be immediate or planned. An immediate response occurs when the stimulus – response rule is currently in the cache. The search-based planning component is needed to learn what new action to incur for the given task. Upon completion the resulting new stimulus-response rule becomes part of the cache.

The technics discussed above also represent the three learning mechanisms of 'caching of inferred beliefs (reactive component), learning from explanations associated with successfully solved problems (search-based planner), and inductive inference for ordering problem solving methods' (Mitchell *et al.*, 1990).

15.4.10 The performance advantages/disadvantages of Theo

The TheoAgent project was designed as a self-improving system, capable of learning and evaluating information without prior knowledge of the environment. The combining of the learning and reactive architecture permits this system to decide on plans of action within subseconds. The Theo system promotes some form of learning throughout its control structure. The cached values serve as a form of rote-learning, while instigating explanation-based reasoning only when forced to plan. The acquiring of slot values permits Theo to adjust the methods of inference based on the accuracy of the available methods. (This form of learing is based on the evaluations of the SE software.)

Other advantages of the TheoAgent architecture are briefly summarized below:

- The accuracy of the EBL system returns a plan in a form (stimulus–response rules) suitable for execution.
- The TheoAgent system 'chooses its own goals based on the sensed world state, goal activation conditions and relative goal priorities' (Mitchell, 1990). With this advantage the system can decrease the number of explicitly designed goals as the number of stimulus–response rules increases.
- Improvements in the reaction time, result from the eager/lazy sensing abilities of TheoAgent. This form of sensing has helped reduce the lower bounds of reaction time since all values are not updated continually.

The self-improving architecture of TheoAgent has numerous advantages that reactive architecture and EBL alone cannot provide. It should be noted this experimental architecture has some flaws.

The disadvantages of the Theo system begin with the limited ability of the knowledge base. The speed and accuracy has only been tested on a limited laboratory setting. The developers are doubtful if the subsecond planning speed could be maintained in a more elaborate environment. Another disadvantage occurs with the planner of the cost of sensing or effector command implementation. The planner might suggest a plan that is not feasible due to lack of knowledge or unknown sensing/effector planning. Therefore, the planner would begin implementing the plan only to adjust for the situation later. The last disadvantage is the lack of inductive learning during the planning phase. The inductive learning will improve the resulting actions created by the system.

15.5 CONCLUSION AND FUTURE WORK

This paper began by exploring the three fundamental components (modeling, planning and program synthesis) of the RTPS. Each phase of the RTPS was followed by a simple example of a block world, that was developed one step at a time. Next we introduced two new architectures for the RTPS called EBL and reactive architecture. Both of these systems provided different perspectives and advantages to the original RTPS; however, separately the systems were lacking certain attributes.

For years, researchers have been developing technics for implementing learning and intelligence; however, until now these methods have had limited success. Recent research has led to a new system, called TheoAgent, that supplemented reactive architecture with an EBL system. The Theo system is a software package influenced by the work of other self-improving systems, such as SOAR (Laird, Newell and Rosenbloom, 1987) and RLL (Greiner and Lenat, 1980). The advantages of the TheoAgent architecture makes this 'uniform architecture for a self-improving system' a unique unifying attempt at several ideologies of artificial intelligence. The representation, indexing, learning and 'self-reflecting' aspects of Theo makes this software function with speed and accuracy.

This comparative study of RTPS has shown the benefits of combining two different approaches to task planning systems. In the future we plan to implement some aspects of a TRPS, that will further utilize the benefits of reactive architecture and perhaps introduce a different ML algorithm based on 'learning from observation'.

REFERENCES

Abramson, B. and Korf, R.E. (1987) *A Model of Two-Player Evaluation Functions.* Proceedings of the 1987 National Conference on Artificial Intelligence. AAAI, Morgan-Kaufmann, San Mateo, CA.

Blythe, J. and Mitchell, T. M. (1989). *On Becoming Reactive.* Proceedings of the Sixth International Machine Learning Workshop. Morgan Kaufmann, San Mateo, CA.

Fu, K.S., Gonzalez, R.C. and Lee, C.S.G. (1987) *Robotics: Control, Sensing, Vision, and Intelligence*, McGraw-Hill, New York.

Kaelbling, L.P. (1986) *An Architecture for Intelligent Reactive Systems.* Proceedings of the 1986 Workshop Reasoning about Actions & Plans, Morgan Kaufmann, San Mateo, CA, pp. 395–410.

Laird, J., Newell, A. and Rosenbloom, P. (1987) SOAR: an architecture for general itelligence. *Artificial Intelligence*, **33**(1), 61–98.

Lee, M.H. (1989) *Intelligent Robotics*, Halsted Press, New York.

Mitchell, T.M. (1990) Learning Robots, in *'90 Proceedings of the 4th Australian Joint Conference on AI*, World Scientific, New Jersey.

Mitchell, T.M., Allen, J., Chalasani, P. *et al.* (1990) Theo: A framework for Self-improving Systems, in *Architectures for Intelligence*, (ed. K. VanLehn), Lawrence Erlbaum, Hillsdale, NJ.

Segre, A.M. (1988) *Machine Learning of Robot Assembly Plans*, Kluwer, Boston.

Toshikazu, T. and Mitchell, T.M. (1991) Embedding learning in a general frame-based architecture, in *Applications of Learning and Planning Methods*, (ed. N.G. Bourbakis), World Scientific, New Jersey.

Vukobratovic, M. *et al.* (1988) *Introduction to Robotics*, Springer-Verlag, New York.

CHAPTER 16

An expert system with an external optimization module for quality control decisions

Alice E. Smith and Cihan H. Dagli

16.1 INTRODUCTION

Artificial intelligence includes several fields of scientific research including natural language processing, artificial neural networks and expert systems. All are concerned with replicating and sometimes improving upon human abilities. For expert systems, the human traits of knowledge storage, rule of thumb invocation and inferential reasoning, are the areas of interest, and interpretation, prediction, design and monitoring are the tasks addressed (Martin and Law, 1988).

Manufacturing quality control is a large field aimed at optimizing quality by preventive and corrective measures; variables in product design, materials, processing, ambient conditions and fabrication are brought under control (Hayes and Romig, 1982). It seeks to balance high quality and the costs of maintaining such quality. Using statistical methods to achieve quality control, statistical quality control was popularized by W.A. Shewhart and his group at Bell Laboratories in the first half of this century, and has been expanded by such notables as W.E. Deming and J.M. Juran until nearly all businesses currently use some form of it.

Engineering a knowledge base for statistical quality control is viable since it is a well structured domain with both analytical and heuristic solutions. A human expert can make very good decisions in little time, whereas a non-expert may need significant training and exposure to reach the same decision-making capability. Statistical quality control fulfills three fundamental criteria for an expert system candidate – relevancy, feasibility and optimality (Leonard-Barton and Sviokla, 1988). Commercial software, such as spreadsheets and statistical packages, can store data and perform the calculations of quality control; this expert system, named Quincy, is designed as a complement to those to assist with decision making and complex analysis.

This chapter will briefly review control charts and acceptance sampling plans, and cite previous expert system applications to quality control. Then, the system structure including reasoning will be presented. Finally, each of the three main areas of Quincy are discussed in more detail and conclusions reached.

16.2 OVERVIEW OF CONTROL CHARTS AND ACCEPTANCE SAMPLING PLANS

Control charts are commonly used in manufacturing environments to analyze process parameters to determine if a controlled process is within or out of control. Some processes which benefit from control chart tracking are filtration, extraction, fermentation, distillation, refining, reaction, pressing, metal cutting, heat treatment, welding, casting, forging, extrusion, injection molding, spraying and soldering (Miller and Walker, 1988). Control charts are plots of sequential points of critical parameters of sample lots. Sample size and sampling interval are generally fixed. Control charts measure location (such as mean or proportion) and variability (such as range or standard deviation). Although computationally simple, control charts are sometimes complex to correctly use because the sample points come from probabilistic distributions and usually require interpretation by a skilled user.

W.A. Shewhart (1931) first proposed the use of control charts, which commonly bear his name as Shewhart control charts. Figure 16.1 shows a portion of a typical control chart with 3σ limits. Through the ensuing years many different formulations became known, such as moving average and range charts, proportion (P) charts, number of defectives (C) charts, cumulative sum (cumsum) charts and control charts with warning limits (Gibra, 1975; Montgomery, 1980; Saniga and Shirland, 1977). However, most manufacturers use versions of the early control charts which track sample mean (X-bar charts) and sample range (R charts) as checks on the process state and the process variability (Saniga and Shirland, 1977). These charts are best suited for processes which typically experience large shifts when out of control, or the penalty for out of control production is not extreme (Montgomery, 1980). One reason these X-bar and R charts may be so popular among all processes is that to select a more appropriate control chart requires greater expertise than is sometimes available.

A difficulty with control charts is the determination of whether a process is actually within control or not. Since sample points are subject to noise due to measurement, human and other factors, they form a non-specified probabilistic distribution. An extreme point may come from a process which is, in fact, under control (a Type I or α error) or a point within boundaries may come from a process which has shifted to out of control (a Type II or β error). A Type I error costs time and money when the process is investigated

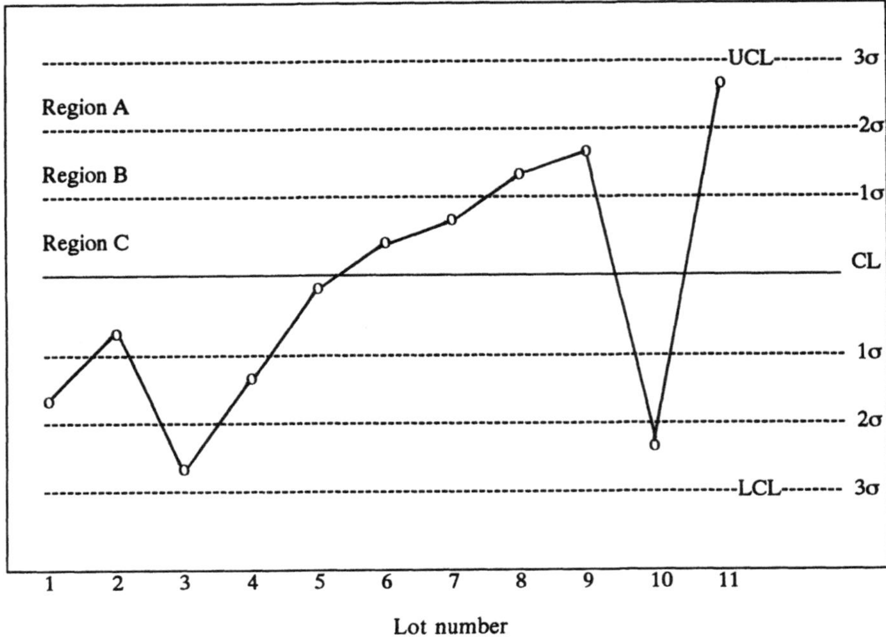

Fig. 16.1. Sample control chart.

to find the cause of the out of control situation (the assignable cause) when, in fact, there is none because the process is not out of control. A Type II error allows a process to continue even though it is out of control. Besides control limits, there are certain patterns of sample points which indicate the process is moving towards, or cycling through, undesirable situations. Again, classifying these points correctly is stochastic, and requires a certain expertise in the process and the concept of control.

Another important decision area in quality control is the selection of acceptance sampling plans. Acceptance sampling can be done by the supplier or buyer. It aims to ensure a certain level of quality while minimizing sampling costs. Choosing an acceptance sampling plan depends on the nature of the product, sampling costs, the relative importance of rejecting good items or accepting bad items, the overall quality level of items, and the sample size and variability. Once a plan has been selected, certain key variables can be calculated to estimate the average quality level, and the cost and effort of sampling.

Some earlier work has been done to relate expert systems to control charts. Most of these have selected proper control methodologies and advised on the analysis of the selected methodologies (Alexander and Jagannathan, 1986; Dybeck, 1987; Eid and Losier, 1990; Evans and Lindsay, 1988; Hosni and Elshennawy, 1988; Scott and ElGomayel, 1987). Quincy, an earlier version of which was introduced in Dagli and Smith (1991), is different from this previous work in two respects. First, it is more comprehensive than earlier expert systems since it includes acceptance sampling and control chart diagnoses along with control chart selection. Second, it integrates with an external optimization module transparently so that the user can obtain optimal sampling parameter values, along with the expert system recommendations, within one consultation.

16.3 THE EXPERT SYSTEM STRUCTURE

One of the reasons for the popularity of knowledge-based systems is the proliferation of sophisticated but developer and user friendly expert system shells. These shells differ from conventional programming by relying on heuristics rather than algorithms, being symbolic instead of numerically oriented and by processing interactively, not sequentially (Harmon and King, 1985). This application was built on version 1.2 of Level5, an expert system shell, from Information Builders Inc. running on a PC under DOS. Level5 for DOS works by compiling an ASCII text file with the suffix '.PRL' into an executable knowledge base with suffix '.KNB'. The ASCII file can be created in any text editor or word processor as long as the format required by the Level5 compiler is followed. Level5 can work with external programs and the expert system described in this chapter uses a C optimization module.

16.3.1 System features and structure

Quincy is rule based, i.e. the primary intelligence lies in its rules. Each rule is *modus ponens* (IF–THEN), and consists of antecedent conditions, known facts and additional user supplied answers resulting in one or more consequents (conclusions and actions). The inference engine (logic system) pursues the selected goal by backward chaining. Backward chaining is efficient when the problem has a well defined goal and means of selection to reach that goal. Quincy employs exhaustive reasoning, which means the inference engine does not stop when one applicable conclusion is reached. Reasoning continues until all possible avenues are explored. Exhaustive reasoning allows more than one conclusion to a given antecedent to be reached, and this, in turn, is used to weigh the value of the different conclusions.

This system has 118 rules, 246 facts, 16 optional explanations, 34 text expansions and 36 displays. Figure 16.2 shows the present and proposed system structure of Quincy. Facts are either attribute (qualitative variables), numeric (numeric variables) or string (literate variables). Quincy employs goal selection, i.e. allows the user to choose one of the three main areas (control chart selection, control interpretation or sampling plan selection)

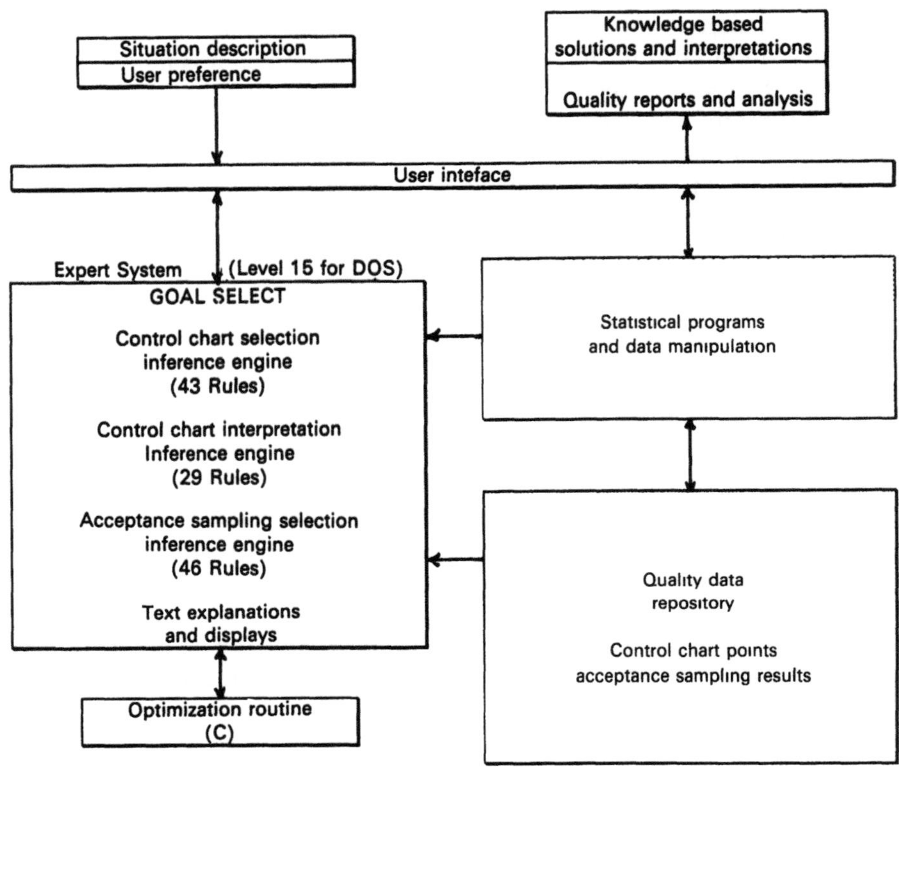

Fig. 16.2. System architecture of Quincy.

for consultation. For any question, a user may respond with 'Unknown' and the system will attempt to proceed with the analysis without that information. Quincy has been designed to minimize user supplied information and is therefore very limited in its ability to reason with unknown responses. At the end of the session, the user may optionally ask for the chain of reasoning utilized, the answers to all questions and the conclusions to all rules. These facilities are contained within the Level5 software shell.

Further enhancing Quincy's usability are explanations, text expansions and displays. Explanations offer optional clarification on questions to the user or elaborate on term definition. They are interactively selected using function keys by the user during the consultation. Text expansions are used for terms which are briefly expressed in the rules. The expansion is automatically shown to the user instead of the brief rule expression during the consultation. Displays are automatically shown too, and are usually used to show the user the conclusions and recommendations of the system. Some Quincy displays also show confidence factors as bar graphs and the calculated values to numeric variables.

Several other concepts have been applied. To handle uncertainty confidence factors for various system generated conclusions are employed. Confidence, or certainty, factors are typically stated in percentages from 0 to 100%, and allow the system to exhaustively pursue reasoning avenues which may not be wholly true and to suggest solutions with varying confidence. In this system, user responses are minimal and are largely based on factual conditions, so use of certainty factors is confined to only a few items, i.e. the control chart selected and the sampling plan selected. These can have multiple attributes, e.g. there can be more than one suggested sampling plan or control chart. Should there be more than one applicable control chart or acceptance sampling plan, Quincy outputs all applicable choices and their confidence factors to the user.

Some basic calculations for acceptance sampling are done within the expert system, as are some error checks. If the user violates the allowable entries, the system outputs an error message and 'loops'. Looping lets the user input the information again, correcting previous errors, without terminating the consultation.

Finally, since expert systems are not designed to act efficiently for storage and manipulation, this shell allows exporting and importing of data to other programs. Quincy interfaces with a C program performing an advanced optimization algorithm, although the data transfer is completely transparent to the user. The C program is QUINCY.EXE. To interface with the expert system shell, QUINCY.EXE uses a C header file, ASCIIPRM.H. Results are returned and displayed to the user by Quincy.

16.3.2 Diskette file descriptions

The diskette included with this book contains a directory, QUINCY, which includes the compiled knowledge base, QUINCY.KNB, and the compiled C optimization module, QUINCY.EXE. To run the knowledge base, Level5 for DOS must be installed on the PC, and QUINCY.KNB and QUINCY.EXE must be placed in a directory called 'PRL'. The QUINCY directory on the diskette also contains three ASCII files:

1. QUINCY.PRL. This is the ASCII version of the expert system. QUINCY.PRL can be inspected by the reader with ease. Annotations are indicated with a '!' at the start of the line. Words in all capitals signify Level5 commands or keywords. The file begins the title screen, then defines all the variables as either attribute, numeric or string. These variables are used to match antecedents and consequents during rule firing. After a few system parameters are set, an outline of the major portions of the knowledge base appears as 'Goals of the system'. Rules

follow, with each rule labeled by a lettered phrase to designate its logical grouping, and by a sequential number to designate the individual rule. After the rules, the automatic text displays (TEXT), the screen displays to the user (DISPLAY) and finally the optional text expansions (EXPAND) are listed. In the displays, the command BAR produces a bar graph of confidence factors.
2. QUINCY.C. This is the ASCII file of the C program for optimization of sampling method when the control selected is X-Bar and R. Nine numeric variables are passed from the expert system during the consultation to the C optimization program as variables inval[0] through inval[8]. Four numeric variables found by iterative search are returned from the C program to the expert system for display to the user. These are labeled in the C program as size, bestfn, bestk and besth (optimal sample size, optimal cost of control, k value for upper and lower control chart limits, and optimal sampling interval).
3. ASCIIPRM.H. This is the ASCII file of the C header file needed by QUINCY.C. It controls the passing of data back and forth from the expert system and the C program.

16.4 THE SYSTEM DOMAIN

The knowledge-based system does not attempt to include those areas better handled by traditional systems, such as data storage, statistical analysis and graphical displays. It does focus on three main areas of decision making:

1. Selection of applicable control charts for various processes.
2. Interpretation of control chart results.
3. Selection of appropriate acceptance sampling plans, with applicable calculations.

Together, these three areas form major decision points of statistical quality control. What is the form of control (control charts for process monitoring and acceptance sampling for components or finished goods), and what is that control expressing? The success of human decisions depend on the expertise of the quality control engineer and time allowed for decision making.

16.4.1 Selecting control charts

Control charts have been used effectively as graphic means to measure the state of a process. The process is tracked over time by mean, variability or range, and is considered out of control when specific limits are reached, such as 3σ. To select a proper control chart, the user must know the objectives of control, the conditions of obtaining control measurements and the applications of the different control charts. The portion of the knowledge-based system which selects control charts contains 43 rules to select one of eleven charts. It also optionally sends data to an external C optimization routine to calculate the optimal sample size, sampling interval, k limits for upper and lower control chart bounds, and the cost of control.

The portion of the knowledge based system which selects control charts could be easily modified to accommodate fewer control chart alternatives, depending on those traditionally used by the particular manufacturer. An initial decision is whether monitoring will be of attributes or variables; attribute data is whether an item should be accepted or rejected whereas variable data are measured on a continuous scale. Attribute data are monitored if time, cost or technological methods preclude measurement (Taguchi, Elsayed and Hsiang, 1989). An example of attribute monitoring is a go or no go test, such as if a scratch exists or whether a color is proper. Measurable product characteristics are weight, dimensions, specific gravity, tensile strength, impurity, resistance and voltage. In some cases, variable data will lend themselves to one sided limits only, e.g. strength having only a lower limit and impurity only an upper limit (Taguchi, Elsayed and Hsiang, 1989). If items to be measured are from a homogeneous sample, e.g. pH of a chemical solution, more than one measurement per sample is redundant and an individuals chart is recommended.

For users of X-bar and σ charts and X-bar and R charts with computer capability a further probe is done to determine if cumulative sum (cumsum) charts would be appropriate. These charts are not true Shewhart control charts but are easily understood plots designed for use on the production line. They detect small process changes more quickly and require smaller sample sizes. Also for X-bar and R charts the user has an option of running an optimization module to determine proper sample size, the minimum cost of control, the k value to determine upper and lower control limits and the optimal sampling interval (Dagli and Stacey, 1988). These parameters further assist the user to precisely design a superior control chart. The optimization routine is a C program, external to the knowledge base, based upon a single assignable cause model approximated by Duncan (1956), Chiu and Wetherill (1974) and Montgomery (1982)

$$E(L)=\frac{a_1+a_2n}{h}+\frac{a_4[h/(1-\beta)-\tau+gn+D]}{1/\lambda+h/(1-\beta)-\tau+gn+D}+\frac{a_3+a_3'\alpha\ \exp^{-\lambda h^*}(1-\exp^{-\lambda h})}{1/\lambda+h/(1-\beta)-\tau+gn+D}$$

where:

$E(L)$	expected cost per hour*
α	probability of false alarm, $\alpha=2\int_k^\infty \phi(z)\,dz$
$\phi(z)\,dz$	standard normal density
k	constant for setting upper and lower control chart bounds*
h	time interval for sampling*
τ	expected time of occurrence of single assignable cause between consecutive samples

$1/\lambda$	expected length of out of control period (based on Poisson distribution)#
β	probability of Type II error
$h/(1-\beta)-\tau$	expected length of out of control period
δ	magnitude of single assignable cause#
n	sample size*
g	time required to take sample#
D	time required to find the assignable cause#
a_1+a_2n	cost of taking the sample#
a_3	cost of finding an assignable cause#
a_3'	cost of investigating a false alarm#
a_4	hourly penalty cost associated with production in the out of control state#

*Is found by optimization routine.
#Entered by user.

16.4.2 Interpreting control chart results

Often control chart results are not obvious and need skillful opinion to determine whether a process is out of control. The sub-module of the system addressing this aspect has 29 rules divided into two main domains: (1) interpreting the general shape of points, and (2) determining the control condition from the number and position of data points. The system distills well known rules of thumb for interpretation of control charts results (Ishikawa, 1976; Messina, 1987), leading the user through applicable questions to arrive at a probable diagnosis.

The shape of sequential control chart points can often indicate vital information about the process. Quincy looks at five common variations – trend, stratification, sine wave, bimodal and shift. After diagnosing one of these common shapes from the description entered by the user, Quincy tells the user typical causes of the shape. This portion of the expert system is modular so that additional shapes and trends can be added easily. It is also intended that this shape analysis, along with the data point position analysis, could be done automatically when Quincy is interfaced with a quality control data repository.

The analysis of data point position begins with the user selecting whether the placement will be specified by halves of the chart (areas on either side of the centerline) or by σ regions A, B and C on either side of the centerline (Fig. 16.1). Once determined, the user is asked for number and location of points commencing with those most likely to indicate an out of control situation. Number of consecutive points falling in certain regions are analyzed via ratios and total numbers to decide if a process is out of control. The user is queried by the system until the system can diagnose whether the consecutive points indicate an out of control situation, or not.

16.4.3 Choosing an acceptance sampling plan

This portion, consisting of 46 rules, focuses on the choice of an acceptance sampling plan and calculates some measurements of the plan. Acceptance sampling based upon statistical properties is intended to reduce inspection costs and improve quality, giving an accept or reject signal for a lot of items. For destructive testing it is mandatory to use a sampling plan. Acceptance sampling can be done at any point during the process; the primary variables are lot size, sample size, the acceptable producer's risk (α risk – rejecting a good lot) and the acceptable consumer's risk (β risk – accepting a bad lot).

The system can recommend a single sample plan, a double sample plan, a multiple plan, a skip lot plan or two types of continuous sampling plans for attribute data. The selection is based upon the sizes of the sample and the lot, the acceptance number, the inspection plan, the proportion defective, whether curtailment is employed, and relative sampling costs versus false acceptance or rejection costs. The system also tests to see if a variable type sample plan would be feasible, and can recommend two kinds of variable sample plans. The final counsel is for an appropriate US Military Standard governing the type of sampling plan selected. These last rules are easily modifiable to accommodate changes.

Once a plan has been recommended, a component of the acceptance sampling rule base calculates the probability of acceptance, the average outgoing quality, the average outgoing quality limit and the average total inspection for the applicable sampling plans. These numeric quantities tell the user the impact of implementing the selected sampling plan, and are based on the acceptance number, the lot and sample sizes, and whether the sample is rectifying, i.e. bad items are replaced with good.

16.5 CONCLUSIONS AND IMPLICATIONS FOR THE FUTURE

Expert systems can augment human skills for structured and semi-structured problems and are particularly effective for small, routine decisions. Tasks which are performed better if there is ample time, the best expert always did them or where consistent decisions are desirable are prime candidates for knowledge engineering (Leonard-Barton and Sviokla, 1988). The primary economic benefit of expert systems are increased productivity by speeding professional and semiprofessional work by factors of tens to hundreds (Feigenbaum, 1990). The shortened time frame for decision making, the ever expanding number of options and the decreasing experience of decision makers are added stimuli for expert systems.

Quality control is a popular focus of new methodologies and implementations as manufacturers and service industries seek to improve their competitive edge. Although statistical quality control is already largely automated for the gathering and processing of data, improperly implemented quality control can yield erroneous and costly results, which may go unnoticed indefinitely. The problem domain, combining diagnoses and decisions based on analytical data, on heuristics and on user choice, is suited for rule-based expert systems.

Quincy is a prototype system which can act either as an intelligent integrator for statistical-based quality control software packages or a stand alone system. It directs the user towards appropriate solutions, analyzes quality data and advises on courses of action. The integration function makes a system like Quincy useful for both experienced and novice quality control engineers. The widespread use of quality control and its similarity from application to application further increase the vaue of an expert system; it can be expanded or altered to adjust to different industrial situations while retaining fundamental expertise.

ACKNOWLEDGMENTS

The authors would like to acknowledge the assistance of Sergio Martinez who wrote the C interface code to link the optimization module with the expert system.

REFERENCES

Alexander, S.M. and Jagannathan, V. (1986) Advisory system for control chart. *Computers & Industrical Engineering*, **10**(3), 171–7.

Chiu, W.K. and Wetherill, G.B. (1974) A simplified scheme for the economic design of x bar charts. *Journal of Quality Technology*, **6**, 63–9.

Dagli, C.H. and Smith, A.E. (1991) A prototype quality control expert system integrated with an optimization module, *Proceedings of the World Congress on Expert Systems*, Orlando, FL, pp. 1959–66.

Dagli, C.H. and Stacey, R., (1988) A prototype expert system for selecting control charts. *International Journal of Production Research*, **26**(5), 987–96.

Duncan, A.J. (1956) The economic design of x bar charts used to maintain current control of a process. *Journal of the American Statistical Association*, **51**, 228–42.

Dybeck, M. (1987) Taking process automation one step further: SPC. *Proceedings of the Sixth Annual Control Engineering Conference*, pp. 643–51.

Eid, M.S. and Losier, G. (1990) QCMS: a quality control management system. *Computers & Industrial Engineering*, **19**(1–4), 495–9.

Evans, J.R. and Lindsay, W.M. (1988) A framework for expert system development in statistical quality control. *Computers & Industrial Engineering*, **14**(3), 335–43.

Feigenbaum, E.A. (1990) Penrose and Feigenbaum on AI: is intelligence manufacturable? *IEEE Spectrum*, **27**(2), 49–50.

Gibra, I.N. (1975) Recent developments in control chart techniques. *Journal of Quality Technology*, **7**(4), 183–92.

Harmon, P. and King, D. (1985) *Expert Systems*, John Wiley & Sons, New York.

Hayes, G.E. and Romig, H.G. (1982) *Modern Quality Control*, Glencoe Publishing, Encino, CA.

Hosni, Y.A. and Elshennawy, A.K. (1988) Knowledge-based quality control. *Computers & Industrial Engineering*, **15**(1–4), 331–7.

Ishikawa, K. (1976) *Guide to Quality Control*, Nordica International Ltd, Hong Kong.

Leonard-Barton, D. and Sviokla, J.J. (1988) Putting expert systems to work. *Harvard Business Review*, **66**(2), 91–8.

Martin, A. and Law, R.K.H. (1988) Expert system for selecting expert system shells. *Information and Software Technology*, **30**(10), 579–86.

Messina, W.S. (1987) *Statistical Quality Control for Manufacturing Managers*, John Wiley & Sons, New York.

Miller, R.K. and Walker, T.C. (1988) *Artificial Intelligence Applications in Manufacturing*, SEAI Publications, Madison, GA.

Montgomery, D.C. (1980) The economic design of control charts: a review and literature survey. *Journal of Quality Technology*, **12**(2), 75–87.

Montgomery, D.C. (1982) Economic design of an x bar control chart. *Journal of Quality Technology*, **14**, 40–3.

Saniga, Erwin, M. and Shirland, L.E. (1977) Quality control in practice... a survey. *Quality Progress*, **10**, 30–3.

Scott, L.L. and ElGomayel, J.I. (1987) Development of a rule based system for statistical process control chart interpretation, in *Quality: Design, Planning, and Control*, (eds R.E. DeVor and S.G. Kapoor), ASME, New York, 73–91.

Shewhart, W.A. (1931) *Economic Control of Quality of Manufactured Product*, Van Nostrand Reinhold, Princeton, NJ.

Taguchi, G., Elsayed, E.A. and Hsiang, T.C. (1989) *Quality Engineering in Production System*, McGraw-Hill, New York.

Index

Paper titles and the corresponding disk directories and knowledge base files

Dir. A Filename: EX.ASC
A knowledge-based system for selection of resource allocation rules and algorithms
G.A. Süer and C.H. Dagli

Dir. B Filename: RULES
Expert system for casting design evaluation
I.C. You, C.N. Chu and R.L. Kashyap

Dir. C Filename: PAPER.TXT
An expert system approach to surface mount pick-and-place machine selection
C.Y. Liu, C.R. Emerson and K. Srihari

Dir. D Filename: TESS-KB.PAP
Knowledge-based surface treatment and coating selection in product design
C.S. Syan

Dir. E Filename: QUINCY.PRL
An expert system with external optimization module for quality control decisions
A.C. Smith and C.H. Dagli

Dir. F Filename: MFGSPVR.KB
An intelligent shop management system for production supervision
G.P. Moynihan

Dir. G Filename: MHT-ES.ASC
Expert system approaches to the selection of materials handling and transfer equipment
J. Rubinowitz and R. Karni

Dir. H Filename: RULEBASE.DOC
Intelligent systems for conceptual design of mechanical products
Q. Wang, M. Rao and J. Zhou

Dir. I Filename: ESRULES
Optimal rule-switching for flow shops with random workloads
J.C. Chen and A. Thesen

Dir. J Filename: FIXPERT.RLS
FIXPERT: A rule based system for workholding device selection of rotational parts
B. Bibanda, P.H. Cohen and C. Tunasar

Dir. K Filename: RULES.APP
A common skeletal framework for knowledge-based solutions to a representative set of manufacturing problems
M. Marefat and P. Banerjee

Dir. L Filename: RULES.TXT
OR/AI rules for the optimal design of manufacturing systems: machine and traffic allocation
Mario van Vliet

Dir. M Filename: APPA, APPB, APPC
A general purpose knowledge-based system and its application to design problems
Setsuo Ohsuga and Jiebo Guan

Dir. N Filename: COLLISIO.LSP, DAF.LSP, KNOW-ACT.LSP, KNOW-BAS.LSP, PREDICAT.LSP
Learning in robotic task planning
Nina M. Berry and Soundar R.T. Kumara

Dir. O Filename: KNOWBASE
A knowledge-based system for scheduling in a flexible manufacturing system
Suranjan De and Anita Lee

www.ingramcontent.com/pod-product-compliance
Ingram Content Group UK Ltd.
Pitfield, Milton Keynes, MK11 3LW, UK
UKHW021530300726
14060UKWH00011B/187
* 9 7 8 0 4 1 2 4 6 6 7 0 0 *